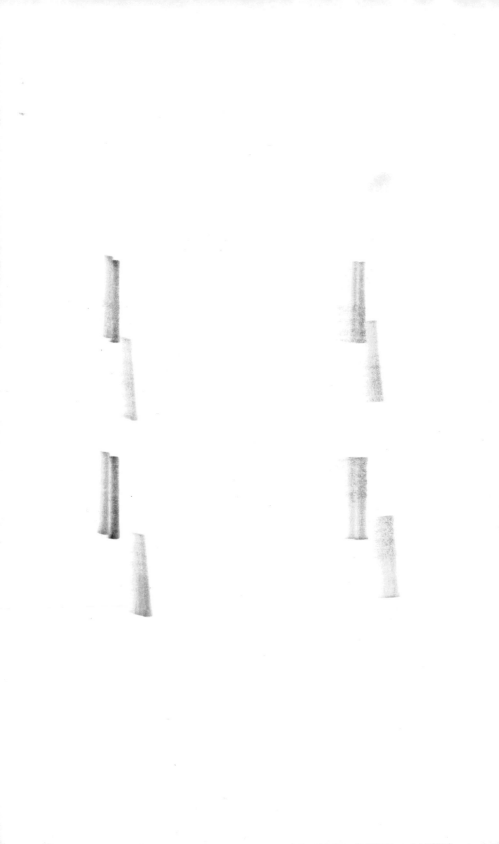

"Dr. Malone's experiences working in basic science and biodefense, along with his incisive analysis, perceptivity, and clarity of thought, make his book a fascinating read that will leave you in wonder and admiration of the breadth and depth of his insights. The forces he understands so well will continue flexing their influence unless deterred by trustworthy leadership and a resistant public."

—**Joseph Ladapo, MD, PhD**, Florida Surgeon General and author of
Transcend Fear: A Blueprint for Mindful Leadership in Public Health

"Reading this book is a bit like having a surrealistic nightmare, but one you don't wake up from. Although factual, informative, and not sentimental, the story is blood-curdling—with a magnitude of scandal that defies any imagination. Robert's deep insights into 'the system' leave us with profound disappointment and disdain for all those involved in this crime against humanity, while offering a realistic perspective to our children and grand-children for a better way forward."

—**Geert Vanden Bossche**, PhD, virologist and vaccine expert,
formerly employed at the Global Alliance for Vaccines and Immunization (GAVI)

"To understand the devastation of our times, to comprehend the scale and depth of the emergency that we face, Robert Malone is the leading person in the world you want as your guide. His scintillating book, filled with candid truths only he would know, is a gift to the world—to you, me, and everyone who seeks to understand. For decades, he has been at the center of the arena, as a scientist, intellectual, and moral force. His credentials are impeccable, even legendary—but just as remarkable is his willingness to speak. He could have been like so many others in his echelon of knowledge; he could have joined the junta of control, played along, or just stayed quiet. Something inside this man said, 'No.' And his moral compass guided him, same as so many other dissidents today. People who care owe a debt of gratitude to Dr. Malone for this literary achievement."

—**Jeffrey Tucker**, author, founder of the Brownstone Institute

"Essential reading for anyone willing to comprehend the madness we have endured during the past few years."

—**Paul Marik, MD**, former professor of medicine at Eastern Virginia Medical School

"Dr. Robert Malone's expertise and knowledge in the fields of vaccinology and infectious disease countermeasure technology is unparalleled. Despite this, he was systemically banned and censored by big tech and the US government for merely sharing his views. His book provides a road map for our nation to reform our crony capitalist society before even more harm results."

—**Andrew G. Huff, PhD, MS**, author of *The Truth About Wuhan:
How I Uncovered the Biggest Lie in History*

"Melding brainpower with compassion and solid values, Robert and Jill Malone stand out among the COVID truth-tellers. They have stayed grounded; they have continued to act with integrity and grace as they led the rest of us toward truth. Now they have the generosity to produce a work like this: a book that tells us exactly where we are, how we got here, and how we can create the world we must now bring into being."

—**Meryl Nass, MD**

"Robert Malone might have been the most influential critical thinker and voice during the corona crisis. He continued to speak out, no matter how much resistance he met."

—**Mattias Desmet**, professor of clinical psychology at Ghent University

"Dr. Robert Malone gives us an essential, captivating, and comprehensive guide to our historical moment, from lockdowns and mRNA vaccines to the administrative state and the game plan to control people via propaganda and groupthink. A scientist who also has deep knowledge of government, history, politics, and psychology, as well as great personal courage, Dr. Malone gives us this page-turning overview of where we are and how to move forward with our humanity intact."

—**Naomi Wolf**, bestselling author of *The Beauty Myth*

"As one of the top vaccinology experts, and the inventor of mRNA technology, Dr. Malone was naturally tapped by the government to help in the early stages of the COVID pandemic. When he began to ask hard questions, the fury of censorship and coordinated personal attacks led him on a journey of self-discovery and awakening. This book reveals the truth of the last two-plus years and exposes how our public health institutions really make the sausage."

—**Ed Dowd,** former Black Rock managing director, author of *Cause Unknown*

"An extraordinary and deeply researched tour through the engineered global brainwashing experiment known as the COVID-19 pandemic. Through hard-hitting, data-driven critiques authored by the brightest medical science thinkers of our time, this book bears witness to the true COVID conspiracy unleashed upon the world. If Western medicine is to be salvaged in the aftermath of this worldwide pandemic fraud, it will require an honest reading of this groundbreaking book that, if properly considered, can change the course of the history of medicine. Dr. Robert Malone and all his co-authors are to be applauded for their courage, determination, and passion, telling the unpopular truth in an age of convenient lies."

—**Mike Adams**, founder, NaturalNews.com and Brighteon.com

"Soon after the madness started, I stumbled across Robert Malone in the forest of the online world. His presence, his voice, what he had to say and why, were balm for my troubled soul. Here in this book is the story that explains why this softly spoken man did what he did. He has been true north for so many who felt utterly lost. I have nothing to offer but gratitude."

—**Neil Oliver**, author and GB News host

"Dr. Malone's critical thinking skills were honed while working with US military intelligence. His precise analysis of the COVID 'science' was not only accurate, but presciently predicted what is now widely accepted. He exposed the lies and fraud of the COVID narrative early on, including during a Joe Rogan podcast that is among the most watched of all time. This book goes further into the truth that attracted those 50 million views."

—**Dr. Joseph Mercola**, founder of Mercola.com,
the most visited natural health site on the internet for the last twenty years

LIES MY GOV'T TOLD ME

AND THE BETTER FUTURE COMING

ROBERT W. MALONE, MD, MS

FOREWORD BY ROBERT F. KENNEDY, JR.

Children's
Health Defense

Skyhorse Publishing

Skyhorse Publishing books may be purchased in bulk at special discounts for sales promotion, corporate gifts, fund-raising, or educational purposes. Special editions can also be created to specifications. For details, contact the Special Sales Department, Skyhorse Publishing, 307 West 36th Street, 11th Floor, New York, NY 10018 or info@ skyhorsepublishing.com.

Skyhorse® and Skyhorse Publishing® are registered trademarks of Skyhorse Publishing, Inc.®, a Delaware corporation.

Visit our website at www.skyhorsepublishing.com.

10 9 8 7 6 5 4 3 2 1

Library of Congress Cataloging-in-Publication Data is available on file.

Print ISBN: 978-1-5107-7324-0
eBook ISBN: 978-1-5107-7325-7

Cover design by Brian Peterson

Printed in the United States of America

A Tribute to My Partner in Everything

The truth of the matter is that both my daily life and this book have been a partnership, and Dr. Jill Glasspool-Malone PhD (Biotechnology and Public Policy) has contributed at least as much as I have to the resulting products. We requested that she be listed as a coauthor, but the (sky)horse had already left the barn—the cover graphics were prepared long ago, initial marketing had already been done, and the project had developed such momentum that we could not turn it around. She often writes under my name, after all these years we still routinely finish each other's sentences, and this book absolutely would not have been completed without her constant effort, intellectual contributions, advice, daily writing and editing for over a year now. Her spirit, ethical compass, and probing mind is interlaced throughout the resulting work. At our wedding so many decades ago, I read the passage from Kahil Gibran's The Prophet concerning marriage, and as I look back, I believe that we have lived to that advice. Over the many years of our partnership, we have truly become both two separate and one together. When you read the word "I" in this book, often it should really read "we," as both have experienced each of these events as one, and the journey of the book and the intellectual insights herein have emerged from our constant shared dialog. Allowing my sole authorship while so freely giving of herself has been her gift to me, but the reader should know that the product has completely been a joint effort, and will please recognize and acknowledge the shy intellectual genius tomboy who has been woven throughout these pages. If you read carefully, you will see her peek out here and there from behind the words and ideas. Thank you, Jill, for all that you do, have done, and have freely given these many, many years. I look forward to many more, continuing to love and protect, and hope that we can return someday soon to our quiet life together of farm, gardening, horses, and dogs, far from the madding crowd.

Contents

FOREWORD
by Robert F. Kennedy Jr.

Lies My Gov't Told Me is an apropos title for a book by a man who knows those lies from the inside, where he spent much of his life, and from the outside, where many powerful people want him to spend the rest of his life.

Reflecting on decades of work in biodefense and vaccinology, Malone writes, "I never really allowed myself to confront the possibility that we might not be the good guys." These pages bring the reader on the journey that opened his eyes and closed so many doors.

For decades, the military industrial intelligence apparatus has routinely taken advantage of catastrophic crises to increase their power and control, and this time, Robert Malone was among the few who stood up in their path. For this, he earned their disdain, and my enduring respect.

Being a highly accomplished and internationally recognized physician-scientist, a pioneer and expert in mRNA and DNA vaccines and therapies, and a researcher and developer of biodefense countermeasures for US Department of Defense contractors, Malone posed a special problem for those in power: He couldn't be easily dismissed or debated. So instead, he was quickly deplatformed and canceled by corporate and social media at the behest of the government, then vilified, marginalized, and lied about. The dust is still settling, and Robert Malone is still standing.

He is supported by other physicians, scientists, scholars, attorneys, and activists who contributed chapters that are woven throughout this book. Readers will be guided through the rabbit hole of falsehoods and misrepresentations that beguiled millions of Americans into accepting mandated vaccines and barely tested drugs, without even the pretense of informed consent. Parents agreed to give mystery injections to their children and babies, yet can't explain the risks or supposed benefits.

Malone envisions a different future, one in which our citizens understand enough to defend their freedoms, medical and otherwise. By giving us all deeper insight into the global pandemic of government deceit and overreach, this meticulously researched book can be an important part of reaching that better future.

INTRODUCTION:
Things Fall Apart; the Center Cannot Hold

Prelude

Before the time of COVID, my wife and I had built a quiet life on a Virginia horse farm.

Both of our homeschooled sons were healthy and happy, had graduated from college, and were married. We had one grandchild. The farm and tractor were mostly paid off. We had homesteaded the place, starting with unimproved rolling hay fields purchased directly from the prior owner—no bank loans necessary. Beginning with an old office trailer, we had built up fences, power, well, septic, barn, and both a main and a guest house over five years. Run-down historic outbuildings were being renovated. Years of experience in rebuilding and landscaping small farms had allowed us to create a working operation, our own park and garden.

Our refuge is located in a sleepy Virginia county with about as many residents as before World War II, an hour and a half south of the traffic and bustle of the nation's capital. Using American political slang, a red county in a purple state, stretching along the western side of the Shenandoah National Park. Internet access is a problem, and television requires a satellite dish. The historic farms of USA founding fathers Thomas Jefferson (Monticello) and James Madison (Montpelier) are only a short drive away. The first Lutheran church built in North America is two miles over the hill as the crow flies. Old established farming families control local politics. Trees pop up if no one mows the grass. Amish and Mennonite communities work nearby farms. Our Portuguese senior stallion was coming along nicely in his dressage training, we had a great string of brood mares, and homebred Australian shepherd dogs were our daily companions. My wife and I planned trips to the Golega Lusitano horse fair in Portugal and a horse competition in Texas. Price and availability of hay was a constant topic. Far from the madding crowd.

Together with Dr. Jill Glasspool, my wife and partner in all things for over forty years, I was maintaining a boutique medical research consulting practice that paid the bills. When we started our lives together, I was working as a short-order cook, farmer, and carpenter; she was a waitress, and we managed to work and pay our way through years and years of university training. This was our fifth small farm rebuild. Our primary challenges at the time consisted of business development, writing, reviewing, and executing contracts, and juggling the very different demands of the consulting business, the farm and gardens, and the horse-breeding operation. Occasionally I was asked to lead an NIH contract study section or review a manuscript for some journal, but that was just about all the contact I still had with the world of academia that I had chosen to leave almost twenty years prior. I had recently picked up a promising new Rockville, Maryland-based client that supported clinical research and regulatory affairs for Chinese pharmaceutical and biotechnology companies seeking to bring their products to the US market. Jill and I were trying to build a more international consulting practice and reduce our dependence on what often seemed like arbitrary and capricious US Government contracts, and we had planned and executed a series of actions toward that goal. It was a quiet and fulfilling life.

The Twin Towers, Pentagon, and anthrax-powder letter attacks had changed both the face of infectious disease research and my professional life as profoundly as had the advent of AIDS at the very beginning of my career. Shortly after the terror attacks, the Norwegian investors in the genetic vaccine company we had helped launch (Inovio) pulled back out of fear of US instability. We were left high and dry with neither clients nor academic appointment, so by necessity I joined a Department of Defense contract management firm called Dynport Vaccine Company (DVC) as assistant director of clinical research. At the time, DVC had recently received the "prime systems contract" for managing all advanced development (clinical and regulatory steps for licensure) for all Department of Defense biodefense-related drugs and vaccines. Little did I know when I took the job that Dynport's majority owner, Dyncorp, ran one of the two main US-based mercenary armies; that the field of "biodefense" was about to explode; that my career path would be transformed forever; and that I would be catapulted into the shadowy realm that exists between academic biotechnology research and US government-funded infectious disease intelligence, surveillance, and threat mitigation.

While employed at DVC, I had the epiphany that if I really wanted to help people, I needed to leave the cloistered, backbiting, and self-aggrandizing reality of academic discovery research and embrace the world of advanced medical product development. The professional culture around me neither wanted nor needed more "academic thought leaders," and the true unmet need was for people who understood both the wild west of discovery research as well as the highly regulated world of advanced development—clinical research, regulatory affairs, project management, and all that goes into making licensed medical products. If I really wanted to help people by enabling development and licensing of lifesaving treatments, I had to forget about the ivory tower world of academics and learn the skills necessary to help companies navigate the world of the Food and Drug Administration and the European Medicines Agency. So that became my new career path, and I threw myself into learning all that was required to meet this need. In the ensuing years I exceeded my goals by winning or managing billions of dollars in US federal contracts doing precisely that.

Over the years before COVID, Jill and I had developed a modest network of friends and professional colleagues scattered across the globe. This network was built from our consulting practice, from when I was working on US Government-funded biodefense and influenza vaccine contracts, as well as my prior days as an academic teaching pathology and molecular biology to medical students while doing bench research, writing papers, filing patents, and getting involved in various biotechnology start-up companies. And we had our horse friends of course. LinkedIn, Facebook, occasionally Twitter, and email correspondence allowed us to stay in touch with all of our friends and colleagues. Social media censorship and shadow banning was something that happened to people who lived in China—I could not imagine that it could happen to me. Jill and I simultaneously lived in two very different worlds that rarely touched each other; one in cutting-edge biotechnology and infectious disease medical countermeasure research, and the other immersed in horses, hay, orchards, farm equipment, construction, and the local feed store.

Somewhere between September and December 2019, a novel coronavirus entered the human population and began spreading like wildfire across the globe, turning my world upside down. Maybe it also transformed your life, too? If someone had written a letter describing my life today to the person I was before this outbreak, the old me would have concluded that the author specialized in (improbable) dystopic fiction and should probably be looking for another line of work.

Looking back, I am struck by how sheltered and naive I was (pre-COVID), and how much my worldview and my role in it have been radically shifted by subsequent events.

Will you take a memory walk with me for a moment?

Until COVID, I thought that free speech was a protected fundamental right guaranteed to all citizens of the United States of America by the Bill of Rights. Having been assigned core texts like *1984, Brave New World, Animal Farm, Lord of the Flies*, and *The Trial and Death of Socrates* in fourth and fifth grade as a "gifted and talented" student in the California school system of the time, I believed there was no way anything like what was written in those books could happen here in the USA during the 21st century. Internet censorship and government-controlled propaganda were unfortunate things that happened to those who lived in the People's Republic of China under totalitarian Communist Party control, but I had been born into a modern Western free society and had the luxury of watching this play out from afar. Social media was a tool that we used to chat with friends, sell horses (Facebook), write about the scientific issues of the day, and look for new biotech clients (LinkedIn).

Trained at one of the top clinically focused medical schools in the United States, Northwestern Feinberg School of Medicine, I believed that physicians were deeply committed to upholding the Hippocratic oath (principle of nonmaleficence), had freedom and responsibility to diagnose and treat patients as individuals, and were guided by a shared core of bioethical principles codified after the Second World War and incorporated into US federal law as the "common rule." At the center of this training was the practice of taking a detailed history and physical exam, beginning with the "chief complaint"—uncovering the real problem that brought the patient to the physician. Patients had medical autonomy; and "informed consent" for any medical procedure was ethically critical. I knew that corporatized (and computer algorithm-driven) medicine was placing ever-heavier burdens on the daily grind required to maintain a clinical practice—an unfortunate reality that practicing physicians and medical care providers had to endure if they elected to work under those systems. But for my colleagues, there was always the option to leave for private practice. One edgy new frontier for clinical practice was direct payment to physicians, practicing in the new world of outpatient surgical centers, and "doc-in-a-box" group practices, somewhere between the local doctor's office of my youth and an emergency room setting—thereby bypassing established hospital networks with their huge costs, kludgy bureaucracies, and massive burden of administrative oversight.

State medical boards were primarily in place to ensure that physicians and allied medical professionals met educational standards, provided patients with a high standard of care, and did not engage in overtly unethical practices or gross misconduct. Examples warranting medical board review or disciplinary actions included violations of the principles of nonmaleficence, beneficence, patient autonomy, or justice; violations of which would occasionally rise to the level of medical malpractice—usually by physicians who had developed a substance addiction. State medical boards were not generally involved in policing off-label prescribing practices of licensed drugs, or in terminating medical licenses unless a medical care provider was clearly mentally compromised or abusing the right to prescribe a medicine. I had never heard of a medical board policing free speech by a physician, whether it involved politics or prescribing practices. One example of disallowed medical practices that would trigger disciplinary action involved prescribing powerful addictive opioids without a compelling medical indication, typically leading to both patient addiction and high physician revenue. But most medical boards seemed hesitant to even discipline that behavior. Other examples involved physician compromise due to personal drug addiction or inappropriate sexual contact with patients resulting in an abuse of the patient-physician relationship. For those situations as well, the usual medical board intervention involved nothing more than a requirement for remedial training with a possible temporary suspension of medical privileges.

The current practice of "hunting physicians" by filing complaints with medical boards to withdraw their license to practice medicine for trying to help their patients with new therapeutic strategies, or for questioning the safety or effectiveness of a current medical intervention, was unheard of. Dissent and discussion within the medical community was a time-honored tradition with a long history of leading to improvements in medical care. Early in my career, I collected old medical texts as a way to remind myself of how far medical science had come, how far we still needed to go, and how frequently the deeply held medical treatment paradigms of different ages had been proven ineffective or even harmful. One practical consequence of these oversight policies was that for the preceding two decades, medical practitioners were consistently ranked the most trusted professionals by the Gallup Honesty and Ethics poll.

A key part of my consulting practice as a Maryland licensed physician and experienced scientist involved my deep experience in clinical research, with years of training in all of the related disciplines combined with three decades of practical experience in academic and industrial bench research, regulatory affairs, and clinical trials. As a requirement for being allowed to serve as a "Principal Investigator" for both federal biomedical research grants, contracts, and human clinical research trials, I had completed extensive and repeated coursework in medical and research ethics. A few years before, I had completed a prestigious fellowship at Harvard Medical School in Global Clinical Research as a Research Scholar, which rounded out my skills and training in clinical trial design, bioethics, epidemiology, clinical data interpretation, regulatory affairs, and biostatistics. During the winter of 2019, I was completing training for board certification in Medical Affairs, the term applied to the discipline of managing all communications between a pharmaceutical company, physicians, and patients, and for insuring compliance with rapidly expanding legal requirements. I was taking this additional training because so much of my consulting practice involved advising executive-level clients on a wide range of issues involving communication and medical affairs. Clients sought me out because of my deep understanding of FDA-compliant clinical research, my prior experience as an entrepreneurial bench researcher with many issued fundamental patents (including the initial DNA and mRNA vaccine patents from my early work while I was in my late 20s), and my extensive experience and understanding of vaccine and biodefense-related medical countermeasure development. And in particular, they valued my willingness to speak freely, forthrightly, and honestly about whatever issues that they wanted me to look into. Apparently, this has become a rare trait in modern business settings—particularly in the pharmaceutical business.

Late in 2019, working with a scientific friend and colleague, our consulting firm had been awarded a modest pilot subcontract from the Department of Defense (DoD) Defense Threat Reduction Agency (DTRA). The objective was to demonstrate the usefulness of combining the latest computer-based drug screening methods with high throughput robotics to test very large libraries of drug candidates and discover inhibitors of organo-phosphate-based biowarfare nerve agents. I had previously helped my colleague develop and win a large Department of Defense contract for building and staffing one of the "advanced development" antibody and vaccine production facilities that were built after the Obama White House had realized that the United States had lost much of its biologic drug manufacturing capacity to Europe, India, and China. The scope of work and approach that the DoD had funded was in large part an extension of a prior start-up company that I had founded called "Atheric Pharmaceuticals," which had been focused on partnering with DTRA and USAMRIID (United States Army Medical Research Institute of Infectious Diseases) to use high throughput robotic screening technology to repurpose drugs for treating diseases caused by viruses such as Zika, Ebola, and Yellow Fever. We had great success in achieving the mission (patents were filed for use of hydroxychloroquine, nitazoxanide, niclosamide, and many others based on our work), but we also learned the hard lesson that the investors had no appetite at that time to fund drug repurposing for emerging infectious diseases and viral biothreats.

The Call

Then everything changed, first for me and our DTRA-funded research group, and then for the world.

I took a call on my cell phone on January 4, 2020, from a fellow physician who had been in Wuhan, China, for some unknown period of time via an academic exchange program with a Chinese university. Dr. Michael Callahan is a brilliant infectious disease and intensive care specialist with both a long history of working at the forefront of biodefense and medical countermeasure development as well as a faculty appointment at Harvard University. Many years before, he had been introduced to me as a CIA employee and key DARPA leader, but his status regarding the CIA as of January 4, 2020, was and remains unknown to me. Michael and I had copublished academic papers in the past (involving the Zika virus outbreak) during my Atheric Pharmaceutical days, and I knew that he was exceptionally well-connected with those who live in the edgy gray zone of global infectious disease outbreaks and the US intelligence community. Of course, he knew that I had previously succeeded in collaborating with leading scientists at USAMRIID, the nation's biodefense epicenter, to identify repurposed drugs active against the Zika virus. Michael called to warn me that there was a new coronavirus on the loose in Wuhan, China, and to recommend that I get my group spun up to apply our tools, skills, and knowledge to address this new biothreat.

And with that fateful call, our quiet lives on our Virginia horse farm were completely transformed.

Jill and I had been at the forefront of so many of these outbreaks in the past: HIV, the Anthrax spore events, influenza virus (multiple times), West Nile, Ebola, Zika, etc. Our initial response to the alert call from Michael Callahan was a reflexive "here we go again," with a topper of "time to get going."

Having a proven ability to make a difference is both a gift and a curse. Chaos reigns early in a potential infectious disease pandemic. As if God's hand were guided by the words of Shakespeare's Mark Antony: "Cry 'Havoc!' and let slip the dogs of war." The onset of war is the proper metaphor, and the fog of war descends over everything. For those at the tip of the spear, it gives rise to an addictive sustained adrenaline rush like no other, coupled with constant risk of going overboard if you lose perspective.

Action

Once again, we got to work. Jill is very local community-oriented, and she poured her heart, mind, and soul into writing a kind of survival manual for those at risk and self-published the book via Amazon. An avid reader, she had become a big fan of self-published books and her Kindle. I threw myself into getting the team assembled for the DTRA project spun up and providing direction by diving into the coronavirus literature and selecting a specific protein target to apply the repurposed drug discovery/computational docking tools to. I helped Jill with her book by collecting and expanding some of the thoughts and comments I had been posting on LinkedIn to create content about the virology and immunology and assisted on editing the text. We worked like demons, side by side, day after day, and she was able to self-publish during the first week in February 2020. Within a mere five weeks, she completed the first edition of *Novel Coronavirus: A Guide for Preparation and Protection*.

Meanwhile, I got my scientific research group motivated, energized, and activated to volunteer their time, skills, knowledge, and abilities to try to discover repurposed drugs able to act as inhibitors of the critical SARS-CoV-2 protein known as the papain-like protease, otherwise known to virology experts as the 3-chymotrypsin-like cysteine protease (3CLpro). When the sequence of the "Wuhan Seafood Market Virus" was uploaded to the NIH sequence database, I applied computer software tools developed at UCSF to model the structure of that protein based on publicly available (previously published) crystal structures of the closely related 3CLpro from the SARS coronavirus. With SARS, this protein had been one of the leading antiviral drug targets, so it was reasonable to apply what had been learned with SARS to this new coronavirus. A specific region (binding pocket) of the protein had already been identified for drug development for the original SARS virus. Digital libraries representing detailed models of all known licensed drugs and nutraceutical compounds were obtained. Different software tools were then used to virtually dock each drug into the binding pocket of the modeled 3CLpro, resulting in a ranked list of possible inhibitors, which we then compared to the known safety profile and pharmaceutical characteristics of the leading drugs. This began a months-long process of testing, refinement, and retesting to optimize a list of drug candidates for further testing as antiviral compounds in the "real world."

Censorship

Jill's book was published February 11, 2020, with a plan to constantly update the editions as more data and information became available. We hoped to create what is known as a "living document" that would be updated as the pandemic evolved. There were no other books available at the time that had been written by medical and scientific professionals. Most people were still unaware of what they were about to be hit with as the virus made its way into Italy, the rest of Europe, and then the United States.

As more and more people became aware of the threat posed by this novel coronavirus, the book began to sell. Although sales were modest, charts from the Kindle Direct Publishing website showed steady increases in February and March. The Amazon reviews were all five stars, and Jill felt a strong sense of pride. Her first book!

Little did we know we were about to encounter the new reality of government-corporate cooperative censorship, which would become a major theme throughout the entire history of the COVID-19 public health event.

Jill's book was censored by Amazon. No explanation, no appeal. When we went to upload the most current edition in March, we received messages stating that Kindle Direct Publishing was "experiencing a temporary delay in publishing some titles." On the phone we were told this was a normal delay due to lack of editors.

We then received multiple messages, stating that the book did not meet "community standards"—which many of us have come to recognize as the standard phrase used to justify censorship in the time of COVID. We spoke with multiple people at KDP, who assured us that the reviewers would speak to us about why, as that was standard Amazon policy. That usually such problems could be worked out.

A few days later, people at Amazon told us by phone that the reviewers would not speak with us and that the book didn't meet community standards. They stated they did not know the reason the book was banned, and they were "very sorry." Multiple phone calls produced the same results. They refused to pass on our wish to speak with a supervisor, and they refused to answer our questions. At no point did we lose our temper or raise our voice. They just refused all inquiries and stated that the reviewers did not wish to speak with us. We could find nothing in the "community standards" statements that applied to anything we had written.

And at that moment, we knew that something very dark was happening, something we had never seen before. Little did we realize that this was just a very early example of what was to become a large movement over the next two years, a global movement involving collusion between government, corporatized legacy media, social media, big technology, big finance, and nongovernmental organizations to completely control and shape all information and thought concerning the public health response to the novel coronavirus.

Publications and social media posts about the coronavirus began being removed from all over the Internet. Although the original intent was to remove books that promoted "snake oil," or were out to make a quick profit, this censorship quickly turned into something far more insidiously dangerous. That is, books that didn't share the messaging of the US government were removed. Amazon represents the biggest bookseller in the USA. When Amazon censors reading material—where does this leave us as a nation? Apparently, the government believed that we as a nation must give up our precious freedoms of free speech and a free press due to declaration of a public health emergency. I will write it as clearly as I can: censorship and its "big brother"—propaganda—is not the answer.

The ramifications of these choices by our government to censor, lie, and obscure will go down in history. If the truth-tellers—scientists, writers, journalists, and authors—are not allowed to document the true story, a revisionist history will emerge. The alternative history being provided by the US government and promulgated by tech giants will allow such outbreaks to occur more easily in the future and allow those who failed us to remain in control of our governmental functions.

In the short span of three months, Jill and I had gone from a peaceful life on our farm, to receiving an alert from an American physician and intelligence operative operating in a region of China that I had never even heard of before, to self-publishing a modest guide for preparing and protecting yourself from the coming wave of infection, to directly experiencing the effects of an emerging Orwellian collusion among an international non-governmental organization (WHO), a US Government (which appeared to have casually cast aside the First Amendment enumerated in the Bill of Rights), and the largest bookseller and retailer in the world.

I am often asked, "What made you decide to speak out about what you saw going on during this 'pandemic?'" I have been told that I have become radicalized (by Steve Bannon, no less!) or "red-pilled" over the ensuing many months. The truth is, my quest to understand how, why, and by whom this global public health event has been weaponized against all of us began with a simple and inexplicable book banning. Many have since reviewed Jill's book looking for some subtle offense and found nothing. The incredible effort and work product of my treasured wife and companion had been taken from her and thrown away with neither rationale nor explanation.

Doctrine

As time went on, it became clear to me that the World Health Organization, as well as senior members of the US Government Department of Health and Human Services, were repeatedly lying to the world. Almost daily, the official "leaders" speaking to the world, using the megaphone of mass media, were substituting their own personal opinions and biases for what was being presented to the general population as fact or data-based information. The phrase

"Follow the Science" became a global joke, compounded by the amazing self-own statement of Dr. Fauci in which he told MSNBC's Chuck Todd during June of 2021 that attacks on him were "attacks on science." I began feeling an almost overwhelming sense of vertigo while struggling to find truth in the middle of this sea of mis(mal?)information, "factcheckers," gaslighting, defamation, and chronic falsehoods. Then, while I was participating with other physicians in an effort to support two of our colleagues who were being threatened with loss of job and medical license for merely voicing concerns about the genetic vaccines and support for early interventions with repurposed licensed drugs, I was presented with an amazing document titled "The Malone Doctrine." The authors told me that they had listened to everything I had said in my various public statements up to that point, had read everything I had written, and had developed a declaration based on the "white space between the lines" of all that I had spoken or written. They asked us to read and sign the declaration that they had prepared.

As Jill and I began to read their work, a smile crept across each of our faces and grew into outright joy. This was a first step toward recovery from the trauma and darkness that so many of us were experiencing. A new dawn. At that moment, we began to see the outlines of a better future coming, a future worth fighting for.

The Malone Doctrine
A Declaration of Independence
From the Decisions of Institutions That Lack Integrity

We the Undersigned:

Demand that all underlying data that contributes to a body of work under consideration must be made available and must remain accessible for analysis.

Proclaim the value of knowledge to society is not determined by any given creator of information. Instead, that it is the beneficiaries of knowledge who assign value to a proposition only through thorough critique and relentless scrutiny.

Establish the free and open exchange of information and establish as a duty the authority to serve as the custodians of all data forming the basis of our decisions.

Require the full disclosure of all sources of funding regarding any citation noted or references made pertaining to any matter under consideration.

Commit to impartiality in consideration of all analytical information and data brought before us and expect the same from all others.

Foster rigorous open debate and scrutiny in consideration of and for any matter of concern.

Shall promptly make the discovery of intellectual dishonesty or professional irresponsibility known to all.

Ensure the health, welfare, and safety of any whistleblower, bringing forth and/or making public an abrogation of the beliefs held herein.

Stand in opposition to censorship and will not accept representations of parties holding within themselves values that conflict with principles of free expression.

Deny no person the right to challenge, debate, petition, redress, examine, or protest with facts and evidence any decision of this body.

Purpose

In one sense this book documents a personal journey, a long effort to get to the bottom of the fundamental questions that have dominated every waking moment of my life ever since. It includes a series of essays composed during late 2021 through 2022, each of which addresses some aspect of the enormity of what we have all experienced. Who is responsible for all the globally coordinated propaganda, information management, mind-control efforts, lies, and mismanagement we have experienced? How has it been globally coordinated, and what can we do to stop this sort of thing from ever happening again? What are the root causes of this incredibly dysfunctional "public health" response that frequently seems to have nothing to do with public health? Has there been a truly nefarious agenda, or is this dysfunction merely the unintended consequence of interactions between separate, random events amplified by incompetence and exacerbated by hubris?

During this journey, I have seen, experienced, and learned so many new things, met so many people, made many new friends, and listened to so many stories. What follows in this volume is an attempt to process and comprehend the incomprehensible human tragedy and horror of what has occurred during this "pandemic," and to find some path forward that could lead to a better future for all of us. A future that will require people who still believe in the core principles that form the bedrock upon which Jill and I have built our lives: acting with integrity, respecting the fundamental dignity of other human beings, and making a commitment to community. The principles that formed the foundation of the American Enlightenment, resulting in the US Constitution and Bill of Rights.

I am firmly committed to a belief that the American experiment in self-governance, forged in another crucible, the tyranny of a mad king, remains relevant today. I reject the twisted logic of those who assert that these principles are obsolete, antiquated, and must be replaced with a system built upon a collectivist and globalist totalitarian vision, a system of government and command-and-control economic activity that have consistently failed every time they have been tried throughout history.

Jill and I have lived our lives as free and honest people. It has not been an easy path to walk, but as we begin to approach the end of our journey, we would have it no other way. This commitment and belief system form the subtext that is woven throughout the following chapters. A commitment to integrity, dignity, and community, tempered with empathy, offered without apology.

PART ONE:
HISTORY AND PHYSICAL EXAM – HOW DID WE GET HERE?

Few are aware that on September 28, 2022, during a World Economic Forum "disinformation panel" discussion, United Nations' global communications representative Melissa Fleming openly stated, "We partnered with Google, for example. If you Google climate change, at the top of your search, you will get all kinds of UN resources. We started this partnership when we were shocked to see that when we'd Googled climate change, we were getting incredibly distorted information right at the top. We are becoming much more proactive. *We own the science and we think that the world should know it, and the platforms themselves also do.* But again, it's a huge, huge challenge that I think all sectors of society need to be very active."

Fleming also stated, "Another really key strategy we had was to deploy influencers […] and they were much more trusted than the United Nations […] We trained scientists around the world and some doctors on TikTok, and we had TikTok working with us."

Moderating the "Tackling Disinformation" panel was the World Economic Forum managing director Adrian Monc. Both Ms. Fleming and Mr. Monc tied these UN and WEF information control strategies to COVID as well as "global warming," with Mr. Monc stating that there has been "professionalization of disinformation" including "COVID-19 state-sponsored actors engaged in that." What does that even mean? That somehow those of us critical of the COVID-19 policies are "state-sponsored" actors? What their statements did reveal is that there has been a group of scientists and physicians who have been trained by the UN and WEF to actively promote "The Science" concerning COVID as "owned" by the UN and WEF, and to do so on a variety of media (corporate and "news" media) channels. The terms typically used for such activities would be "controlled opposition" and "agents provocateurs." Or just plain "propaganda" and "propagandists."

Almost everyone, whether or not they have accepted an inoculation labeled as a vaccine, has been infected by one or more of the SARS-CoV-2 variants at some point. Each has their own story and experience, and each of these stories are facets of individual and collective truth that transcend all attempts by media, governments, nongovernmental organizations, pharmaceutical companies, and other stakeholders to manage and manipulate the coronavirus narrative to advance a wide range of agendas. For some, the tide of events has cost their lives or those of friends and loved ones. For others they have destroyed their businesses or livelihoods. And for a small subset, particularly those dissidents who have raised alarms about the many breaches of fundamental medical ethics, human rights, freedom of speech, clinical research, and regulatory norms and guidance, it has cost them reputations and careers. Vocal dissident medical professionals have been bombarded by withering and highly coordinated attacks in their places of employment, by their medical licensing boards, on social media, and in a bewilderingly globally coordinated array of corporatized legacy mass-media outlets.

How to begin to capture and make sense out of the breadth and depth of the global human tragedy known as COVID-19? The concentration of such immense power to control information and understanding in so few individuals and organizations is unprecedented in human history. Those in power not only promoted their story, but effectively crushed dissent, along with the medical ethics and the civil liberties norms that so many of us had taken for granted.

Humans perceive and interpret the world by comparing the information that they receive through their senses to internal models of reality. Our conscious mind does not directly know reality. It holds a model of what it believes to be true, and then compares incoming information to this model. Psychological experiments involving hypnosis have demonstrated that if our internal models of reality are shaped to deny the possibility of an existing object, we will actually not be able to "see" that which is demonstrably present in the stream of photons that our eyes detect or the audio waves that our ears hear. In other words, we can only see that which we believe exists, that which is consistent with our own personal model of reality.

The key challenge for any person who seeks to make sense out of the confusing and often mesmerizing flow of information bombarding us during the COVIDcrisis is to develop an extended internal model of the world that can help their own mind process all of this. Unless steeped in the world of biowarfare, pathogen bioengineering, psychological operations, and the "intelligence community" (as I have been), it is normal for humans to instinctively recoil from the possibility that SARS-CoV-2 is an engineered pathogen, that the COVIDcrisis could have been exploited to advance the economic and political interests of a small group of people, or that there may be those who support the concept of global depopulation or culling of "useless eaters." For most of us, such possibilities are so far from our internal models of the world (and of Judeo-Christian ethics) that we immediately, reflexively reject them.

This book is designed to help you to recognize that the coronavirus narrative that has been so actively promoted over the last three years is not the only model for understanding the present and predicting the future, but rather one of many alternative models, one that is being heavily promoted by people and organizations who have an angle and vast resources. People and organizations with a conflict of interest, one way or another. Furthermore, this book is intended to serve as a first draft of an alternative dissenting version of history, as a recitation of the lies and harms that have been inflicted on all of us, and a means to help you make sense out of the bewildering array of lived events. My hope is that it will also help us all process our collective experiences and will help us to derive lessons and identify actions that we might take to move toward a better future, informed by this global experience that we have all shared.

I believe that this sense of cognitive dissonance, of psychological pain, that often occurs when encountering facts or ideas that are different from the ones we have relied upon in the past (and have previously employed to make sense of the stream of the present) can be a signpost pointing toward an opportunity for personal growth. However, one thing that we have become acutely and very personally aware of is that there seems to be a movement in modern society to avoid information, theories, or opinions that trigger cognitive dissonance and the associated psychological pain. Often associated with terms such as "cancel culture," "virtue signaling," and "wokeism," this movement appears to have manifested as a belief system that holds that both individuals as well as the collective body politic have a fundamental right to intellectual protection, to not encounter unpleasant thoughts, information, or ideas that are inconsistent with their internal model of reality. These are the intellectual roots that nurture censorship, denialism, and the weaponized gaslighting, defamation, and slander that many have experienced, as well as the idea that anything that causes individuals to lose faith in their government constitutes domestic terrorism and should be treated as such. There is a long and rich human history of punishment by death for such dissident thought crimes. I suggest that these behaviors and actions are among the ugliest manifestations of the unpleasant tribal human tendency to reject those who are willing to speak inconvenient truths, and that this tendency has always been behind the dark reactionary aspect of common processes by which scientific and medical knowledge advance. Awareness of this phenomenon is not something just recently discovered. It extends back even before Galileo Galilei and the Roman Catholic Inquisition to at least the fourth century BC, and probably further beyond that into the mists of time.

About 2,400 years ago, the Athenian philosopher Plato (student of Socrates, mentor of Aristotle) described the Allegory of the Cave, writing while using the voice of his martyred mentor Socrates. Socrates is most famous for his powerful approach for avoiding hubris during logic-based reasoning, beginning all philosophical and logical quests for truth with the position that "The only true wisdom is in knowing you know nothing."

The setting for the Allegory of the Cave is a hypothetical dark cavern inhabited by a group of prisoners who are all bound hand and foot facing the same wall. The prisoners have been there since birth; this is the only reality that they know. Behind them is a burning fire maintained by the rulers of the cave. The rulers have different objects and puppets that they hold up so that the prisoners can see the shadows cast by the objects as they interrupt the light of the fire, and the rulers make sounds and generate echoes for the prisoners to hear. These rulers of the cave are the puppet masters, able to control the reality that the prisoners are able to experience. The prisoners accept this shadow reality and do not question it.

One day, one of the prisoners gets loose. His chains break, and in a confused state he stands for the first time, looks around, and sees the fire. Lying on the ground next to the fire he sees the puppets and objects that correspond to the shadows on the wall. In a great leap of insight, he concludes that the shadows came from these objects, and that the puppets and fire represent a greater reality than that which he had previously known. Outside the cave, he sees color, sun, and trees, and he is filled with joy.

In the hope of enlightening his friends, he returns to the cave. He explains the new reality that he has experienced, but they cannot even begin to understand what he is trying to describe. The cave is all they have ever known. They have no way of knowing that they are, in fact, imprisoned. But they do notice that he is different now, his eyes look different, and he has trouble seeing, naming, and interpreting the shadows. They laugh at him, and all agree that leaving the cave is a fool's errand. Then, they threaten to kill their brother and anyone else who dares to leave the cave, break their bonds, shatter their reality.

This ancient parable presents a dilemma that I also address in this book. For those emancipated from the confines of their old perception of reality, it is natural to hope to share observations and experiences about a new reality, despite the vast difference from the approved narrative. These people, and perhaps you are one of them, have already begun to question what they are being told by the puppet masters. For those who do not accept the official story, the first challenge is learning how to communicate something we believe is essential and vital to the health and well-being of family, friends, and the world at large. The second challenge is how to avoid being treated as a dangerous threat by everyone else still captivated by shadows on the wall.

Physicians and other medical practitioners are constantly encountering things that do not make sense. The good ones become a kind of detective, specializing in interpreting the shadows on the walls of the cave that they know best. Most of the rest become masters of naming the shadows. A very few are occasionally able to see outside the cave. But almost inevitably these few are initially rejected, defamed, and ridiculed by their peers. Yet they often persist, armed with conviction that they have seen a new reality, and the knowledge of how other dissenters who came before helped advance the common good. But it is neither easy nor pleasant to enlighten their fellow prisoners, many of whom will never accept that there is something more than the shadows to which they have become attached and familiar.

This book follows the basic process that physicians are taught to use when encountering a patient. A well-trained and experienced physician begins by trying to make sense out of what has brought the patient to seek care, a process that begins by getting the patient to speak about why they have come to the physician seeking treatment (the chief complaint), gathering information both as a history in the patient's own words as well as results from a physical examination and laboratory tests. This information is then compared to the many models of disease that the physician holds in their head (and sometimes in books or computers), and a hypothesis is developed that seeks to answer the question "What are the causes of this particular patient's complaints and symptoms?" The resulting diagnostic hypothesis may be challenged and supported by performing additional examination or tests. A treatment plan is then developed based on the working model (hypothesis) for what is causing the patient to have a complaint or what appears to be a particular disease. The treatment plan is implemented, and after a period of time the physician and patient come back together to see if the treatment has been effective or if the hypothesis needs to be modified or rejected.

In the case of the current work, we have assembled a number of personal stories that we hope will help the reader start to see underlying patterns and problems. These chapters are essentially personal histories that describe the chief complaints of different people from all over the world who have been impacted by the COVIDcrisis. Think of these as case studies, from which observations and hypotheses about the diagnosis of "what has caused us this pain" during the COVIDcrisis can be derived. Then there are essays developed during the course of these events that strive to comprehend and make sense of the events and forces that have caused these various complaints and symptoms. Finally, there are the chapters that have been most difficult for me to write, the treatment plans. The collected thoughts and ideas that, if implemented, offer hope for recovery and prevention of future global calamities akin to that which we are now (hopefully) emerging from.

These case histories illuminate only a fraction of the tragic collective human suffering we have all endured. And the treatment plans proposed are only a starting point for a broader plan. I neither pretend to have the answers, nor to understand the full "truth" of what we all have experienced. If we can achieve one thing only, it will be in helping others awaken to the possibility that the models of reality with which we have become familiar and attached just may be deleterious to our health. If, with this book, we can open your "Overton window" just a bit more, perhaps individuals like you, like me and Jill, and like the contributing authors in this volume can help create a better future for our children and grandchildren.

But don't be surprised if you find yourself wanting to avert your eyes or don a pair of sunglasses. Cognitive dissonance hurts when you first venture out of the cave and encounter the bright light of the sun.

CHAPTER 1
How I Got Red-Pilled, and the Gradual Reveal (TNI, WEF)

Who is Robert Malone? Husband, father, and grandfather for starters. Still happily married to my high school sweetheart. Carpenter, small farmer, equestrian. There have been periods in my life where I was desperately poor, and other periods when I have been comfortably middle class. Together with Jill I manage a forty-acre horse farm in the Virginia foothills of the Shenandoah Mountains. Like all of us, I do not really fit into any one category, although there have been many attempts to stereotype me by various media outlets over the last couple of years.

I am an internationally recognized scientist/physician, and the original inventor of mRNA and DNA vaccination (resulting in nine issued patents with a priority date of 1989) as well as mRNA- and DNA-based gene therapy [1–8]. I am also an inventor or early adopter of multiple nonviral DNA and RNA/mRNA platform delivery technologies. I hold numerous fundamental domestic and foreign patents in the fields of gene delivery, delivery formulations, and vaccines. I have been working in the fields of advanced clinical development and vaccinology for almost forty years. My Google Scholar ranking is 50, which is the ranking of an outstanding full professor.

In short, I have spent much of my career working on vaccine development. I have also had extensive experience in drug repurposing for infectious disease outbreaks. My contributions to science and industry are outstanding. I am proud of my contributions. My friendships and connections with professional colleagues have persisted for years.

So, when I am defamed by the *New York Times*, *Washington Post*, *The Atlantic*, or others, I know that there is more driving their character assassination attempts than efforts to report actual truth. These attacks are not about "me" personally, but rather about me speaking outside of the approved government and WHO/WEF narrative concerning COVID-19 policies. It is about me criticizing the government, the vaccine clinical trial failings, the pharmaceutical companies, the significant adverse events and their cover-up, the amazingly counterproductive pandemic public health policies and about how the government and WHO has mishandled this pandemic from the very beginning. It is about advocating early on that we have multidrug, multistage lifesaving treatments that could have saved so many lives that have been lost, treatments that are used every day in hospitals around the country for related conditions as well as for COVID. These attacks are also about me supporting the position of the Great Barrington Declaration, which basically stated that we should have focused our risk mitigation efforts on the elderly, and that the US should not have vaccinated healthy, normal children (who do not die of COVID) with an experimental vaccine. Finally, it is about the 18,000 signatories of the Global COVID Summit declaration that ratified that Declaration.

A Freedom of Information Act (FOIA) request by Blaze Media has revealed that the Department of Health & Human Services (HHS) through the CDC has spent one billion tax dollars on propaganda to push the safety and effectiveness of these vaccines and to stop "misinformation." The money was given to ABC, NBC, CBS, MSNBC, the *Washington Post*, and the *Los Angeles Times* (the mainstream media), who have not disclosed that their articles and journalists were funded by taxpayers. This campaign was a national push to improve public "trust," using fear-based articles to threaten the population, promote the safety and efficacy of the gene therapy based COVID-19 vaccines, and defame those deemed as critical of the endeavor. For instance, the *Los Angeles Times*' "experts" advised how to persuade skeptical friends and relatives to get vaccinated. Furthermore, the CDC produced a series of non-peer-reviewed articles that promoted the vaccine. They used these articles to push the narrative of "safe and effective," and to discredit the vast number of peer-reviewed journal articles demonstrating the significant adverse events associated with the SARS-CoV-2 genetic vaccines.

The Gates Foundation has also trained, employed, and given press association memberships to reporters, especially in fields of health, education, and global development, where Gates wants the most influence. He has paid more than $319 million to control the mainstream media—*The*

Atlantic, NPR, BBC, PBS—and foreign media organizations like *The Daily Telegraph, The Financial Times,* and Al Jazeera. Intelligence agencies were also used in this global campaign to eradicate antivaccine messaging. In addition, the Chan Zuckerberg Initiative paid out vast sums to magazines and journals, such as the *Atlantic Monthly,* to smear those who criticized how the government handled vaccine development and production, as well as the vaccine itself.

People sometimes ask me what has brought me to the point of daily podcasts, interviews, op-eds, advocacy with legislators, and building a twitter feed of almost a half million people (before it was deleted) and then to build a 400,000-follower GETTR feed and a Substack daily publication that has a subscriber list of over 200,000 and is read by about 500,000–700,000 people a day.

It started with my own experiences and concerns regarding the safety and bioethics of how the COVID-19 genetic vaccines were developed and forced upon the world, and then expanded as I discovered the many short-cuts, database issues, obfuscation, and, frankly, lies told in the development of the spike protein-based genetic vaccines for SARS-CoV-2. My commitment to public truth-telling was accelerated by my professional and personal experiences in identifying, developing, and trying to publish peer-reviewed academic papers focused on drug repurposing for the early treatment of COVID, advocating for the rights of physicians to practice medicine, and witnessing close colleagues encounter similar roadblocks to advancing repurposed drug treatments.

Finally, as unethical mandates for administering experimental vaccines to adults and children began to be pushed by governments, my research exposed what I believe is authoritarian control by governments in coordination with large global corporations (big finance, big pharmaceutical, big media, and big technology). This discovery influenced and then eventually transformed my worldview. As the slow reveal of the vast array of adverse events associated with these vaccines has occurred, I have been shocked by the governmental response of actively hiding and obfuscating the data. This culminated in both the shocking revelation that the CDC has been hiding the majority of data about the vaccines [9], and the further complicity of the CDC trying to stop the release of the clinical trial data as well as the postvaccination Pfizer study data from the public. Due to a FOIA request for access to these documents, the CDC went so far as to ask the courts for the papers to be sealed for fifty-five years.

I have always been taught and believed that vaccines must be developed in conjunction with lifesaving treatments for an emerging infectious disease or a pandemic. To reiterate: I am a vaccinologist. I invented the core mRNA vaccine technology platform. I have spent much of my career working on vaccine development. I have also had extensive experience in drug repurposing for infectious disease outbreaks. I am not an "antivaxxer" in any way, shape, or form. But I do believe that the shortcuts that the US Government (USG) has taken to bring the mRNA and the adenovirus vaccines to market for this pandemic have been detrimental and contrary to globally accepted standards for developing and regulating safe and effective licensed products.

I used to believe that the FDA, NIH, and CDC were working for the citizens of the United States, not Big Pharma. I thought that if we could just repurpose already known, safe drugs for emerging infectious diseases, we could quickly find ways to reduce the COVID death rate. I thought that drug and vaccine development were regulated by the federal government for the common good. What I have learned over the last two years is that regulatory capture of the federal government has warped and shaped the work of Congress and federal agencies to such an extent that they no longer represent what is in the best interests of the nation, the world, and humanity. The more I have expressed data-based concerns about what is happening with the vaccines and the USG and WHO responses, the more I have been censored, defamed, and slandered with various forms of character assassination by big tech and corporate-controlled legacy media (which, in fact, are being paid by the CDC to do so). I am not alone in being targeted. During the COVID-19 pandemic, mainstream media has attacked and censored other prominent physicians/scientists who dissent, on scientific grounds, to the approved government narrative. That narrative instructed physicians to send their newly diagnosed COVID patients home and wait until they get better or become so sick that they can't breathe and their lips have turned blue. Only then are patients allowed to go to the emergency room. Never in the history of medicine have doctors given out this type of advice. It is medical malpractice.

From there, journalists took to hunting down physicians who gave early treatment and exposed them. Once physicians are exposed in the media, medical licensing boards have been encouraged to investigate and remove medical licenses from physicians who don't comply with federal "guidance," that guidance being to let people become so sick in their homes that their chances of death are much higher even if they are hospitalized.

The harassment, censorship, and defamation have developed into a standardized process. Government agencies, hospitals, medical boards, and mass media companies have deployed this technique worldwide for suppressing physicians who are guilty of the "sin" of treating patients with lifesaving drugs in an outpatient setting. These lifesaving treatments use standard therapies and FDA licensed drugs with extensive safety data. These treatments involve common-sense solutions that physicians developed in the field by a combination of knowledge, insight, and trial and error using well-established medical practices. For this "sin," our government, hospitals, medical boards, and corporate legacy media have persecuted these medical-freedom heroes. All this has resulted in physicians bullied, licenses imperiled, and, most tragically, many lives lost due to lack of lifesaving early treatment.

What is happening "is not right, it is not proper, and it is not fair."

I'm not alone, as you will see.

CHAPTER 2
Children on the Back of a Mad Elephant

By Gavin de Becker

Gavin de Becker is considered the leading security specialist in the United States. His security and consulting firm, Gavin de Becker and Associates, protects government agencies, public figures, corporations, and universities. Through his work keeping some of the world's most prominent people safe, de Becker has gained a singular perspective on fear, threat assessment, and preparedness in the face of threats. He has earned three Presidential appointments and is a bestselling author of *The Gift of Fear*. In this essay, de Becker looks at fear in the context of the COVID situation and propaganda surrounding it.

* * * * *

I've spent a long career studying risk, danger, safety, and fear.

I've sat across the table and seen fear in the eyes of public figures who were stalked and threatened—and I've seen the same fear in the eyes of assassins, convicted murderers, soldiers, rape victims, battered women, and police officers. I've discussed fear with a president who was shot at, with another who was hit, with the widow of one who was killed, with an athlete who was stabbed at a sports event, with an iconic public figure who was attacked by an assassin, and with children who grew up surrounded by danger. The fear I've seen has worn a thousand faces, but when unmasked it is the same as yours and mine. Occasional fear and anxiety are features of human beings, just as co-opting fear and using it to advantage is a feature of some human beings.

Throughout history, fear has been used to persuade and control populations. When those in power tell us about the next enemy or danger, it's our—often shirked—responsibility to fully understand what it is we are being encouraged to fear. (An odd phrase, en*couraged* to fear.)

Not all citizens are willing to tease out what's relevant to our safety from the long menu of things we are encouraged to dread. A brief inventory of fears promoted during the past few decades tells a clear story: unidentified external enemies; identified external enemies like the Russians, the Chinese, Gaddafi, Saddam, and Bin Laden; Middle Eastern extremists, home-grown extremists, illegal immigrants and legal immigrants; communists, communism, terrorists, and terrorism; Mad Cow disease, flesh-eating disease, and killer bees; Bird Flu, the seasonal flu, Swine Flu (1976), Swine Flu (2007), AIDS, West Nile Virus, Ebola, Anthrax; and last but not least, Y2K.

And in case you haven't heard, COVID-19. And Monkeypox.

To be clear, all these things harm some people. Should any of these things ever have become the central issue of concern for every American? Depends who you ask. Ask a politician, and the answer is *"Yes please."* Ask a government, and the answer is *"Yes please"*—backed up by force. Ask the news media, and the answer is *"Yes please and keep 'em coming."* Ask me and the answer is *"No."*

In order to succeed at separating the bullets from the blanks, in order to decide which fears are warranted, which are worth investment of our energy, time, and attention, it's helpful to first understand what fear is. There are two broad categories:

1. True fear is a signal *in the presence of danger*. It is meant to be brief and unignorable. True fear is always based on something we perceive in or near our environment—something we see, hear, smell, taste, feel.
2. Unwarranted fear is based on something in our imagination or memory.

How to tell the difference: You are at the airport and suddenly feel fear about the flight you're about to board. That fear is almost certainly based upon something in your imagination or your memory, a news story about a plane crash, for example. That is unwarranted fear.

But if your fear is based on seeing the disheveled pilots stumble out of the airport bar stinking of whiskey and making their way onto the plane, that's true fear.

Those who benefit from our fears know that the most frightening place is our imagination, and they work to populate our imaginations with all variety of unfamiliar risks that only they can fully understand, only they can lessen, and sometimes only they can even see. The fears that are easiest to exploit are a bit mysterious, because barking at us about the real dangers in our lives just won't cause enough alarm:

> INJURIES AT HOME! Every week, more than a million Americans rush to hospitals due to falls, cuts, and other serious injuries. In the next week, those injuries will kill more than 3,000 of you!

All true, by the way, only not as scary as an invisible virus.

When presented with some new risk that's hard to conjure and understand, many people ask, What's the worst-case scenario? *Doctor, what's the worst-case scenario?* Death. *Officer, what's the worst-case scenario?* Murder. *Captain, what's the worst-case scenario?* Fiery crash.

A worst-case scenario is a theoretical sequence of events intentionally devised to be as bad as possible, the word scenario coming from scene, as in a play or movie. Worst-case scenarios are creative exercises, not predictions of likely events.

Most worst-case scenarios enter the stream of discussion specifically because they are unlikely, specifically because they are at the far end of possibility, and usually because the worst-case outcome is not coming. Anthony Fauci has shown this again and again during his half-century elevating fear of real and concocted viral outbreaks—HIV/AIDS in 1983, West Nile Virus in 2001/2, SARS in 2003, bird flu in 2005, swine flu in 2009, dengue in 2012, MERS in 2014, Ebola in 2014/16, Zika in 2015/16, and COVID-19 in 2020.

Even way back when he was promoting fear of AIDS, Fauci had already perfected his method of *ad-fear-tising*, using remote, unlikely, far-fetched, and improbable possibilities to frighten people. He terrified tens of millions into wrongly believing they were at personal risk of getting AIDS when they were not. Looking at just one of Fauci's old interviews (has anyone ever done more interviews?), you'll immediately recognize his special and awful style. I've highlighted the conditional language and cunning caveats that let him say almost anything about anything:

> The long incubation period of this disease we *may be* starting to see, as we're seeing *virtually*, as the months go by, other groups that *can* be involved, and seeing it in children is really quite disturbing. *If* the close contact of the child is a household contact, *perhaps* there will be a *certain number* of individuals who are just living with and in close contact with someone with AIDS *or at risk of* AIDS who *does not necessarily* have to have intimate sexual contact or share a needle, but just the ordinary close contact that one sees in normal interpersonal relationships. Now that *may be* farfetched *in a sense* that there have been no cases recognized *as yet* in which individuals have had merely casual contact, close *or albeit* with an individual with AIDS who *for example* have gotten AIDS. *For example*, there have been no cases *yet* reported of hospital personnel, who have *fairly* close contact with patients with AIDS. There have been no case reports of them getting AIDS; but the *jury is still out* on that because the situation is constantly *evolving* and the incubation period is so long, as you know. It's a *mean of about* fourteen months, *ranging from* six to eighteen months. So what medical researchers and public health service officials *will be*—are *concerned with* is what *we felt* were the confines of transmissibility now going to be *loosening up and broadening up* so that *something less than truly* intimate contact *can* give transmission of this disease.

Translated into English, those 250 rambling and tricky words can be boiled down to just twelve words of truth:

> There have been no cases of AIDS spread by ordinary close contact.

But the message people understandably took away from Fauci's fear-bomb was quite different: *You can catch this disease by less than intimate contact.*

Despite a history of untruths at the center of his pronouncements, the Fauci of today is a world-class expert at frightening the public, exaggerating the severity of contagions, and always focusing on the terrible outcomes that could, maybe, perhaps, conceivably occur, over time, at some point in the future, unless we do exactly what he tells us to do, and even, apparently, after we do exactly what he tells us to do, because after all, the situation is always evolving and the jury is still out and transmissibility can be expected to loosen up and broaden up and tick up, which all remains to be seen. Perhaps.

By design, the human mind pounces on anything that can seem relevant to survival. We're built to entertain every thought of danger that's put in front of us, to turn it over, to look at it from every angle. The more enormous a lethal danger might be and the more people it might harm, the more fascinating. But for us to be fascinated by something, it has to be made accessible to our minds. The Earth coming out of its orbit and spinning off into a collision with Jupiter is too hard for us to get our minds around, but the idea that a virus could kill us (all)—that idea has been made to appear plausible by repetition, promotion, and outright advertising.

Alarming words are dispatched by Fauci like soldiers under strict orders: Cause anxiety that cannot be ignored. Surprisingly, their deployment isn't entirely bad news. It's bad, of course, that someone wants to scare you, but warnings always mean that at least for now, the terrible outcome isn't happening to you.

Though you wouldn't know it by the reaction they frequently earn, whatever power resides in Fauci's words is derived from the fear instilled in the target (you and me). How one responds to a fear-bomb determines whether it will be an effective instrument or mere words. Thus, it is the listener and not the speaker—we and not the government—who decides how powerful the words will be.

Our social world relies on investing some words with credibility while discounting others. A belief that the city will tow the car if we leave it here encourages us to look for a parking space unencumbered by that particular threat. The disbelief when our joking spouse threatens to kill us if we are late to dinner allows us to stay in the marriage. And finally, knowing that worst-case scenarios are, at the end of the day, scenarios can help us place them in context with everything else in our lives.

I noted above that all governments in world history have used fear to persuade and control their populations. A few quick examples demonstrate that the object of fear is never as significant as the efforts to exploit the fear.

America 1917–1918: President Woodrow Wilson fervently did anything he could to create support for America to enter WWI, ironic since he had just gained reelection on the slogan "He kept us out of war." Wilson created an Orwellian police state with a robust propaganda campaign called the *Committee on Public Information*. Sound familiar?

Then the government enacted the Espionage Act of 1917 and the Sedition Law of 1918, leading to citizens spying on their neighbors, students reporting on teachers, and organizations—such as the American Protective League—pledging to defend their country from undesirable citizens. Sound familiar?

During the Wilson administration...

- Journalists and others were imprisoned for speaking out
- Newspapers and periodicals were shut down
- 1,500 citizens were arrested for opposing the war
- Others were lynched by vigilante mobs

People understandably feared losing their jobs, being ostracized, being arrested—and similar campaigns have followed through the generations:

- If you don't speak out against communism, you must be a communist
- If you don't speak out against racism, you must be a racist
- If you don't support a war, you must be a traitor
- If you don't support the fight against terror, you must be a terrorist sympathizer
- If you don't support mass vaccination, you must be against public health

Soviet Union 1930: Stalin instituted a series of purges against church leaders, ordinary citizens, and even his own his military officers. The secret police and their network of informants created a crippling climate of fear that enabled Stalin to gain complete control over "truth." Sound familiar?

The best defense against being the next person arrested was to inform on someone else, with the result being that 20 million Russians were sent to the Gulag. At least half died there. An interesting example of just how intense the fear became: members of the audience at one of Stalin's speeches were so scared to be the first to stop clapping that the applause went on for more than ten minutes. The manager of a paper factory was the first person to sit down, and that night he was arrested and sentenced to years in prison. Eventually, a light was installed at Stalin's speeches; when it flashed, everyone could stop clapping.

Rome 390 BC: Gallic tribes marched from Gaul (France) over the Alps and sacked Rome. For hundreds of years thereafter, Romans were constantly reminded that the enemy could invade at any moment.

Eventually, Caesar conquered Gaul (52 BC), bragging in his commentaries that of the three million people there, he killed a million and enslaved another million. With Gaul conquered, Caesar needed a new enemy to induce fear. He delivered—by turning his army on Rome itself, eventually crossing the Rubicon River and defeating Pompeii in a civil war. After that, Caesar declared himself dictator for life.

Other examples include exploiting the fear of a slave uprising (Spartacus), the post-911 fears used to expand government control (DHS, TSA, the Patriot Act, legal torture, etc.), today's fear of the virus, then the unvaccinated, then the variant of the virus, then the next variant, then Monkeypox. These last few were used to force social distancing, face coverings, vaccine mandates, restrictions on visiting relatives in hospitals, business closures, church closures, school closures, censorship of doctors and scientists who favored early treatment or opposed mass vaccination, medical board investigations of doctors, delicensing doctors, firing doctors, competing media companies joining together to support government positions, travel restrictions, vaccine passports, mass firings (34,000 healthcare workers in New York alone), expanded travel requirements, and whatever else is coming.

Was all this done to address the virus, or is all this the most recent incarnation of what powerful governments have always done?

A note on censorship: while it is nothing new, we have not in our lifetimes seen this level of censorship in America. It reminds me of a little-known piece of history: King Charles banned coffeehouses in Britain because they became centers for spirited political discussion and sharing news and ideas the King didn't want expressed. In his own words, "by occasion of the meetings of such persons therein, diverse False, Malicious and Scandalous Reports are devised and spread abroad, to the Defamation of His Majesty's Government."

In other words, *misinformation.*

King Charles ordered local officials to deny licenses to businesses that sold "Coffee, Chocolet, Sherbett or Tea, as they will answer the contrary at their utmost perils." During the early COVID lockdowns, our coffeehouses weren't prohibited just from selling coffee and chocolate; they weren't allowed to sell *anything.* Plus parks and even beaches were closed, visited by citizens "at their utmost perils."

Ultimately, God didn't save the King, and he soon allowed coffeehouses to sell coffee again. Similarly, America allowed businesses to open again, once it was clearly established that elected and unelected officials at every level of government could do whatever they wanted to do.

Though it began as a mysterious disease we were told could kill any of us, we've learned much more since we first heard the word COVID. Unfortunately, many people are stuck on the first story: over 60 years old = Death.

Politicians and media and government encourage us to go to war with death, but it's good to remember that life is a sexually transmitted, always fatal condition. We don't want to live encamped in a thousand precautions, ever-mindful of the newest frightening study and the latest emergency-concocted drug, ever-alert to a thousand unlikely risks as if the alertness would make any difference whatsoever to death. With a billion dollars of marketing, COVID became conflated with death, though they are not the same thing. Not even close.

Let's quickly put COVID-19 into perspective, using information my firm reported to our clients in 2020, at the very start of the pandemic:

ASSESSMENT: RISK OF DEATH FROM COVID-19

The average age of death attributed to COVID is 79.5 years old (later moved to 81 years old).

Even among hospitalized COVID patients who are 90-years and older, *nearly 90% have survived.*

Different hospitals, states, cities, and jurisdictions gather and report statistics differently, and because the interpretation of statistics is fertile (play)ground for politicians, we also assessed data from overseas, and data from various US states.

Massachusetts, for example, counted people "*who have tested positive and who have died.*" It's a nuanced and intelligent phrase that doesn't automatically assume every person who died *with* COVID died *from* COVID.

We also reviewed daily reports from Italy's National Institute of Health and learned that *almost 100% of the patients whose deaths were attributed to COVID were already struggling with chronic fatal illness, in most cases between two and three other fatal conditions.* (It took more than a year before CDC finally acknowledged that patients in the US whose deaths were attributed to COVID had also been diagnosed with, on average, 2.6 other fatal diseases, now 3.7 other fatal diseases.)

The ISS Italian National Health Institute of March 17, 2022 shows that more than 99% of those whose deaths were attributed to COVID were already sick:

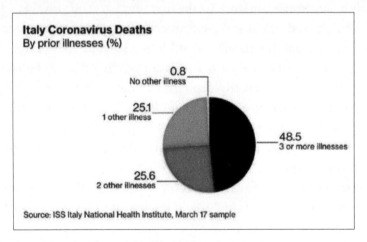

Italy Coronavirus Deaths
By prior illnesses (%)

0.8
No other illness

25.1
1 other illness

48.5
3 or more illnesses

25.6
2 other illnesses

Source: ISS Italy National Health Institute, March 17 sample

South Korea reported that as many as *99% of active cases in the general population did not require any medical treatment,* dramatically affirmed in this Reuters story [11]:

"In four U.S. state prisons, nearly 3,300 inmates test positive for coronavirus—*96% without symptoms.*"

Worth repeating: *96% without symptoms.*

Consider that the third leading cause of death in America is Medical Error (e.g., too much of that drug, or too little or too late of this drug, or the wrong drug altogether, or pressure too strong on a ventilator). A Johns Hopkins study concluded that "more than 250,000 people in the United States die every year because of medical mistakes, making it the third leading cause of death after heart disease and cancer." Other studies have placed this number at 450,000 per year.

In normal times, medical error would account for more than 5000 deaths *per week.* But since early 2020, medical error hasn't killed a single person, if we rely upon news stories and the CDC. It's all COVID all the time. In normal times, medical errors were the bane of hospital legal departments, but not during the pandemic. Anomalous deaths of people who had at some point tested positive for COVID were not scrutinized, investigated, debated, or litigated over—and there were no autopsies. Everything was automatically a COVID death. And even if we were to accept every single one of those as having been caused by COVID, the stats would remain:

Almost all of the patients whose deaths were attributed to COVID-19 were elderly people already struggling with more than two fatal illnesses.

The majority of those whose deaths were attributed to COVID were people living (dying) *in nursing homes*, as was the case for more than 70% of all deaths attributed to COVID in Canada, for example. So, if you are very old, or very sick, or already living in a nursing home, COVID might be quite serious for you—just like every health challenge for people in that situation can be quite serious (a cold, a fall, a flu, being startled, choking, etc.).

But if you are not in those two categories, and you aren't living in a nursing home, here are the hoops you'd have to jump through to die:

1.   Already be very unhealthy
2.   Get the virus

3.   Have any symptoms at all (most people don't)
4.   Feel sick but avoid medical care (maybe deterred by news stories)
5.   Finally feel sick enough to go to the hospital
6.   Be admitted to the hospital (only 10% of symptomatic people are admitted, so you have a 90% chance of being sent home)
7.   End up in critical care (only 12% of hospitalized patients do, meaning you have an 88% chance of not ending up in critical care)

Now, imagine a person 55 to 60 years old lost on all those odds and ended up in the hospit*al*. He still has a *99.4% likelihood of surviving.* And this was in the early months when there was less known about treatment, and when ventilators were still widely [mis]used.

If we could find stocks or a game of chance with more than 99% chance of success, we'd all jump at it.

So that was the situation my firm reported to our clients in early 2020. But that was not the story our government and news media told us, was it?

Of course not, because the fear citizens feel is immensely valuable—in dollars, in policy, control, and power.

To be clear, I do not oppose considering risk and creating strategies for reducing risk. What I oppose is wasting time and energy, and everything to which we give energy takes energy away from something else. Accordingly, we are wisest to put our resources where they'll be most likely to return some benefit.

You already live your life according to that equation, deciding where to put your cautionary resources at home, for example. Though intruders could land a helicopter on your roof and core through the ceiling, you've decided that entry via the front door is more likely.

If there's an emergency phone list in your home, the names and numbers reflect your family's assessment of likely hazards. The phone number for the US Nuclear Emergency Search Team is not on your refrigerator door. You also have a list in your head of things you want to avoid or prepare for. You base the list on experience, logic, new information, and intuition. That list has limits—because it has to.

Conversely, worst-case scenarios promoted by governments have no limits. Wherever their imaginations can travel, your mind can take you there. But the trip is voluntary. Even when news media and Pharma and big corporations are urging you on, even when your own government is urging you to take that trip ("for your own good"), you don't have to.

When everybody is discussing something, it's easy to assume the thing is likely to happen, but that's not true. What's true is that reality is warped when news media, politicians, pharma, and governments are all closely aligned, as is the case today with COVID, and whatever viruses follow. Simply put, the national dialogue being promoted now does not match reality. And this has happened before, or more accurately, it's never not happened.

In 1997, then-Secretary of Defense William Cohen appeared on ABC News and held up a five-pound bag of sugar, threatening that "This amount of Anthrax could be spread over a city—let's say the size of Washington. It would destroy at least half the population of that city. If you had even more amounts—" Let me interrupt Mr. Cohen for a moment and recall that he also said, "One small particle of Anthrax would produce death within five days." With that kind of inaccuracy and exaggeration, every anthrax scenario promoted by government involved the death of hundreds of thousands or even "millions," as Cohen was intoning when interviewer Cokie Roberts actually said to him, "Would you put that bag down please."

We never heard an anthrax scenario that went like this:

> Somebody will put anthrax spores in envelopes and send them to companies in a few East Coast cities. About 100 people will be exposed to the bacteria, 30 will get sick, and be successfully treated. Five others will die. Hundreds of times fewer Americans will die from this anthrax attack than from insect stings in the same period.

The reality of anthrax ended up looking like the paragraph above, and nothing like the scenarios that were promoted by government officials.

To be clear, I'm not saying that bad things don't happen. I am deeply involved every day in managing —and imagining—bad things that happen or might happen. Rather, I'm saying that the popular worst-case scenarios are just that: popular—and they remain popular as long as news media companies promote them. If a terrible thing actually happens, it moves from our imaginations to our reality, moves from a scary possibility to something we can assess and manage. So far, none of the much-discussed catastrophes has wiped us all out, though each imagined catastrophe was used to erode more of our freedom.

Today, we are challenged to accurately decide which is the bigger risk: Is it COVID, or is the new unchecked power gathered up in the name of COVID?

Whether you feel governments are sincerely trying to protect you from COVID, or you feel they're using fear to gather up new powers, they now have control technologies that every despot in history would have envied.

* * *

Way back in 1918, Randolph Bourne famously wrote, "War is the health of the State." In his unfinished letter to the American people, he expressed concern about the State's sudden acquisition of greater power and undue control of individuals. It used to be that in times of peace, "the sense of the State almost fades out of the consciousness of men," but unfortunately, that is no longer the case. Since 2020, we have had to engage with the State a hundred times a day, as we presented a government card to get into a restaurant, school, or airplane; when we went outside, when we failed to wear a face covering, when we had relatives over for the holidays (often violating some emergency order to do so), when we traveled, when we visited loved ones in a hospital, and more often when we didn't visit loved ones in a hospital. You get the idea.

And Bourne got the idea:

"Every individual citizen who in peace times had no function to perform by which he could imagine himself an expression or living fragment of the State becomes an active amateur agent . . . reporting spies and disloyalists . . . propagating such measures as are considered necessary by officialdom." Sounds familiar.

Bourne described "irresistible forces for uniformity, for passionate cooperation with the government in coercing into obedience the minority groups and individuals which lack the larger herd sense." Sounds familiar.

"By an ingenious mixture of cajolery, agitation, intimidation, the herd is licked into shape, into an effective mechanical unity . . . under a most indescribable confusion of democratic pride and personal fear they submit to the destruction of their livelihood if not their lives, in a way that would formerly have seemed to them so obnoxious as to be incredible." I wish that didn't sound familiar, but it does.

Bourne's most famous quote can be paraphrased to better fit the present moment: "Disease is the health of the State." After some time, it will be war again, and then something else. Pessimistic, I know, and also realistic.

Speaking of pessimism, Bourne wrote, "It is difficult to see how the child on the back of a mad elephant is to be any more effective in stopping the beast than is the child who tries to stop him from the ground."

Today, more and more people are recognizing reality and trying to stop the mad elephant, so many that there might soon be good reason for optimism. Maybe we're there already and just don't know it yet. I hope so.

CHAPTER 3

The Extraordinary Story of a Truth Warrior Persecuted for Advocating and Providing Lifesaving Treatments

By Meryl Nass

I first met Meryl when she came to our farm to work on a strategic plan for Children's Health Defense in the summer of 2021. We quickly became friends, as we both enjoy conversing intensely on bioethics and vaccines. Meryl has a quick wit, with an intellectual curiosity that cuts through even the toughest of defenses. Her boundless enthusiasm for both her patients and for the practice of medicine is combined with bravery that few can match. This made her one of the few physicians willing to take on the establishment regarding early treatment for COVID-19 and vaccines. She's always willing to engage in a conversation about science or medicine, and she does not back down to authority. She has most definitely been radicalized by her experiences with the US government, and her writing below reflects that. I admire her ability to get to the heart of any problem and to say things as she sees them. She has a long history of confronting the issues and, most important, put her own career on the line for what is right.

Meryl has been persecuted by her medical board for her stance on giving lifesaving treatments to her patients, and for being willing to talk about those treatments to the press. For her "crimes" of saving lives, her medical board suspended her license to practice medicine in the state of Maine. This, despite the fact that no patient actually filed a complaint against her. Furthermore, they required that she undergo a neuropsychological exam, with a psychologist of the board's choosing, before she could have a hearing. I do not know of any state medical board in the USA behaving in this manner before. It is truly draconian and most likely illegal. Meryl is taking the fight to the courts. As far as I am concerned, she is a true medical-freedom warrior.

Meryl has allowed me to modify and print her essay on the suppression of early treatments for COVID-19.

* * * * *

The Extraordinary Story of How Patient Access to COVID Treatments Was Denied, Eventually Involving Witch Hunts of Physicians Who Dared to Treat Patients

By Dr. Meryl Nass

I have an unusual professional background. My day job is as an internal medicine physician. But I also have a strong background in biological warfare. I am the first person to have proven that an epidemic (actually an epizootic, in which people catch a disease from animals) was due to biological warfare, way back in 1992. I did this by examining every aspect of the outbreak and showing that none of them conformed with what would be expected from a natural event. This happened during the Rhodesian civil war, and it was a form of low intensity warfare. Anthrax was spread to kill cattle mostly. It was designed to impoverish and starve the black population, who provided support to a guerrilla movement. There was actually an official "food control" program being carried out by the Rhodesian white minority during the guerrilla war.

I am a really good problem solver. In 1993, Cuba was suffering from an epidemic of blindness and peripheral neuropathy. Asked to investigate, I discovered (as did a few others) that the illness was due to cyanide, coupled with nutritional deficiencies that inhibited the body's natural processes for detoxifying the cyanide.

I also have a compulsive streak regarding "First do no harm." When Defense Secretary Cohen announced in November 1997 that all members of the armed services would be receiving anthrax vaccinations, my ears perked up. I knew the vaccine had never been shown to actually work for inhalation anthrax. I also knew there had been a congressional hearing in which it had been suggested that soldiers who received anthrax vaccinations were more likely to develop Gulf War syndrome than those who had not. So, I wrote a very short paper about this, finishing it in half a day, for an email mailing list I was on. Unexpectedly the paper went viral. I was soon recognized as an expert on anthrax and anthrax vaccine (basically because it was such an arcane area that almost no one else knew anything about it).

I really hadn't expected it, but the anthrax vaccine started causing grievous injuries in a considerable minority of those vaccinated. I was contacted by thousands of ill soldiers. I wound up helping to lead a coalition of service members and their families trying to stop anthrax vaccine mandates. There were a dozen congressional hearings that looked into the vaccine and the vaccination program. It almost was cancelled, but when the anthrax letters appeared, the military anthrax vaccine program roared back to life. This gave me a profound experience in how the system of government works, and how federal agencies knowingly create fraudulent scientific studies to fulfill their "mission." The same shaming and punishing of vaccine refusers went on then. Even though almost everyone in the military knew how bad the anthrax vaccines were, giving in and getting jabbed became a biological loyalty oath. You had to take it or be docked a month's pay, be given extra duties, or even be court-martialed. Some soldiers were held down and vaccinated. Nothing was allowed to stop the program, even though the vaccine wasn't safe and probably didn't work.

My colleagues (some of the most amazing people on this planet) organized a dream team of attorneys, conducted immense research, worked closely with members of Congress, and eventually brought suit against the vaccine. In 2004 we won! DC District Court Judge Emmett Sullivan threw out the anthrax vaccine license because the vaccine had never met the FDA requirements and had skated through an FDA review, probably as a special favor the FDA gave to DOD.

I learned then that DOD did not care about Congress, public opinion, or bad press—and they tried to ignore Judge Sullivan too. Almost as soon as he pulled the vaccine's license, DOD slapped an Emergency Use Authorization on the vaccine. And DOD attempted to mandate anthrax vaccinations again. Our team went back to court, and Judge Sullivan told the Defense Department, in no uncertain terms, that while soldiers may risk their lives fighting for the US, they could not be forced to risk their lives as guinea pigs for an experimental vaccine.

While there were more shenanigans to come, I learned an important lesson: it is possible to finally end a grievous injustice in the courts. I also learned the win might not last. You see, the government has an army of lawyers and have unlimited funds. They will fight forever if necessary. While you, on the other hand, are spending tons of time and money to try and prosecute a case. Justice can be achieved sometimes, but the costs are high, and victory may be brief.

In 2005, FDA rubberstamped the anthrax vaccine license. There was still no evidence of whether it worked, and plenty of evidence it was not safe. No matter. The courts, when we appealed said FDA had deference. What that meant is even if the vaccine falls far short of FDA's standards for licensure, it doesn't matter. FDA doesn't have to obey its own rules. You do; it doesn't.

After that I investigated the 2009 swine flu pandemic and the vaccines rushed out for it. I learned that pandemics are like wars: when there are a lot of experimental drugs and vaccines or vaccine components sitting on a shelf, you grab the opportunity to try them out when there is an emergency. This happened during the Gulf War. Swine Flu. Ebola. Zika.

You see, it is very expensive to test a new drug or vaccine in a randomized clinical trial. It generally costs thousands of dollars per human subject. You have to test the product in animals first, you need 3 human trial phases, and the entire process takes many years.

Not so in an emergency. Patients become free human subjects. Regulation gets condensed to almost nothing. Billionaires are minted.

First Came the Chloroquine and Hydroxychloroquine Suppression

Then came the "Novel Coronavirus," now named SARS-CoV-2 or COVID-19, in 2020. As usual, I tried to find solutions. I discovered that the chloroquine drugs had been tested against SARS and MERS (successfully!) and the history behind the logic for using these cheap, re-purposed drugs is compelling.

The drug looked very promising for both prevention and treatment of the first SARS virus. Chloroquine is an interesting drug; it has been used for many decades to both prevent and to treat malaria. It is used as an anti-inflammatory against rheumatoid arthritis, it is used as an anti-parasitic by changing the body's pH for malaria and other parasitic diseases and it has antiviral properties. There appear to be multiple mechanisms of action by which Chloroquine acts as an anti-viral, and one of the leading ones is that Chloroquine increases the pH of the lysosomes and the late endosome (endosome uptake being the way that many viruses enter and infect cells), causing the impaired release of viruses from the lysosome or the endosome. This makes the virus unable to release its genetic material into the cell and replicate. Chloroquine also acts as a zinc ionophore that allows extracellular zinc to enter the cell and inhibit viral RNA-dependent RNA polymerase [12]. So, it is no surprise that this drug would be considered a viable anti-viral treatment against beta-coronaviruses, such as SARS-CoV-2.

In 2005, five CDC (US government) scientists published a paper, along with three Canadian government scientists in the Journal of Virology, showing that chloroquine was an effective drug against SARS coronaviruses [13]. The CDC paper is entitled *"Chloroquine is a potent inhibitor of SARS coronavirus infection and spread"* and concludes with the following quote: "chloroquine has strong antiviral effects on SARS-CoV infection... suggesting both prophylactic and therapeutic advantage." A similar study was conducted in 2004 by a group of European scientists [14].

In 2014, scientists working at the National Institute of Allergy and Infectious Diseases (NIAID), showed the same results. Not only did chloroquine work *in vitro* against the MERS coronavirus, but *dozens of existing drugs*, which could have been tested in patients as soon as the pandemic started, were also effective against SARS and MERS coronaviruses. The study was published in the journal "Antimicrobial Agents and Chemotherapy" and was called "Repurposing of Clinically Developed Drugs for Treatment of Middle East Respiratory Syndrome Coronavirus Infection." [15]. The NIAID authors wrote:

> Here we found that *66 of the screened drugs were effective* at inhibiting either MERS-CoV or SARS-CoV infection in vitro and that 27 of these compounds were effective against both MERS-CoV

and SARS-CoV. These data demonstrate the efficiency of screening approved or clinically developed drugs for identification of potential therapeutic options for emerging viral diseases, and also *provide an expedited approach for supporting off-label use of approved therapeutics.*

Just in case you think these papers were flukes, two unrelated groups of European scientists found essentially the same thing. The 2014 European paper entitled "Screening of an FDA-approved compound library identifies four small-molecule inhibitors of Middle East respiratory syndrome coronavirus replication in cell culture" was published back-to-back with the NIAID paper above [16].

I have to repeat myself, because the information is so shocking and I don't want you to miss it: our governments already knew of options for treating COVID before it appeared, but instead of immediately trying these already identified, safe, cheap, and available repurposed drugs, and offering early treatments, they did everything they could to stop people obtaining the chloroquine drugs. I have written two reports on this topic, one is called "WHO and UK trials use potentially lethal hydroxychloroquine dose—according to WHO consultant" [17] and the other is "Even worse than 'Recovery,' potentially lethal hydroxychloroquine study in patients near death" [18]. They are about how patients were administered borderline lethal doses of hydroxychloroquine to give the drug a black eye. In my opinion, these are medical crimes against humanity. Yet this has never been investigated by mainstream media, bioethicists or regulatory or licensing agencies.

In 2020, I compiled a list of over 50 ways authorities and pharma companies in multiple countries stopped the use of the chloroquine drugs for COVID [19]. This was (and is) a stunning collection, which has been widely read and reproduced on many websites. It is astounding to learn that all the US (and many international) public health agencies took many different actions to *increase* deaths and destruction from COVID and prolong the pandemic by suppressing information on life saving early treatments. Taking hydroxychloroquine for COVID was equated to drinking bleach, and "avoiding the Trump drug" served as a great cover story. But here's the kicker: the authorities knew all about chloroquine and other effective treatments for COVID *before* there was a COVID19 [20], as well as early data showing efficacy against COVID in 2020 [21]. Chloroquine was first used as an effective anti-viral against HIV and its anti-viral properties are well documented in the peer reviewed literature. This is because they had figured it out for the 2003 SARS epidemic and the 2012 MERS epidemic, both caused by related coronaviruses [22], but as documented below, it was hushed it up. This has to be investigated and justice obtained, to prevent such crimes from happening to patients ever again.

The "Why?" and "How could this be?" requires people to take a huge leap in order to understand the world we live in. Many don't have the fortitude to dissect their world view and rebuild it in accord with the facts that have spilled out over the last two years. But I am about to present some more facts that I hope you can assimilate into your understanding of the world. It might require a stiff drink, or perhaps some chocolate. Whatever it takes, read on, as it might save your life or someone else's.

Ivermectin

Ivermectin had not been identified in the studies I mentioned above as a potentially useful coronavirus drug. But some people knew it was likely to work in early 2020. For instance, the French MedInCell company, supported by Bill Gates, was working on an injectable version (which would make it patentable) of Ivermectin for COVID, issuing a press release about this on April 6, 2020 and an informational paper on April 23, 2020.

> On March 29, 2020, researchers from Monash University in Melbourne, Australia published results from a laboratory cell study showing that Ivermectin can kill the coronavirus in less than 48 hours.6 Studies have been carried out by research institutes for the past few months to assess the effectiveness of treatment using Ivermectin on hospitalized patients with Covid-19. MedinCell published last January data showing that Ivermectin can be formulated with our BEPO® technology as long acting for varying doses and durations of up to several months [23].

There was a brief run on the veterinary drug at this time in the US, according to an FDA warning issued on April 10, 2020, indicating some people knew it might be an effective COVID treatment and were acquiring veterinary versions [24]. But there was not a lot of buzz and sales did not take off at that time. Here is what FDA said on April 10, 2020:

> FDA is concerned about the health of consumers who may self-medicate by taking Ivermectin products intended for animals, thinking they can be a substitute for Ivermectin intended for humans . . . Please help us protect public health by alerting FDA of anyone claiming to have a product to prevent or cure COVID-19 and to help safeguard human and animal health by reporting any of these products.

In December 2020, a full eight months later, Ron Johnson held a Senate hearing that was focused on Ivermectin's benefits for COVID. Intensive care specialist Dr. Pierre Kory, originally a New Yorker, gave a particularly compelling speech. People began paying attention to the drug. YouTube then removed Kory's speech, censoring a Senate hearing!

I think the authorities were initially scared to repeat the same tricks with Ivermectin that they had used to beat down the chloroquine drugs. And because Ivermectin has efficacy in both late early stages in the disease and is not toxic at several times the normal dose, some of the tricks used against chloroquine (giving it too late in the disease course or overdosing patients) simply would not work with Ivermectin.

But then Ivermectin's popularity started exploding. CDC published a report in late August showing that Ivermectin prescriptions had quadrupled in a month, and the drug was now selling at 25 times the pre-COVID rate [25].

An article in *Business Insider* exclaims: "More than 88,000 prescriptions for the drug were filled by pharmacies in the week ending August 13, the CDC said in a report published August 26 [26]." Apparently, the prescription sales of Ivermectin terrified the powers-that-be. What if the pandemic got wiped out with Ivermectin? Would that be the end of vaccine mandates, boosters, vaccine passports, and digital IDs? The end of the Great Reset? Something had to be done, and fast. It had to be big. It had to be effective. They couldn't simply take the drug off the market; that would require a long process and a paper trail. What to do? There was probably only one option:

Scare the pants off the doctors. Loss of license is the very worst thing you can do to a doctor. Threaten their licenses and they will immediately fall into line. You can't get a prescription if there is no doctor to write it.

This method of going after the licenses of physicians who prescribe Ivermectin has already been tried and tested in the Philippines with great success [27].

The powers-that-be could also scare the pharmacies at the same time. This required stealth and cunning—there couldn't be a paper trail. Intimidation was required, backed by a one-two punch. They would actually be suspending doctors' (and maybe pharmacists') licenses. They could couple that with a huge media offensive, and threats from an industry of medical "nonprofits." You suddenly invent "misinformation" as a medical crime, studiously failing to define it. You make people think the legal prescribing of Ivermectin and hydroxychloroquine is a crime, even though off-label prescribing is entirely legal and customary under the federal Food, Drug and Cosmetic Act. Did Fauci give the order? Walensky? Acting FDA Commissioner Woodcock? It was probably some combination, plus the public relations professionals managing the messaging and the media.

This all seems so implausible. Yet here we are. This is actually what happened.

1. Senator Ben Ray Lujan (D, NM) and several other Senators introduced the "Health Misinformation Act" in July 2021 because "misinformation was putting lives at risk," he said [28]. A huge supporter of COVID vaccinations, the 49-year-old Senator suffered a stroke on February 1, 2022.
2. The pharmacies suddenly could not get Ivermectin from their wholesalers. No reason was given except 'supply and demand.' But it seemed the supply was cut off everywhere. Ivermectin was dribbled out by the wholesalers, a few pills a week per pharmacy, not enough to supply even one prescription weekly. Some powerful entity presumably ordered the wholesalers to make the drug (practically) unavailable. With no shortages announced. I called the main manufacturer in the US, Edenbridge, and was told they were producing plenty. (Editor's note: I was a personal friend of the CEO of Edenbridge, and he had also informed me that there was plenty of supply).

Hydroxychloroquine had been restricted in a variety of ways, determined by each state, since early 2020. It had also been restricted by certain manufacturers and pharmacy chains in 2020. Suddenly, in September 2021, it too became considerably harder than it already was to obtain.

3. In late August, CDC sent out a message on its emergency network about Ivermectin, but the urgent warning contained only 2 examples of anyone having a problem with the drug: one person overdosed on an animal version and one overdosed on Ivermectin bought on the internet [25]. This should not have been news. However, pharmacists and doctors read between the lines and knew this was code for "verboten." Almost all stopped dispensing Ivermectin at that time. It should be of interest to everyone that our health agencies now speak in coded messages to doctors and pharmacies, presumably to avoid putting their threats on paper and being accountable for them. What a way for government to do business.

4. Also in August 2021, various "nonprofit" medical organizations started issuing warnings, in concert, regarding doctors prescribing Ivermectin or hydroxychloroquine, and spreading misinformation, especially about COVID vaccines. These organizations included the Federation of State Medical Boards, the American Medical Association, the American Pharmacy Association, and several specialty Boards. Here is an example of the AMA's language:

Spreading Falsehoods

The COVID-19 pandemic continues to spawn falsehoods that are spread by a whole host of people such as political leaders, media figures, internet influencers, and even some health professionals—including by licensed physicians.

The words and actions of this last group may well be the most egregious of all because they undermine the trust at the center of the patient-physician relationship, and because they are directly responsible for people's health. A handful of doctors spreading disinformation have fostered belief in scientifically unvalidated and potentially dangerous "cures" for COVID-19 while increasing vaccine hesitancy and driving the politicization of the pandemic to new heights, threatening the public health countermeasures taken to end it [29].

These organizations have told doctors they could lose their licenses or board certifications for such "crimes." Mind you, none of these so-called nonprofit organizations has any regulatory authority. Nor do I believe they have any authority to claw back a Board Certification. They were blowing smoke. And they were probably paid to do so. Who paid?

5. Over the course of 3 days at the end of August 2021, national media reported on 4 doctors in 3 states whose Boards were investigating them for the use of Ivermectin. In Hawaii, the board *really* wanted to make an example by going after the state's chief medical officer, who had had the courage to treat COVID patients.

6. The Federation of State Medical Boards (FSMB) is an organization that assists 71 state and territorial medical boards with policies, training, etc. Members pay dues and the organization accepts donations. It has its own foundation, too. Its President earns close to $1,000,000/year, not bad for a backwater administrative job at an organization headquartered in Euless, Texas. After the FSMB instructed its members that misinformation was a crime, somewhere between 8 and 15 of its member boards began to take action. (Media have reported that 8, 12 or 15 boards of its 71 member Boards did so, according to the FSMB, which is closely monitoring the results of its calumny.)

7. On February 7, 2022 the Department of Homeland Security issued its own dire warning about the spread of misinformation, disinformation and a neologism, malinformation [30].

> The United States remains in a heightened threat environment fueled by several factors, including an online environment filled with false or misleading narratives and conspiracy theories, and other forms of mis- dis- and mal-information (MDM) introduced and/or amplified by foreign and domestic threat actors. These threat actors seek to exacerbate societal friction to sow discord and undermine public trust in government institutions to encourage unrest, which could potentially inspire acts of violence. Mass casualty attacks and other acts of targeted violence conducted by lone offenders and small groups acting in furtherance of ideological beliefs and/or personal grievances pose an ongoing threat to the nation.

Thus, it appears that Misinformation and Disinformation have been selected to play an important role in a newly developing narrative, as the Pandemic restrictions and older narrative comes to an end.

8. I presume the majority of the 71 Medical Boards' attorneys knew something about the Constitution, knew that every American has an inalienable right to freedom of speech, and simply ignored the FSMB's exhortation to go after misinformation spreaders. The Maine Board, however, went along. Three doctors in Maine have recently had their licenses suspended or threatened for writing waivers for COVID vaccines, spreading misinformation, and/or prescribing Ivermectin and hydroxychloroquine. (All three of which are legal activities for doctors.) But Boards have broad powers to intervene in the practice of medicine, and their members are shielded from liability as agents of the state. And so they went after a chronic Lyme doctor several years ago, who found, as expected, that it would be too onerous to fight back, and he gave up his license.

Finally, this is what my state (Maine) medical licensing board claims about me:

> The board noted that Ivermectin isn't Food and Drug Administration "authorized or approved" as a treatment for COVID-19 in the suspension order.
>
> The board said that her continuing to practice as a physician "constitutes an immediate jeopardy to the health and physical safety of the public who might receive her medical services, and that it is necessary to immediately suspend her ability to practice medicine in order to adequately respond to this risk.

I am 71 years old, and my medical practice was set up as a service to provide care during the pandemic that was otherwise very hard to get, so everyone could access COVID drugs who wanted them. My fee was $60 per patient for all the COVID care they needed.

I am sure the Board had calculated that given all the above, I would not challenge the Board's suspension and would simply surrender my license, since it would probably cost hundreds of thousands of dollars to fight the Board's actions in court.

However, I was surprised to find that on the day my license was suspended, there was massive national publicity about my case. The story was on the AP wire, covered from the *San Francisco Chronicle* to the *Miami Herald*. And for some reason, it was not behind the usual paywall. *The Hill*, *Newsweek*, the *Daily Beast*, and many other publications all ran hit pieces about me.

I gathered that my situation was bigger than just a Maine renegade Medical Board: I had been selected to serve as an example to physicians nationwide who might be prescribing early treatment for COVID.

Once I realized I was to be made an example of, to assist with a national fear campaign followed by a purge of doctors who think independently, I decided to fight back. Fortunately, Children's Health Defense is helping with my legal expenses, which is what allows me to mount a strong attack against the bulldozing of free speech, patient autonomy and the doctor-patient relationship. There is a lot riding on the outcome, and we have only just begun to fight for medical freedom and the physician's right to practice medicine.

CHAPTER 4
The Anatomy of a Career-Ending Sham Peer Review

By Paul Marik

My father died of sepsis in 2018. I was at his bedside, and despite being a physician, I had no idea how little could be done to treat sepsis and was truly appalled at the lack of treatments available. For some statistics, sepsis accounts for 20 percent of all deaths globally, almost 270,000 Americans die of sepsis each year, and one in three hospital patient deaths in the USA is from sepsis. Furthermore, there is evidence that once a patient has had sepsis, their immune system becomes more compromised. Sepsis may be the most underreported acute health issue of our time.

When one reads pharma-sponsored webpages regarding sepsis, they typically state with a ferocious confidence that scientists have not yet developed a medicine that specifically targets the aggressive immune response seen with sepsis, despite a vast research effort to do so. Fortunately, this is not the case—it is just that the treatments are low-cost and do not bring dollars to the pharmaceutical industry, as the therapies are off-patent (such as corticosteroids) or involve nonpharmaceutical treatments.

This brings me to my first experience with Dr. Marik. As I was frantically searching around for anything that could save my father's life, my wife brought to my attention the work of Dr. Paul Marik in Norfolk, Virginia. Dr. Marik had pioneered the use of vitamin C to treat sepsis and when used early and at appropriate doses can be a lifesaver [31]. His work published in a number of newspaper articles, and we quickly began synthesizing information on his lifesaving protocols to bring to the attention of my father's physicians. Unfortunately, my father passed shortly thereafter. But I have always remembered the brilliant detective work of Dr. Marik.

So, when I first met Dr. Marik in Tennessee at a COVID Early Treatment Summit in December of 2021, I knew exactly what a great mind, physician, and researcher I was engaging with. Dr. Marik has over 500 peer-reviewed publications, he has an H-index of 105 (productivity and citation impact of the publications: extremely high), over 48,000 citations of his work, and until recently, he was Professor of Medicine (Deans' Endowed Chair), Chief of Pulmonary and Critical Care Medicine, at East Virginia Medical School.

Dr. Marik, South African by birth, brings a quiet confidence and demeanor into his discussions and carries himself with dignity. I am proud to call him my friend. Here is his story, which helps address the question of why so many physicians have been silent regarding the failure of the medical-pharmaceutical industrial complex to enable optimal treatment of patients both outside and inside our hospitals.

* * * * *

The Anatomy of a Career-Ending Sham Peer Review
By Paul Marik, MD

In 1986, the United States Congress enacted the Healthcare Quality Improvement Act (HCQIA), which granted immunity to hospitals and reviewers participating in "good faith" peer review of physicians and dentists. These reviews were envisioned to be vehicles by which it could be determined if any actions or recommendations against a physician should become necessary on the measures of incompetence, unprofessional conduct, or behaviors that impact the doctors' clinical privileges. However, of late, HCQIA has resulted in many unforeseen consequences, not the least of which is the rise of "sham peer reviews" [32]—and the consignment of guiltless, lifesaving, preeminent physicians into obscurity.

What is "Sham" Peer Review?

Sham peer review is an adverse action taken in bad faith by a hospital for purposes other than the furtherance of quality healthcare. It is a process that is disguised to look like legitimate peer review [32–37]. But sham peer review is not objectively reasonable, precisely because it is not performed to advance the quality healthcare (violation of safe harbor provision 1; see below) [38].

A sham peer review happens when the hospital invents some pretext on which to attack the physician and acts to disguise the adverse action against the targeted physician by conducting such a review—*where the truth and the facts do not matter, because the process is contrived to be rigged, and the outcome is predetermined* [33].

Over the years, sham peer reviews have unfortunately become fairly well known. Hospitals in the United States have mounted these proceedings for at least four decades to rid themselves of physicians who "get in their way." Often, they are doctors who don't "follow the party line" and for whom they consider "disruptive" [32]. Hospital officials are resistant to physicians who bring patient safety or care-quality concerns to their attention. Some hospitals retaliate against these whistleblowers by instigating these sham peer reviews.

Consider this: In the criminal justice system, accused serial murders, rapists, child molesters, drug dealers, and thieves are entitled to due process and are presumed innocent until proven guilty. Unfortunately, accused physicians in the hospital sham peer review process are presumed "guilty." They are frequently afforded limited (if any) due process; and they are subsequently dismissed from the hospital [33].

How Sham Peer Review Works

Hospitals that use sham peer review bring trumped-up, fabricated, and thoroughly false charges against the targeted physician. Although no court of law would permit depriving an accused person of files or records needed to defend himself, as it is fundamentally unfair and in violation of due process, hospitals that employ sham peer review frequently refuse to provide records required to the physician under review [33]. Based on these totally erroneous and phony charges, the physician's *hospital privileges are summarily suspended.* The physician is usually given fourteen days to respond in writing to the sham charges. The charges and the physician's response are then supposedly shared with the Medical Executive Committee (MEC). The physician then meets with the Medical Executive Committee. The physician is usually denied legal representation (which is unlawful), and the meeting takes the form of a kangaroo court.

Though the concocted accusation(s) are contemptible, the MEC is usually (and inexplicably) not given either the complaint or the physician's response. As the hospital has no legitimate case against the targeted physician, the Chair of the MEC and his/her coconspirators will frequently abruptly change course and focus instead on behavioral accusations [33]. The hospital then accuses the physician of a pattern of unprofessional behavior, yet once again these accusations have no supportive evidence [33]. Why do they do this? Because accusations involving behavior or conduct are much easier for a hospital to prosecute, since typically the only "evidence" required is the accusation itself—not who made the accusation, when it was made, or a copy of the complaint. *In the end, the physician is, of course, found guilty of being disruptive, with his/her privileges revoked and being reported to the state board of medicine and the National Practitioner Data Bank (NPDB)—an action which effectively ends the physician's career.* The hospital attorney usually cites "peer review privilege," so as to prevent the plaintiff physician from discovering and revealing what really happened, in secret, behind closed doors at the hospital. The suspension of the physician's hospital privileges is extended beyond thirty days, at which time the hospital reports the physician to the NPDB. *Even if the charges are subsequently proven to be fraudulent, it is nearly impossible to remove the suspension of privileges from the NPBD.*

In summary, sham peer review is a perversion of the process intended to protect patients and colleagues from ill, incompetent, unethical, dangerous, and unprofessional practitioners. Sham peer review is an illegal, unethical, immoral, and highly virulent process. Participating in sham peer review is a violation of codes of professional conduct. Participants disrespect professional colleagues, engage in vested self-regulation, and promote discriminatory standards of professionalism [35].

Hospitals Don't Have Unrestricted and Unlimited Immunity.

The immunity provided by HCQIA has been abused by hospitals and physicians to harm "disruptive" physicians (i.e., whistleblowers). However, immunity under HCQIA can only be asserted if four safe-harbor provisions are met as set out in 42 U.S.C. §11112 [32, 38]. These four provisions stipulate that the professional review action was taken:

1. In the reasonable belief that the action was in furtherance of quality healthcare,
2. After a reasonable effort to obtain the facts of the matter,
3. After adequate notice and hearing procedures are afforded to the physician involved or after such other procedures as are fair to the physician under the circumstances, and
4. In the reasonable belief that the action was warranted by the facts known after such reasonable effort to obtain facts and after meeting the requirements as outlined in provision 3.

In addition, the statute requires that the physician must be advised of the hearing procedure, including a list of witnesses that will be called, notice that the physician may call and cross examine witnesses, and that he/she may present evidence. HCQIA establishes the right of a physician to representation by an attorney at peer-review hearings [39].

Not only is sham peer review not objectively reasonable, but since the basis for it is often completely fraudulent and done for some purpose other than the advancement of quality healthcare, sham peer review does not qualify as a "professional review action" under the definition provided in the HCQIA and is therefore NOT PROTECTED by immunity. This assertion is supported by case law. As summarized in the 10th Circuit Court in Brown vs Presbyterian Healthcare Services, the court rejected the defendant's assertion of immunity, as they failed to meet the safe-harbor provisions. In the case of Poliner vs Presbyterian Hospital of Dallas, the jury awarded damages in the amount of $366,211,159.30 to Dr. Poliner [40]. The jury found that Defendant's actions were not immune from civil liability under the federal or state peer-review statutes. The jury found in favor of Plaintiffs on all of their claims, including breach of contract, defamation, business disparagement, tortious interference with a contract, and intentional infliction of emotional distress [41]. Further, the jury found the defendants violated medical staff bylaws, and that they found that defendants failed to comply with the reasonableness standards of HCQIA. The jury further found that Defendants had acted maliciously and without justification or privilege.

My Case

As will be described below, the "peer-review" process that I was subjected to by Sentara Norfolk General Hospital (SNGH)/Sentara Healthcare system did not meet the four safe-harbor provisions as set out in in 42 U.S.C. §11112, and therefore SNGH cannot claim peer review immunity. I was not advised of the hearing procedure (which took place on December 2, 2021), and I was provided neither a list of witnesses (who were anonymous) nor the ability to cross-examine the "witnesses." I was not provided with any evidence supporting any of the claims, nor was I permitted to provide any evidence to refute the claims. In addition, and most important, I was forbidden from having legal representation at the sham hearing. Furthermore, what is truly astonishing is that despite the multiple accusations and the claim that *"significant numbers of individuals who reached out to leadership with substantially similar information; all of which who are individuals who have nothing to gain from reporting such concerns,"* I was never provided with documentation of a single complaint. In summary, I was not afforded the "specific procedural requirements of a peer-review process that will be entitled to immunity" as set forth by the HCQIA [32].

In my case, Sentara Norfolk General Hospital followed the exact playbook as outlined above. On November 9th I filed a suit against Sentara Healthcare System for instituting a policy preventing myself and other physicians from administering proven, safe, "off-label" FDA-approved, lifesaving therapeutics for the treatment of COVID-19. The case was heard in the Circuit Court of the City of Norfolk on November 18th. As agreed by all parties, I was scheduled to work in the ICU at SNGH the weekend starting Saturday November 20th. When I arrived at work on the 20th, I found an envelope (with no postmarks) on my desk containing a letter from the Sentara Medical Staff Office. The letter was marked overnight delivery and by email (which was never sent). The letter signed by *the* President of Medical

Staff, Sentara Hospitals, and *the* President, Sentara Norfolk General Hospital. The letter stated the following:

> The purpose of this letter is to inform you that a series of events have recently been reported to Hospital Administration and Medical Staff leadership at Sentara Norfolk General Hospital that have caused significant concern about your ability to conduct yourself in a professional and cooperative manner in the Hospital, which is essential for the provision of safe and competent patient care.

They went on to accuse me of the following falsified charges *(see letter of response)*:

1. That I refused to participate in rounds with residents.
2. That I instructed nurses to crush medication and put them into a feeding tube despite the pharmacy warning against this practice.
3. It was reported that I instructed residents to tell patients' families to hide Ivermectin in Twinkies and sneak it up to patients in the hospital.
4. Nurses have also reported that I have forced them to give a medication, even though doing so might cause the patient to have an anaphylactic reaction.
5. We were also informed that you have been approaching ICU families and telling them that it would be "child abuse" for them to vaccinate their children against COVID19.
6. Additionally, it was reported that you informed patients that "your hands were tied" and that there was nothing more you could do for them. *(Even though my hands were in fact tied. because the hospital barred me from offering them the medicines that would help them.)*
7. It was reported that you have continued to start patients on plasma exchange protocols, despite the fact that such treatment is no longer recommended, and in fact may be dangerous.

It is clear that a third-grader could come up with more credible accusations. Furthermore, they did not provide any substantive evidence to support these outrageous claims, nor did they provide me with any patient details in order for me to refute these claims. The letter went on to state:

> Based on these incidents, Medical Staff leadership has determined that your behavior causes such concern that there are grounds to impose a precautionary suspension of your Medical Staff appointment and clinical privileges in accordance with Section 6.D.1 of the Medical Staff Credentials Policy based upon a conclusion that failure to take such action "may result in imminent danger to the health and/or safety of any individual." This precautionary suspension is effective immediately.

These statements are profoundly offensive and wholly unfounded. Sentara's own data indicate that starting in July 2020, which is when I lost administrative control of the General Intensive Care Unit (GICU), the mortality rate in the GICU had *DOUBLED*. Clearly, my presence in the hospital was beneficial to patients (and their families), nurses, and residents.

Although the Hospital/MEC claim this action is unrelated to my legal case *(see below)* and that the timing is purely coincidental, it is categorically clear that this sham peer review process was nothing more than a corrupt retaliation. Furthermore, as noted by Judge Lanetti, overseeing my legal case:

> Of note, Marik alleges that he received a letter from Sentara when he reported to work on November 20, 2021, stating that, as of November 18, 2021-the date of the Hearing-his hospital privileges at Sentara Norfolk Hospital had been suspended (Pl.'s Nov. 22, 2021, Letter 1.).
>
> Although Sentara maintains that Marik "would not be disciplined for discussing his protocol as a treatment alternative with his patients," it does not dispute that the suspension is related to Marik's care of his COVID-19 patients. (Def.'s Nov. 22, 2021, Letter 1.)

Further, the Sentara letter stated, *"If you wish, you may provide a written statement to the Medical Executive Committee in advance of the meeting as well.*

Please send any such written statement to me by Monday, November 29th by fax and I will ensure that the Committee members receive it prior to the meeting on December 2nd 2021." As requested, I submitted my response to these false allegations on the 29th of November, 2021. Despite the assurance cited above, I have been informed that members of the MEC received neither the original complaint nor my response.

The Sentara letter also stated:

> Pursuant to Section 6.D.2. of the Medical Staff Credentials Policy, the Medical Executive Committee must meet to review the matter resulting in a precautionary suspension within 14 days. That meeting has been scheduled for *December 2, 2021* and you are required to attend. These proceedings are confidential; there will be no legal counsel.

I attended the MEC meeting on December 2nd, which in reality was a kangaroo court. I was confronted by about twenty-five angry people (who did not introduce themselves) including the president of the medical staff, the chief medical officer, the chair of the Peer Review Committee, and a number of department chairs. As the sham peer review "blueprint" outlined above, the chair did not want to discuss the charges that led to this meeting. Rather, she and the chair of the Peer Review Committee raised new allegations of "unprofessional behavior." When I asked about specific instances of unprofessional behavior and who had generated these complaints, the answers were not forthcoming. In regard to my "unprofessional" behavior, those with whom I have worked and the patients and families to whom I have ministered know that during the course of my entire career, I have prided myself on my professional conduct and being courteous and polite to students, residents, nurses, patients, and their families. In my career spanning over thirty-five years, during which time I have published more scientific articles in medical journals on critical care than any other physician in America, I have NEVER been sued, NEVER had a single patient compliant, and without exception, the evaluations of myself by students, residents, fellows, and the nursing staff have been consistently outstanding. I have never had a complaint lodged alleging unprofessional behavior. Clearly, these fictitious claims were a continuation of the original untruthful, phony accusations.

Following the kangaroo court, on December 6th I received a letter once again from the president of the medical staff stating the following:

> When the MEC met with you on December 2, it was felt that your behavior was consistent with the concerns that have been raised most recently. Your demeanor was extremely hostile, and you appeared angry and defiant. You vehemently denied that your behavior has ever been questionable and accused the MEC of retaliating against you for filing suit against Sentara Health-care, despite the fact that the matter does not involve Medical Staff leadership and was not the subject of this inquiry.

An individual witness to these proceedings has confided in me and has stated that "This was pure retaliation, and the accusations are all fabricated." (He/she will testify to such under oath.) The letter continued:

> Overall, the MEC felt that your categorical denial that any of the reported concerns that were described to you in our previous correspondence had occurred was not credible given the significant number of individuals who reached out to leadership with substantially similar information; all of which who are individuals who have nothing to gain from reporting such concerns and the vast majority of which have expressed being fearful about the possibility of retaliation. Given the above, the MEC was unable to resolve the concerns raised and *remains extremely concerned about the continued risk of imminent danger to the health and/or safety of patients, families, medical staff, and employees of the Hospital.* Ultimately, the MEC considered your behavior to be unacceptable and determined that not only are these circumstances concerning on their face, but they also appear to be further evidence of your lack of professionalism when interacting with colleagues and Hospital staff. For this reason, the MEC voted to initiate a formal investigation into this matter pursuant to Section 6.C.2 of the Medical Staff Credentials Policy and to continue your precautionary suspension in place pending the outcome of this review process.

I received a follow-up letter from the president of the medical staff dated December 23rd, 2021. This letter served as "*formal notice of the meeting that the ad hoc Investigating Committee has scheduled with you pursuant to Section 6.C.3 (d) of the Medical Staff Credentials Policy.*" The meeting with the ad hoc committee was scheduled for January 17th, 2022. Furthermore, this letter outlined fourteen accusations against me, including the seven previous charges against me with additional new implausible charges that had been invented since the first accusations of November 18th. My response to these absurd, bogus accusations is outlined in a letter dated December 30th, 2021.

It was perfectly clear that Sentara Health System would continue this sham process, continue with their lies and spurious allegations, and deny me any semblance of due process, with the ultimate goal of revoking my hospital privileges and thereby ending my career. Furthermore, as I understand, my suspension of hospital privileges has been reported to the NPDB. Based on this reality I felt I had no option but to resign from my position as tenured professor of Medicine at Eastern Virginia Medical School (EVMS) effective December 31, 2021. As Sentara Health System had achieved their goal of ending my career, they cancelled the ad hoc committee meeting, which now served no purpose.

In summary, the sham peer review assault perpetrated by Sentara Healthcare System was immoral, unethical, illegal, and unconscionable. It represents evil in its most vile form. Sentara has "acted maliciously and without justification or privilege." The actions of Sentara Healthcare System are, however, in keeping with what one can only intuit as a total disrespect for the sanctity of human life. Hundreds of patients have died as a result of their contempt for science as witnessed by their unconscionable ban of lifesaving COVID-19 therapeutics within the hospital, and their self-serving financial and political interests.

CHAPTER 5
Treating Patients and Fighting for Medical Freedom

By Pierre Kory

Dr. Pierre Kory, MD, MPA, is a specialist in pulmonary diseases, internal medicine, and critical-care medicine. He serves as the president and chief medical officer of the Front-Line COVID-19 Critical Care Alliance. Formerly, Dr. Kory worked as the chief of the critical care service and medical director of the Trauma and Life Support Center at the University of Wisconsin.

During the pandemic, Dr. Kory led ICUs in multiple hotspots and authored several peer-reviewed papers on COVID illness and treatment. He has testified on two occasions to the US Senate on the medical evidence supporting the use of early treatment for COVID-19.

In this essay, Dr. Kory distills lessons that could lead us to a better destination than where we ended up in this pandemic.

* * * * *

After two years like 2021–2022, it's important to take a moment to reflect and distill lessons that may help us change course toward a happier destiny. There's a lot I could say about what has been happening, given the state of our country and our medical community, but I will focus on what I see as the four major takeaways to guide us forward:

1. Do no harm does not mean do nothing

Many healthcare professionals in the US immediately adopted an approach of not trying any treatments until large, expensive, and prolonged randomized trials could be performed, so they could have the security of knowing their treatments were recommended by powerful health agencies.

Some of us, meanwhile, got down to the business of medicine, studying the mechanisms of this novel disease and then formulating treatment approaches using readily available medicines with known properties that could counteract these mechanisms. We did everything possible to give patients the best chance of coming out of this disease alive and free from harm. I'm proud to be in this camp because the results speak for themselves.

The "maverick" doctors in the US who took the above-mentioned path experienced both resistance and punishment from our administrative leaders and government managers, while other countries and regions around the world adopted similar approaches with outstanding results.

I frequently cite the example of Uttar Pradesh, one of India's largest states with a population two-thirds the size of the US. With a careful door-to-door surveillance strategy in combination with a prevention and early treatment regime using Ivermectin, Uttar Pradesh effectively eliminated COVID-19 from their state of 241 million people. The history books will (I hope) rightly recognize their efforts as one of the most successful public health interventions ever [42, 43].

The Brazilian city of Itajai is another great example. The city offered Ivermectin preventively to the entire city's population; 60 percent of the population (133,051 people) agreed to take it every two weeks for six months. The city's health service collected data on the entire population prospectively and found that Ivermectin users had a 70 percent lower mortality rate, and a 67 percent lower hospitalization rate, while the citywide COVID mortality fell from 6.8 percent to 1.8 percent during the program [44].

Similar results have been seen in places like Mexico, Peru, Argentina, the Philippines, Japan, and elsewhere. But in North America, Europe, and Australia, organized, deep-pocketed, and highly effective opposition to such programs led to some of the highest case fatality rates in the world. What will the history books say about that?

2. Treating COVID is about more than Ivermectin

It's easy to think of me as "the Ivermectin doctor," but that's only because the drug is so effective in all phases of COVID-19 that it forms the core therapeutic in the protocols developed by the organization I cofounded, the Front Line COVID-19 Critical Care Alliance (FLCCC), which develops and supports Prevention & Treatment Protocols for COVID-19. There are, however, a whole host of other compounds that work to treat COVID-19, either on their own or in combination. All FLCCC doctors and many, many more physicians throughout the United States and the world pride themselves on these combination protocols, which were carefully constructed to work in synergy. These are multidrug, multistage protocols and not based on one drug or product and can be found on our website [45].

Our group has also developed a treatment protocol that was created for the hospitalized patients based on the core therapies of methylprednisolone, ascorbic acid, thiamine, heparin, and nonantiviral cointerventions (MATH+). There is a scientific and clinical rationale behind MATH+ based on published in-vitro, preclinical, and clinical data in support of each medicine, with a special emphasis of studies supporting their use in the treatment of patients with viral syndromes and COVID-19 specifically [46].

In addition, it's important to recognize that colleagues such as Dr. Paul Marik and Dr. Umberto Meduri, along with myself, were early advocates for the use of steroids to treat COVID patients, a practice initially discouraged by federal health officials, but that has since become the standard of care worldwide [47]. We also had success treating patients with fluvoxamine, a widely used generic antidepressant, in addition to steroids and a number of other repurposed medicines. This protocol contributed to halving deaths in Dr. Marik's hospital [47]. But the hospital stepped in and banned these medicines, largely restricting their doctors to using only remdesivir; we know it doesn't work in late-phase COVID, and worse, the best studies show it actually may be harming patients.

We now know there are a whole host of compounds that work to treat COVID-19, either on their own or in combination. To show how much the US has lost its way in responding to COVID-19, remdesivir is given to nearly every hospitalized patient at a cost of $3,000 per dose. There are "Narco" states and there are "Pharma" states, and the US has clearly fallen into the latter category. We must fight to free ourselves from this oppression.

Regarding fluvoxamine, the FLCCC incorporated it into our treatment protocol on April 27, 2021, with great effect. That practice was affirmed in October 2021, when a large, double-blind randomized controlled trial, published in the *Lancet*, found fluvoxamine reduced COVID-19 mortality rates by up to 91 percent and hospitalizations by two-thirds in those who adhered to the prescribed regimen [48]. This news reinforces the logic of safe, inexpensive, repurposed generic medicines to help get this pandemic under control. Yet the NIH continues to avoid recommending this medicine to treat Americans.

It appears these negligent behaviors at the federal level are finally being resisted at the state and local level. An example is now coming from Florida, where the state's surgeon general, Dr. Joseph Ladapo, recently launched a public service campaign promoting a healthy lifestyle, better nutrition, and early treatment for COVID-19 using many of the compounds in our protocol, including fluvoxamine. It is encouraging to see this kind of move in the country's third most populous state. Here's hoping more states follow Florida's lead in the coming months.

3. Many doctors are too cowardly to speak out

My faith in a lot of things has been weakened since the pandemic began; however, I still believe most doctors go into the profession because they want to help others. I don't think any doctor wants to see their patients suffer needlessly. So, I really shake my head when I see so many doctors standing by and watching, or even participating in the pharmaceutical industry's war on repurposed drugs, dutifully executed via health agencies, medical societies, and state medical boards that scare doctors with bulletins and memos full of threats and fraudulent guidance against using some of the world's safest (and unfortunately for the industry, most highly effective but unprofitable) medicines. The horrific consequences of their decades-long war against repurposed drugs are clearer than ever before. It must stop. Doctors must resist more effectively, and more cohesively.

I really shake my head when I see so many doctors standing by and watching, or even participating in, the pharmaceutical industry's war on repurposed drugs.

We could have put an end to this pandemic and saved countless lives if many more physicians had spoken up in their individual institutions, prioritizing early treatment approaches guided by the precautionary principle and sound risk-benefit decision making. Instead, physician leaders in countless institutions allowed public health agencies and institutions to implement a rigid, top-down approach to treatment, threatening physicians with loss of their livelihoods if they didn't follow their preordained protocols. The physicians' cowardice in staying silent, while patients suffered and died all around them month after month—just to ensure they could stay employed or maintain peaceful relationships with their peers and superiors—is a sad reflection on our medical community. This has led to terrifying outcomes, just as history books will record.

Thankfully, there is a growing number of courageous and outspoken doctors and nurses who are increasingly rising up to do what they are duty-bound to do, and I am honored to count myself among them. These are the people who give me hope and inspire me to keep fighting for the truth no matter how difficult it is sometimes.

4. The powers that be can't keep the truth hidden for long

Here's where the hope shines through the doom and gloom. People are seeing what's happening, and they're getting sick of it. Word is getting out. More and more people are questioning the many misguided policies leading to results more obviously disastrous by the day.

Every week, thousands of people tune in to our FLCCC weekly webinar. Around the world more than twenty countries, representing almost one-fifth of the Earth's population, now use Ivermectin. My Twitter following has grown to over 200,000 people! I don't say this to gloat, but rather to point out that people are hungry for common-sense information they can trust. And I am so proud to be surrounded by a group of pragmatic, caring, thoughtful physicians whose goal is to do just that: use common-sense approaches to fight this pandemic.

A growing number of state attorneys general, including in Nebraska, Louisiana, South Carolina, Oklahoma, and now Tennessee, are moving to protect physicians' ability to use off-label prescribing in the treatment of COVID-19. In an encouraging public statement, Oklahoma's Attorney General John O'Connor said his office would not allow medical boards to prevent doctors prescribing Ivermectin or hydroxychloroquine to treat COVID-19 [49].

I stand behind doctors who believe it is in their patients' best interests to receive Ivermectin and hydroxychloroquine.

This is a huge win for doctors and patients. Just like our long-standing advocacy for early treatment of COVID, the FLCCC has advocated for public officials to let doctors be doctors since the beginning of the pandemic.

CHAPTER 6
Beware the Fact-Checkers
By Leonard C. Goodman

As the late US senator Daniel Patrick Moynihan once said, "People are entitled to their own opinions, but not their own facts." This essay provides a great illustration of how "fact-checkers" are not checking facts, but rather have been acting to police opinion and thought. This is the Orwellian reality that the US Intelligence community advocates for America, and that the corporate legacy media gladly endorses.

Leonard C. Goodman is a Chicago criminal defense attorney and coowner of the for-profit arm of the *Chicago Reader*. This article was first published in the ScheerPost. Republished here by permission of the author.

* * * * *

Beware of the Fact-Checkers

A case study in how allegedly neutral analysts hired by publications or social media can effectively cancel good-faith questions and opinions because they challenge dominant narratives.

By Leonard Goodman / Original to ScheerPost
April 13, 2022

Opinion columnists are familiar with the traditional role of the fact-checker. Prior to publication, an editor checks accuracy of quotes and the sources for factual assertions. Erroneous or unsupported assertions are removed or revised.

But times have changed. Today, an entire fact-checker industry has emerged to check your opinions, making sure you have not strayed beyond acceptable limits for public discourse. These professional fact-checkers are often brought in after publication of a controversial article, opinion piece or podcast to quell a controversy. Acting more like business consultants, they help media platforms large and small stay on the right side of government officials and corporate sponsors.

COVID-19 has been a boon to the fact-checking industry. Big outfits like Politifact and Factcheck.org have special divisions just to police COVID "misinformation." Like the Ministry of Truth imagined by George Orwell in his epic novel, "1984," these outfits will tell you what you can and can't say about the lockdowns, masks, and the mRNA vaccines manufactured by Pfizer and Moderna.

I got a window into the world of professional fact-checkers last November after I published an op-ed for the *Chicago Reader* called, "Vaxxing our Kids, Why I'm not rushing to get my six-year-old the COVID-19 vaccine." In it, I considered the arguments for and against the official policy to vaccinate every child. And I apparently crossed a line by including opinions held by a significant number of prominent scientists and physicians who believe healthy children don't need the vaccine because their risk of severe COVID is minuscule, the vaccine may do more damage than good to children, and it does little to stop the spread of COVID.

Vaxxing our Kids was my 21st column for the *Chicago Reader*. Founded in 1971, the free and freaky *Chicago Reader* has a long history of taking on centers of powers and inviting controversy, including articles exposing the Chicago Police department's systematic use of brutal torture to extract confessions from murder suspects (1990–2007), the Catholic Church's role in covering up allegations of child molestation by priests (1991), and the Israeli government's mistreatment of Palestinians in the Gaza Strip (2002–05).

In 2018, the *Chicago Reader* was insolvent and faced dissolution. I partnered with a Chicago real estate developer to purchase the *Reader* for $1. We assumed its debt and helped pay its operating expenses with the intention to transition the paper to not-for-profit status as the best way to assure its survival into the future.

In 2019, I began writing a semi-regular opinion column for the *Reader*. Taking advantage of its fifty-year history of providing a space for dissent, I focused on subjects that would not be welcome in mainstream papers, such as the connection between convicted pedophile Jeffrey Epstein and U.S. Intelligence, the persecution of Julian Assange and Chelsea Manning, the Obama Foundation's move to privatize 20 acres of historic public parkland on Chicago's South Shore and cut down a thousand trees in order to build a 235-foot-high museum tower on the shores of Lake Michigan, and the collaboration between corporate-friendly federal judges in the Southern District of New York and the Chevron Corporation to punish a lawyer who is trying to make Chevron pay for its deliberate destruction of a large section of the Ecuadorian Amazon Rainforest. Scheerpost copublished several of these columns.

But apparently expressing concerns about giving my six-year-old daughter an mRNA vaccine that was not tested on humans until 2020, and that has been approved only for "emergency use" in kids, took me into forbidden territory. Like all my columns, Vaxxing our Kids was submitted on deadline, fact-checked and edited. At publication, my editor thanked me for taking on the difficult topic and pronounced my research to be "bulletproof." She predicted that the piece would be controversial, but that many parents of young children would appreciate hearing a different point of view. This prediction was accurate. Vaxxing our Kids received 772 likes on Twitter and 323 retweets even though the *Reader* did not support the column. Dozens of parents reached out to tell me that they too were struggling with the decision whether to give their young child an mRNA vaccine and were grateful for information that could not be found in other media. On the other side, a small but angry group of readers and pro-pharma operatives lashed out, demanding that Vaxxing our Kids be taken down off the Reader website and that I be fired as a columnist.

Scheerpost co-published Vaxxing our Kids. But the way Scheerpost and the *Chicago Reader* handled the exact same content, and the ensuing controversy could not have been more different. Scheerpost put the column front and center on its website and invited readers to comment and debate. Last I checked, there were 105 on-line comments and a robust debate, for and against the policy of mass vaccination of children. Many of the posters on Scheerpost shared knowledge, research and expertise on the questions raised in the op-ed, a shining example of how the First Amendment is supposed to work.

The *Chicago Reader* took a different approach. Rather than embrace the controversy and welcome a debate over an important issue of public health, the *Reader* let "the mob ha[ve] the final edit" as one journalist remarked in the *Chicago Tribune*.[6] After disabling all comments on its website, *Reader* management hired an external and anonymous "fact-checker" to rewrite my column and issue a report with nine points of disagreement, later expanded to fifteen points of disagreement. The publisher offered me two options: either remove the column from the *Reader* website, or replace it with the new version that was "extensively modified" by the fact-checker, to be followed by the fact-checker report. I asked to publish a rebuttal to the fact-checker report and was told: "As for rebuttal: Your side is the actual column. The rebuttal is not a 'side' it is a fact-checker's report."

At this point, the *Reader*'s board got involved to protest management's handling of the controversy over the opinion column. The board passed a resolution demanding that the *Reader* guarantee a space for dissenting views before it transitions to not-for-profit status. Management has dug in and refused to engage with the board's demands, leading to a stalemate which threatens the future of the *Reader*.

I accept that it is theoretically possible that I could publish an opinion column that, although extensively researched, edited and fact-checked pre-publication, could be so riddled with factual errors that it needed to be either taken down or extensively modified. On the other hand, I have written more than thirty op-eds for a half dozen publications and never once had to correct a single factual assertion after publication. So it seems highly unlikely that there could be fifteen factual errors in Vaxxing our Kids.

Also, a careful examination of the fact-checker report reveals it to be highly dubious. Most of the items in the report begin with a declaration that a sentence in my column is "untrue" or "misleading," followed by a convoluted word salad that winds up by conceding that what I wrote is 100% accurate. The remaining items in the report are just disagreements with the opinions of the experts that I accurately quote in the column.

For example, item number one in the report takes issue with the following sentence of my column: "Moreover, by not advertising their vaccines by name, Pfizer-BioNTech and other drugmakers are not obliged, under current FDA regulations, to list the risks and side effects of the vaccine." The fact-checker report pronounces this sentence to be both "untrue" and "misinformation." The report then confirms that, "Vaccine manufacturers have not advertised their vaccines at all" and then adds, "If Pfizer begins to advertise its vaccine, which received FDA approval earlier this year, it will have to follow regulations and list side effects." In other words, the report confirms that what I wrote is 100% accurate but nevertheless labels it "misinformation."

Items two and three assert that it was "misleading" for me to criticize the FDA for going "to court to resist a FOIA request seeking the data it relied on to license the Pfizer COVID-19 vaccine." But here again, the fact-checker concedes, in convoluted fashion, that what I wrote is 100% true—the FDA did in fact go to court to resist a FOIA request for the "raw data underpinning the trials." So how is what I wrote misleading? According to the fact-checker, I should have credited the FDA's explanation that, because of "its small department of ten FOIA officers (who are already handling hundreds of other requests)," it needed 55 years (until 2076) to go through the documents and redact "patient information and trade secrets."

In other words, in the age of the fact-checker, an opinion columnist is required to credit the official word of government bureaucrats, even when those bureaucrats are clearly lying, as they were in this case. How do I know they were lying? In early January, about a month after Vaxxing our Kids was published, a federal judge in Texas ordered the FDA to release all the data it relied on to license Pfizer's COVID-19 vaccine at a pace of 55,000 pages a month, rejecting the FDA's argument that its short-staffed FOIA office only had the bandwidth to review and release 500 pages a month. The FDA has so far complied with the court order. And in March, as was widely reported in the media, the first batch of vaccine-trial data was released revealing that Pfizer was aware of 1,291 adverse side effects from its vaccine when it applied for FDA approval.

Most of the other items in the fact-checker report criticize me for accurately quoting opinions that the fact-checker disagrees with. For example, my column cites recent statements from Mexico's health minister, Jorge Alcocer Varela, "who recommends against vaccinating children, warning that COVID-19 vaccines could inhibit the development of children's immune systems." The fact-checker asserts: "There is no evidence that this is the case with COVID vaccines or any other vaccines." But the mRNA vaccines have only been given to children for about a year. No one knows for sure what the long-term effects will be. Dr. Alcocer Varela believes this vaccine could hinder the learning of a child's immune system. He may be right, and he may be wrong. But he is entitled to his opinion. And considering his credentials as an immunologist, researcher, teacher, healthcare professional and government official, parents like myself have a right to consider his views in making healthcare decisions for our young children. People who disagree with Dr. Alcocer Varela are also entitled to express their views in opposition. That is the way free speech is supposed to work.

I got additional insight into the anonymous fact-checker report after a journalist from the Poynter Institute wrote an article weighing in on the controversy at the *Chicago Reader*. The Poynter Institute is a self-appointed leader in "accountability journalism" through its International Fact-Checking Network. The Poynter journalist wrote that "Goodman's column [Vaxxing our Kids] received backlash from readers and staff due to inaccuracies and misleading statements within the piece." In an email, I demanded that the journalist identify these "inaccuracies and misleading statements within the piece." She responded in part that my article cited the views of Dr. Robert Malone; but an article at Politifact.com explains "why he cannot be considered a 'reputable' source on the COVID-19 vaccines."

In other words, Dr. Robert Malone has been cancelled by Politifact. Therefore, op-ed columnists are not permitted to cite Dr. Malone's views even though he is one of the original inventors of the mRNA vaccine technology and scores of people around the world are interested in what he has to say.

I should also point out that the Poynter Institute owns Politifact.

I wrote twenty columns for the *Chicago Reader*, most of which expose connections between government officials and their corporate partners. But it was only after I questioned the official narrative on COVID vaccines that the *Reader* felt compelled to bring in the professional fact-checkers to justify censoring my opinions and cancelling me as columnist.

I suspect that the real objection to Vaxxing our Kids has nothing to do with factual errors. Rather, the piece may have stumbled onto some uncomfortable truths about our official policy to vaccinate every child in America for a virus that poses almost no risk to healthy children. Perhaps the bigger concern was the following excerpt from my column that escaped entirely the fact-checker's red pen:

"This year, Pfizer has banked on selling 115 million pediatric doses to the U.S. government and expects to earn $36 billion in vaccine revenue. Congress is so in the pocket of Big Pharma that it's against the law for our government to negotiate bulk pricing for drugs, meaning taxpayers must pay retail."

That kind of money flowing to a corporate partner makes it hard for government officials to focus on the science.

Moreover, data now becoming available shows the vaccine to have been ineffective in kids. As recently reported by NBC News, "Two doses of the Pfizer-BioNTech COVID-19 vaccine offer almost no protection against coronavirus infection in kids ages 5 to 11, according to new data posted online—a finding that may have consequences for parents and their vaccinated children." Also, more than 17,000 doctors and scientists recently signed onto a declaration that "healthy children shall not be subject to forced vaccination."

These developments, coupled with the court-ordered release of the data from Pfizer's vaccine trials showing more than a thousand undisclosed side effects, may explain the mad scramble to shut down dissenting voices.

As a WWII Air Force pilot was reported to have said: "If you're taking flak, you're over the target."

Since the age of Socrates, truth has been discovered through reasoned debate and discourse. As the places in media to host that debate keep disappearing, some brave board members at the *Chicago Reader* are fighting to rescue the paper from the dark forces of censorship and to preserve its fifty-year tradition of embracing dissenting views.

Below is the original article that prompted the debate with the Chicago Reader and fact-checkers, published in the Reader on November 24, 2021

Vaxxing Our Kids: Why I'm Not Rushing to Get My Six-Year-Old the COVID-19 Vaccine
by Leonard C. Goodman

Like many Americans, I have concerns about giving my six-year-old a new vaccine that was not tested on humans until last year.

As a father of a young child, I am pressured to get my daughter vaccinated for COVID-19. And like many Americans, I have concerns about giving my six-year-old a new vaccine that was not tested on humans until last year, and that has been approved only for "emergency use" in kids. The feverish hype by government officials, mainstream media outlets, and Big Pharma, and the systematic demonization and censorship of public figures who raise questions about the campaign, provide further cause for concern.

This year, Pfizer has banked on selling 115 million pediatric doses to the U.S. government and expects to earn $36 billion in vaccine revenue. Congress is so in the pocket of Big Pharma that it's against the law for our government to negotiate bulk pricing for drugs, meaning taxpayers must pay retail. Corporate news and entertainment programs are routinely sponsored by Pfizer, which spent $55 million on social media advertising in 2020. Even late-night comedians like Jimmy Kimmel, who has called for denying ICU beds to unvaccinated people, have been paid by Big Pharma to promote the COVID-19 vaccine.

It is thus not surprising that most of the information reported in the press about vaccine safety and efficacy appears to come directly from Pfizer press releases. This recent headline from NBC News is typical: "Pfizer says its Covid vaccine is safe and effective for children ages 5 to 11." Moreover, by not advertising their vaccines by name, Pfizer-BioNTech and other drugmakers are not obliged, under current FDA regulations, to list the risks and side effects of the vaccine.

Most Americans are vaguely aware that COVID vaccines carry some potential risks, such as heart inflammation, known as myocarditis, seen most often in young males. But no actual data from the vaccine trials has been provided to the public. After promising "full transparency" with regard to COVID-19 vaccines, the FDA recently went to court to resist a FOIA request seeking the data it relied on to license the Pfizer COVID-19 vaccine, declaring that it would not release the data in full until the year 2076—not exactly a confidence-building measure.

Also troubling is a recent report in the British Medical Journal, a peer-reviewed medical publication, which found that the research company used by Pfizer falsified data, unblinded patients, employed inadequately trained vaccinators, and was slow to follow up on adverse events reported in Pfizer's pivotal phase III trial. The whistleblower, Brook Jackson, repeatedly notified her bosses of these problems, then emailed a complaint to the FDA and was fired that same day. If this scandal was ever mentioned in the corporate press, it was with a headline like this from CBS News: "Report questioning Pfizer trial shouldn't undermine confidence in vaccines."

On the other hand, the initial rollout of the vaccine appeared to be a home run. Reported numbers of new infections went down, and oppressive lockdown rules were lifted. Our bars, restaurants, and gyms opened up. Plus, my own experience getting the vaccine was positive, as I wrote about in an earlier column for the *Reader*. Is it possible that this time, the corporate media and government got it right? Is the mass vaccination of everyone, including kids, really the solution to our long COVID nightmare? I have tried my best to look objectively at the available evidence in order to make the best decision for my daughter. In this column, I share my findings.

The first thing I discovered is that the risk of COVID to healthy kids is extremely low. Or as the *New York Times*'s David Leonhardt recently put it, unless your child has preexisting conditions or a compromised immune system, the danger of severe COVID is "so low as to be difficult to quantify." This raises the question: If the risk for kids is so low, what is the emergency that justifies mass vaccination of children without waiting for proper testing trials of the vaccine?

The argument made most often is that we must vaccinate our kids to protect others. However, while most adults perceive children as little germ factories, the data suggests that kids are at low risk to spread COVID. Reports from Sweden, where schools and preschools were kept open, and kids and teachers went unmasked without social distancing, show a very low incidence of severe COVID-19 among schoolchildren or their teachers during the SARS-CoV-2 pandemic.

I was also surprised to learn that there are reputable scientists opposed to mass vaccination, such as Dr. Robert Malone, an original inventor of the mRNA vaccine technology behind the COVID vaccines. As Malone explains, the mRNA vaccine contains a spike protein, similar to the virus, that stimulates your immune system to produce antibodies to fight COVID. He describes the vaccine as "leaky," meaning it is only about 50 percent effective in preventing infection and spread.

Malone warns that overuse of a leaky vaccine during an outbreak risks generating mutant viruses that will overwhelm the vaccine, making it less effective for those who really need it. "The more people you vaccinate, the more vaccine-resistant mutations you get, and in the vaccine 'arms race,' the more need for ever more potent boosters." Thus, Malone recommends vaccinating only the most vulnerable—primarily the elderly and individuals with significant comorbidities such as lung and heart disease or diabetes—and not healthy children.

If these views sound unfamiliar, it's likely because Malone and other critics of mass vaccination have faced heavy suppression on social media and vicious attacks from corporate media outlets.

Meanwhile the U.S. mainstream press has ignored recent statements by Mexico's health minister, Jorge Alcocer Varela, who recommends against vaccinating children, warning that COVID-19 vaccines could inhibit the development of children's immune systems. "Children have a wonderful immune system compared to the later phases . . . of their life," he explained, warning that "hindering" the "learning" of a child's immune system—the "cells that defend us our whole lives"—with a "completely inorganic structure" such as a vaccine runs counter to public health.

A recent Harvard study provides further evidence that while vaccines protect us against serious COVID illness and deaths, they alone are not very good at stopping the spread of the disease. The study looked at COVID numbers in 68 countries and 2,947 counties in the United States during late August and early September. It found that the countries and counties with the highest vaccination rates had higher rates of new COVID-19 cases per one million people. And suggested other measures, like mask wearing and social distancing, in addition to vaccination.

In place of mass vaccination, Malone recommends early intervention with therapeutics shown to be effective against COVID, including Ivermectin. In contrast, the corporate press has shamelessly attacked early treatments, and especially Ivermectin, which it calls a veterinary drug, in reference to the fact that it is used to treat both animals and humans, along with many other drugs, including antibiotics and pain pills.

In October, popular podcaster Joe Rogan announced on his program that he had contracted the virus and took Ivermectin, prescribed by a doctor, along with other therapeutics including monoclonal antibodies, and that he only had "one bad day" with the virus. CNN ridiculed Rogan for taking "horse dewormer." On his show, Rogan grilled CNN medical expert Sanjay Gupta. "Why would they lie [at your network] and say that's horse dewormer? I can afford people medicine." Rogan pointed out that the developers of Ivermectin won the Nobel Prize in 2015 for the drug's use in human beings.

Why indeed is CNN and much of the mainstream press lying about Ivermectin, a drug that has been used by literally billions of people to treat tropical diseases, and has been shown to be safe and effective in treating COVID in countries such as Mexico, India, Japan, and Peru? First, in order for there to be an emergency use authorization for the vaccines, there has to be no treatment for a disease. Thus, any potential treatments must be disparaged. That is, of course, until Pfizer releases its antiviral drug, PF-07321332.

Second, Ivermectin is off patent, meaning Big Pharma can't make a profit on it. It has been made available to poor people around the world at pennies a dose. In contrast, Pfizer's COVID pill will be priced at more than $500 per course.

At this point, you can guess the end of the story. The final straw for me is the apparent lack of durability of the COVID vaccines. Recent data indicates that the limited protection from the vaccine lasts only four to six months. Since COVID is not going away, is it Pfizer's plan to artificially boost my daughter's immune system every four to six months for the rest of her life?

We have been kept in the dark about vaccine safety and efficacy by our government and its partners in Big Pharma, who tell us they have looked at the science and it supports vaccinating our children against a virus that presents them with only the most miniscule risk of serious illness. As a parent, I will demand more answers before simply taking their word.

CHAPTER 7
Mass Formation and the Psychology of Totalitarianism

By Mattias Desmet

As many of you know, I have spent time researching and speaking about mass formation (psychosis) theory. Although the roots of the theory can be traced back to the Allegory of the Cave described in Plato's Republic (and in the introduction to this book), most of what I have learned has come from Dr. Mattias Desmet. During the COVIDcrisis, Mattias realized that this form of mass hypnosis, of the madness of crowds, can account for the strange phenomenon of about 20–30% of the population in the Western world becoming entranced with the Noble Lies and dominant narrative concerning the safety and effectiveness of the genetic vaccines, and both propagated and enforced by politicians, science bureaucrats, pharmaceutical companies, and legacy media.

What has been clearly observable with the mass hypnosis is that a large fraction of the population is completely unable to process new scientific data and facts demonstrating that they have been misled about the effectiveness and adverse impacts of mandatory mask use, lockdowns, and genetic vaccines that cause people's bodies to make large amounts of biologically active coronavirus Spike protein.

These hypnotized by this process are unable to recognize the lies and misrepresentations they are being bombarded with on a daily basis and actively attack anyone who has the temerity to share information with them that contradicts the propaganda that they have come to embrace. And for those whose families and social networks have been torn apart by this process, and who find that close relatives and friends have ghosted them because they question the officially endorsed "truth" and are actually following the scientific literature, this can be a source of deep anguish, sorrow, and psychological pain. At times when I have spoken about this theory to large groups, I have looked out over the audience and seen grown men with tears streaming down their faces. So many families and interpersonal relationship have been deeply damaged, all too often completely torn apart, during the COVIDcrisis. I believe that one of the most important aspects of Dr. Desmet's profound insights into this phenomenon is that it can help people to understand and (in some cases) to forgive their neighbors, peers, and family members who have become hypnotized by the propaganda, thought, and information control that they have been subjected to during the COVIDcrisis. Here is the story of how Dr. Desmet broke free of his own hypnosis during the time of COVID and realized that the academic research area that had been the focus of his life's work had influenced both his own thinking as well as that of much of the world.

* * * * *

From Our Rationalist View on Man to the World to Mass-Formation

Dr. Mattias Desmet: professor of clinical psychology, Ghent University

At the end of February 2020, the global village began to shake on its foundations. The world was presented with a foreboding crisis, the consequences of which were incalculable. In a matter of weeks, everyone was gripped by the story of a virus—a story that was undoubtedly based on facts. But on which ones? We caught a first glimpse of "the facts" via footage from China. A virus forced the Chinese government to take the most draconian measures. Entire cities were quarantined, new hospitals were built hastily, and individuals in white suits disinfected public spaces. Here and there, rumors emerged that the totalitarian Chinese government was overreacting and that the new virus was no worse than the flu. Opposite opinions were also floating around: that it must be much worse than it looked, because otherwise no government would take such radical measures. At that point, everything still felt far removed from our shores, and we assumed that the story did not allow us to gauge the full extent of the facts.

Until the moment that the virus arrived in Europe. We then began recording infections and deaths for ourselves. We saw images of over-crowded emergency rooms in Italy, convoys of army vehicles transporting corpses, morgues full of coffins. The renowned scientists at Imperial College confidently predicted that without the most drastic measures, the virus would claim tens of millions of lives. In Bergamo, sirens blared day and night, silencing any voice in a public space that dared to doubt the emerging narrative. From then on, story and facts seemed to merge, and uncertainty gave way to certainty.

The unimaginable became reality: we witnessed the abrupt pivot of nearly every country on Earth to follow China's example and place huge populations of people under de facto house arrest, a situation for which the term "lockdown" was coined. An eerie silence descended—ominous and liberating at the same time. The sky without airplanes, traffic arteries without vehicles; dust settling on the standstill of billions of people's individual pursuits and desires. In India, the air became so pure that, for the first time in thirty years, in some places the Himalayas became once more visible against the horizon.

It didn't stop there. We also saw a remarkable transfer of power. Expert virologists were called upon as Orwell's pigs—the smartest animals on the farm—to replace the unreliable politicians. They would run the animal farm with accurate ("scientific") information. But these experts soon turned out to have quite a few common, human flaws. In their statistics and graphs they made mistakes that even "ordinary" people would not easily make. It went so far that, at one point, they counted *all* deaths as corona deaths, including people who had died of, say, heart attacks.

Nor did they live up to their promises. These experts pledged that the Gates to Freedom would reopen after two doses of the vaccine, but then they contrived the need for a third. Like Orwell's pigs, they changed the rules overnight. First, the animals had to comply with the measures because the number of sick people could not exceed the capacity of the healthcare system (flatten the curve). But one day, everyone woke up to discover writing on the walls stating that the measures were being extended because the virus had to be eradicated (crush the curve). Eventually, the rules changed so often that only the pigs seemed to know them. And even the pigs weren't so sure.

Some people began to nurture suspicions. How is it possible that these experts make mistakes that even laymen wouldn't make? Aren't they scientists, the kind of people who took us to the moon and gave us the Internet? They can't be that stupid, can they? What is their endgame? Their recommendations take us farther down the road in the same direction: with each new step, we lose more of our freedoms, until we reach a final destination where human beings are reduced to QR codes in a large technocratic medical experiment.

That's how most people eventually became certain. Very certain. But of diametrically opposed viewpoints. Some people became certain that we were dealing with a killer virus that would kill millions. Others became certain that it was nothing more than the seasonal flu. Still others became certain that the virus did not even exist and that we were dealing with a worldwide conspiracy. And there were also a few who continued to tolerate uncertainty and kept asking themselves: how can we adequately understand what is going on?

***

In the beginning of the coronavirus crisis, I found myself making a choice— I would speak out. Before the crisis, I frequently lectured at university and I presented at academic conferences worldwide. When the crisis started, I intuitively decided that I would speak out in public space, this time not addressing the academic world, but society in general. I would speak out and try to bring to people's attention that there was something dangerous out there, not "the virus" itself so much as the fear and technocratic–totalitarian social dynamics it was stirring up.

I was in a good position to warn of the psychological risks of the corona narrative. I could draw on my knowledge of individual psychological processes (I am a lecturing professor at Ghent University, Belgium); my PhD on the dramatically poor quality of academic research, which taught me that we can never take "science" for granted; my master's degree in statistics that allowed me to see through statistical deception and illusions; my knowledge of mass psychology; my philosophical explorations of the limits and destructive psychological effects of the mechanist-rationalist view on man and the world; and last but not least, my investigations into the effects of speech on the human being and the quintessential importance of "Truth Speech" in particular.

In the first week of the crisis, March 2020, I published an opinion paper titled "The Fear of the Virus Is More Dangerous Than the Virus Itself." I had analyzed the statistics and mathematical models on which the coronavirus narrative was based and immediately saw that they all dramatically overrated the dangerousness of the virus. A few months later, by the end of May 2020, this impression had been confirmed beyond the shadow of a doubt. There were no countries, including those that didn't go into lockdown, in which the virus claimed the enormous number of casualties the models predicted it would. Sweden was perhaps the best example. According to the models, at least 60,000 people would die if the country didn't go into lockdown. It didn't, and only 6,000 people died.

As much as I (and others) tried to bring this to the attention of society, it didn't have much effect. People continued to go along with the narrative. That was the moment when I decided to focus on something else, namely, on the psychological processes that were at work in society and that could explain how people can become so radically blind and continue to buy into a narrative so utterly absurd. It took me a few months to realize that what was going on in society was a worldwide process of *mass formation*.

In the summer of 2020, I wrote an opinion paper about this phenomenon, which soon became well known in Holland and Belgium. About one year later (summer 2021), Reiner Fuellmich invited me onto *Corona Ausschuss*, a weekly livestream discussion between lawyers and both experts and witnesses about the coronavirus crisis, to explain mass formation. From there, my theory spread to the rest of Europe and the United States, where it was picked up by such people as Dr. Robert Malone, Dr. Peter McCullough, Michael Yeadon, Eric Clapton, and Robert F. Kennedy Jr. After Robert

Malone talked about mass formation on the Joe Rogan Experience, the term became a buzz word and for a few days was the most searched-for term on Twitter. Since then, my theory has met with enthusiasm, but also with harsh criticism.

What is mass formation actually? It's a specific kind of group formation that makes people radically blind to everything that goes against what the group believes in. In this way, they take the most absurd beliefs for granted. To give one example, during the Iran revolution in 1979, a mass formation emerged, and people started to believe that the portrait of their leader—Ayatollah Khomeini—was visible on the surface of the moon. Each time there was a full moon in the sky, people in the street would point at it, showing one another where exactly Khomeini's face could be seen.

A second characteristic of an individual in the grip of mass formation is that they become willing to radically sacrifice individual interest for the sake of the collective. The communist leaders who were sentenced to death by Stalin—usually innocent of the charges against them—accepted their sentences, sometimes with statements such as "If that is what I can do for the communist party, I will do it with pleasure."

Third, individuals in mass formation become radically intolerant of dissonant voices. In the ultimate stage of the mass formation, they will typically commit atrocities toward those who do not go along with the masses. And even more characteristic: they will do so as if it were their ethical duty. To refer to the revolution in Iran again: I've spoken with an Iranian woman who had seen with her own eyes how a mother reported her son to the state and hung the noose with her own hands around his neck when he was on the scaffold. And after he was killed, she claimed to be a heroine for doing what she did.

Those are the effects of mass formation. Such processes can emerge in different ways. It can emerge spontaneously (as happened in Nazi Germany), or it can be intentionally provoked through indoctrination and propaganda (as happened in the Soviet Union). But if it is not constantly supported by indoctrination and propaganda disseminated through mass media, it will usually be short-lived and will not develop into a full-fledged totalitarian state. Whether it initially emerged spontaneously or was provoked intentionally from the beginning, no mass formation, however, can continue to exist for any length of time unless it is constantly fed by indoctrination and propaganda disseminated through mass media. If this happens, mass formation becomes the basis of an entirely new kind of state that emerged for the first time in the beginning of the twentieth century: the totalitarian state. This kind of state has an extremely destructive impact on the population because it doesn't only control public and political space—as classical dictatorships do—but also private space. It can do the latter because it has a huge secret police at its disposal: this part of the population that is in the grip of the mass formation and that fanatically believes in the narratives distributed by the elite through mass media. In this way, totalitarianism is always based on "a diabolic pact between the masses and the elite" (see Arendt, *The Origins of Totalitarianism*).

I second an intuition articulated by Hannah Arendt in 1951: a new totalitarianism is emerging in our society. Not a communist or fascist totalitarianism, but a technocratic totalitarianism. A kind of totalitarianism that is not led by "a gang leader" such as Stalin or Hitler, but by dull bureaucrats and technocrats. As always, a certain part of the population will resist and won't fall prey to the mass formation. If this part of the population makes the right choices, it will ultimately be victorious. If it makes the wrong choices, it will perish. To see what the right choices are, we have to start from a profound and accurate analysis of the nature of the phenomenon of mass formation. If we do so, we will clearly see what the right choices are, both at strategic and at the ethical levels. That's what my book *The Psychology of Totalitarianism* presents: a historical–psychological analysis of the rise of the masses throughout the last few hundreds of years as it led to the emergence of totalitarianism.

***

The COVID crisis did not come out of the blue. It fits into a series of increasingly desperate and self-destructive societal responses to objects of fear: terrorists, global warming, coronavirus. Whenever a new object of fear arises in society, there is only one response: increased control. Meanwhile, human beings can only tolerate a certain amount of control. Coercive control leads to fear, and fear leads to more coercive control. In this way, society falls victim to a vicious cycle that leads inevitably to totalitarianism (i.e., extreme government control) and ends in the radical destruction of both the psychological and physical integrity of human beings.

We have to consider the current fear and psychological discomfort to be a problem in itself, a problem that cannot be reduced to a virus or any other "object of threat." Our fear originates on a completely different level—that of the failure of the Grand Narrative of our society. This is the narrative of mechanistic science, in which man is reduced to a biological organism. A narrative that ignores the psychological, spiritual, and ethical dimensions of human beings and thereby has a devastating effect at the level of human relationships. Something in this narrative causes man to become isolated from his fellow man, and from nature. Something in it causes man to stop *resonating* with the world around him. Something in it turns human beings into *atomized subjects*. It is precisely this atomized subject that, according to Hannah Arendt, is the elementary building block of the totalitarian state.

At the level of the population, the mechanist ideology created the conditions that make people vulnerable to mass formation. It disconnected people from their natural and social environment, created experiences of radical absence of meaning and purpose in life, and led to extremely high levels of so-called "free-floating" anxiety, frustration, and aggression, meaning anxiety, frustration, and aggression that is not connected with a mental representation; anxiety, frustration, and aggression in which people don't know what they feel anxious, frustrated, and aggressive about. It is in this state that people become vulnerable to mass formation.

The mechanist ideology also had a specific effect at the level of the "elite"—it changed their psychological characteristics. Before the Enlightenment, society was led by noblemen and clergy (the "ancien régime"). This elite imposed its will on the masses in an overt way through its authority. This authority was granted by the religious Grand Narratives that held a firm grip on people's minds. As the religious narratives lost their grip and modern democratic ideology emerged, this changed. The leaders now had to be *elected* by the masses. And in order to be elected by the masses, they had to find out what the masses wanted and more or less give it to them. Hence, the leaders actually became *followers*.

This problem was met in a rather predictable but pernicious way. If the masses cannot be commanded, they have to be *manipulated*. That's where modern indoctrination and propaganda was born, as it is described in the works of people such as Lippman, Trotter, and Bernays. We will go through the work of the founding fathers of propaganda in order to fully grasp the societal function and impact of propaganda on society. Indoctrination and propaganda are usually associated with totalitarian states such as the Soviet Union, Nazi Germany, or the People's Republic of China. But it is easy to show that from the beginning of the twentieth century, indoctrination and propaganda were also constantly used in virtually every "democratic" state worldwide. Besides these two, we will describe other techniques of mass manipulation, such as brainwashing and psychological warfare.

In modern times, the explosive proliferation of mass surveillance technology led to new and previously unimaginable means for the manipulation of the masses. And emerging technological advances promise a completely new set of manipulation techniques, where the mind is materially manipulated through technological devices inserted in the human body and brain. At least that's the plan. It's not clear yet to what extent the mind will cooperate.

***

Totalitarianism is not a historical coincidence. It is the logical consequence of mechanistic thinking and the delusional belief in the omnipotence of human rationality. As such, totalitarianism is a defining feature of the Enlightenment tradition. Several authors have postulated this, but it hasn't yet been subjected to a psychological analysis. I decided to try to fill this gap, which is why wrote *The Psychology of Totalitarianism*. It analyzes the psychology of totalitarianism and situates it within the broader context of the social phenomena of which it forms a part.

It is not my aim with the book to focus on that which is usually associated with totalitarianism—concentration camps, indoctrination, propaganda—but rather the broader cultural–historical processes from which totalitarianism emerges. This approach allows us to focus on what matters most: the conditions that surround us in our daily lives, from which totalitarianism takes root, grows, and thrives.

Ultimately, the text explores the possibilities of finding a way out of the current cultural impasse in which we appear to be stuck. The escalating social crises of the early twenty-first century are the manifestation of an underlying psychological and ideological upheaval—a shift of the tectonic plates on which a worldview rests. We are experiencing the moment in which an old ideology rears up in power, one last time, before collapsing. Each attempt to remediate the current social problems, whatever they may be, on the basis of the old ideology will only make things worse. One cannot solve a problem using the same mind-set that created it. The solution to our fear and uncertainty does not lie in the increase of (technological) control. The real task facing us as individuals and as a society is to envision a new view of humankind and the world, to find a new foundation for our identity, to formulate new principles for living together with others, and to reclaim a timely human capacity—Truth Speech.

Corona mass formation as a societal symptom

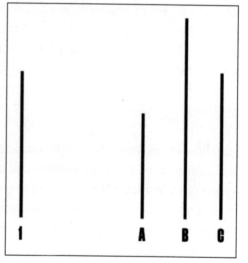

Take a good look at this figure. Which of the line segments A, B, and C has the same length as line segment 1? That was the question Asch asked the participants of his experiment on peer pressure. Each group of eight subjects contained seven Asch employees. They all replied without blinking, "line segment B."

The eighth participant—the only real test subject—gave mostly the same answer as his predecessors. Only 25% consistently expressed what even a blind person can see: not line segment B, but line segment C, is the same length as line segment 1.

After the experiment, some test subjects said that they did know the right answer but were afraid to argue with the group. More interestingly, others admitted that under pressure from the group they had begun to question their own judgment and took the absurd group judgment as true.

We have to face it: even in the COVID crisis, public opinion is in the grip of absurd judgments. The best-known example is, of course, that the reported number of corona deaths in residential care centers was far too high because all deaths were counted, but many other reported figures, such as infection rate and reproduction rate, were also unrealistic.

However wrong it may be, such messages determine public opinion. They are brought up by experts, often on national television, which makes them seem widely accepted. As in Asch's experiment, this is enough for many people to prove their correctness: "Surely it can't be that everyone is wrong," "They wouldn't say if nothing is wrong," etc.

A number of questions arise here: Why is a message carried by a crowd, even if it is wrong, so convincingly? How can intelligent people—the experts—send these questionable messages out into the world? What dangers are associated with such massive psychological phenomena, and how should we deal with them as a society?

Mass formation often arises in a social climate steeped in unease, fear, and meaninglessness (see, e.g., the 300 million doses of antidepressants per year in Belgium and the burnout epidemic). In such an atmosphere, the population is extremely sensitive to stories that identify the cause of their fear and thus create a common enemy—the virus—that must then be "destroyed."

This provides psychological benefits. First, the fear that was previously indeterminate in society is now becoming very concrete and therefore more mentally manageable. Second, in the common struggle with "the enemy," the disintegrating society regains minimal cohesion, energy and meaning; the fight against corona becomes a mission fraught with pathos and group heroism.

In more extreme cases, this puts society in a kind of intoxication that also occurs in a crowd that sings together or chants slogans (e.g., in a football stadium). The voice of the individual thereby dissolves into the overwhelmingly vibrating group voice; the individual feels carried by the crowd and "inherits" its blistering energy. What exactly is sung does not matter; what matters is that they sing it together. Asch's experiment shows the cognitive variant of this: what one thinks does not matter; what matters is that one thinks it together.

As Gustave Le Bon, a French sociologist, noted around 1900, the effect of mass formation resembles that of hypnosis. In both cases, a scary story sucks all the attention, and the field of consciousness narrows. Compare it with the circle of light of a lamp that shrinks and makes everything that falls outside it disappear into the darkness (see figure).

In the Corona crisis you can see an illustration of this phenomenon in this simple example: victims who fall due to the "mitigation" measures such as lockdowns and quarantine (e.g., deaths due to emotional and physical neglect in residential care centers, non-corona patients whose treatment was postponed, victims of aggression indoors) are given little attention and empathy relative to those with death and disability attributed to infection with SARS-CoV-2. There are no daily statistics, case reports, testimonials from family members, etc., that record these indirect damages from the "public health" policies. These victims fall outside the circle of light.

This lack of empathy should not be confused with vulgar selfishness. Le Bon noted that both mass formation and hypnosis allow individuals to radically ignore their selfish strivings, yes, even their own pain. With a simple hypnotic procedure, patients can be anesthetized to such an extent that incisions can be made during surgery without problems. Likewise, during the corona crisis, much of the population is curiously willing to accept measures that "cut" deeply into their pleasure, freedom, and prosperity.

But there is also an important difference between mass formation and hypnosis. In hypnosis, only the field of consciousness of the hypnotized is narrowed; the one who speaks the hypnotic story (the hypnotist) is "awake." In mass formation, the person who articulates the story—in this crisis the expert—is also mentally in the grip of the story. In fact, the virologist's field of attention has narrowed even more than that of the population through his training (which is one-sidedly focused on viruses) and the secondary benefits the story brings him (excessive prestige, authority, research funding, etc.). This explains the surprising finding that experts make mistakes that a layman would not easily make (a phenomenon sometimes referred to as "expert blindness").

Those who fanatically trust the experts and those who completely distrust them (and see conspirators in them) may make the same mistake here: they attribute to the experts too absolute knowledge (and power), the first group in a positive sense, the second in negative. The actual masters of the situation are not the experts, but the stories and their underlying ideologies; the stories own everyone and don't belong to anyone; everyone plays a part in it, nobody knows the full script (not even all-American hero Bill Gates).

Mass formation ensures that the shared social story becomes immune to criticism and confirms itself absurdly. For example: In a paradoxical way, the victims who fall due to the measures (e.g., because of loneliness in residential care centers) are used as an argument for the measures. They are innocently added to the general excess mortality and thus used to justify the measures.

The UN warned that famines as a result of the lockdowns could soon cause millions of victims. We run the risk that these will also be incorrectly counted among the corona victims and that the fear and therefore support for stricter measures will increase exponentially. In this way, society can end up in a vicious circle: the stricter the measures, the more victims; the more victims, the stricter the measures.

Don't underestimate what this could lead to in the future. The idea of housing-infected individuals in isolation centers is still considered a "disproportionate" measure. But insofar as society remains mentally glued to a scary virological story, all it takes is an increase in fear to consider this too "necessary for public health."

In combination with the manipulability of corona tests and a feudal redistribution of power (governors and mayors gain unseen power, due to the impasse of national politics), you see what appears on the horizon: arbitrarily picking up, isolating, and "treating contaminated" people. Social systems that tend toward totalitarianism use different discourses, but they all do it about the same.

The mass psychological dynamics that arise around the real core of the corona epidemic exhibit all the characteristics of a psychological symptom and must be analyzed as such. Just like an individual symptom, it has a signaling function. It refers to an underlying social problem, which we described above as a lack of meaning and associated epidemic anxiety and depression.

This can be felt in the workplace, among other things. Now that the lockdown and the accompanying leave (which did not really feel like leave) are almost over, we have to slowly return to the old work regime. Many of us will again be confronted with the experience described in the bestseller *Bullshit Jobs*: the working day seems to be a succession of obligations that one has to fulfill quickly without knowing who actually benefits.

As someone recently told me, you would almost long for another lockdown. For quite a few people this seems like the only way to escape the grueling rat race and at least experience some sense of meaning and connection with the other in the fight against the virus. For example, in the current mass formation we find another characteristic of psychological symptoms: they are attempts to solve the underlying problem, which are harmful in the long term. This is the task we face: to find meaning and connection in life without the need for a war with a virus. Where in our Western worldview is there an opening that offers the prospect of a meaningful existence as a human being?

CHAPTER 8

Shocking Increases in All-Cause Mortality Coinciding with COVID Vaccine Mandates

By Ed Dowd

It was during a local group dinner and fundraiser for the cause, held on the island of Maui, that Ed Dowd and his colleagues first introduced themselves to me. To my surprise, they indicated that they had authored a document they had named "The Malone Doctrine." Taken aback, as neither Jill nor I had contributed in any way, I asked why they were using my name. Ed and his colleagues told me that "we have read and listened to everything you have said and written during COVID, and this is what is written in the white spaces between every line." The wisdom and clarity they captured in that document has become a guiding light for many since it was published, and their "Malone Doctrine" provides the basis for every statement regarding integrity that I have made since that time. But where did this profoundly prescient document come from? A dedicated team, which included a senior building inspector, some young, hard-working idealists, and an experienced hedge fund manager—Ed Dowd. Their doctrine was written to address what they saw as a fundamental societal breakdown in commitment to integrity. Not just for the US Department of Health and Human Services, but for virtually every "vertical" sector throughout society, government, and particularly business.

One day, Ed brought me emerging data concerning all-cause mortality as revealed by senior-life insurance executives. These data cut through the fog of corporate media misrepresentations vaccine-associated adverse event data. These data—coming from a source other than the government—raised profound questions about the "safe and effective vaccines" narrative.

* * * * *

Central Banks, Global Debt, and COVID

My journey into the world of fraud began long before COVID-19. I have spent the majority of my career on Wall Street at firms like HSBC Inc., Donaldson Lufkin & Jenrette, and BlackRock, where I learned about fixed income, currency, and equity markets. My knowledge of global capital markets is very deep. I witnessed the DotCom fraud as well as the Mortgage fraud that was detailed in the movie *The Big Short* from the inside. The key takeaway from the DotCom disaster is that the easy money from the Federal Reserve was responsible for driving the speculation, which in turn led to a gigantic misappropriation of capital, which ended up driving tremendous losses, fraudulent behavior, and theft. Without easy money none of this would have happened. When it comes to the Mortgage fraud, it's important to remember that not one banker went to jail for these crimes; the banks were bailed out, but the homeowners and others who were the financial victims were not.

Thus began the rise of global central-bank dominance. This process was facilitated by unprecedented cooperation between the US Federal Reserve, the EU Central Bank, and the Bank of Japan in the form of money printing and debt purchases. The global governments of the world went into deficit spending to make up for the catastrophic wealth destruction, job losses, and demand destruction, while the banks and investment funds that caused the problem emerged as the dominant global power brokers.

The Most Recent Cycle: Central Bank and Political Fraud with COVID As the Cover-Up

The free-market system (as we previously understood it) ended on March 5th, 2009, when the Fed began its historic bailout of the banking system. The twelve years since then have seen an unprecedented growth in global debt to keep the patient known as the global economy on life support. Crony capitalists and those closest to the money-printing machines have seen their wealth grow, while the rest of the citizenry have been lucky if they were able to march in place and not give up ground economically. Washington DC's power and wealth have increased mightily since that crisis unfolded. The percent of GDP that the government now commands and controls is 40% thanks to the COVID crisis and is up from low double digits forty years ago.

Knowing that all cycles end and that the growth in global debt was unsustainable, many of us in the financial community wondered what the end of this cycle would look like. How would it manifest, we pondered? Political instability? Currency wars? Sovereign debt defaults? In 2019 we saw signs of global growth slowing and a repossession crisis when overnight lending rates spiked in the fall. Corporate credit spreads began to wobble a bit. It looked like we were nearing the end of this cycle. Then the COVID-19 crisis hit, and the central banks had an excuse to print the largest amount of money in the history of the Federal Reserve, with a 65% increase in the money supply from 2019 to 2020.

As a seasoned investment fund manager, my suspicions were triggered by this surge in the money supply. My suspicions were confirmed when I saw Saint Louis Federal Reserve President James Bullard being interviewed about how to reopen the economy on the April 5, 2020, edition of *Face the Nation*. I began suspecting that COVID-19 was being exploited as cover for a new global financial collapse. Bullard indicated that we had new technologies that could test people; those with a negative test could wear immunity badges, and new surveillance technologies would be deployed that could track them. I was blown away. Why was a Federal Reserve president weighing in on public health? I speculated that once a vaccine was introduced, governments would begin to implement vaccine passports. For raising this concern during 2020, I was labeled a conspiracy theorist.

Fund managers such as myself typically operate based on various models they develop to explain long- and short-term political and economic trends. I developed a working thesis that COVID-19 would be used as an excuse to control travel and clamp down on global riots once the debt collapse began in earnest. The collapse of the world's economies would be blamed on COVID-19, and ensuing "safety" measures would be put in place as a system of control and compliance—for our own good. Also, continued virus evolution and outbreaks could be used as additional excuses to print and inject more money into the collapsing economy by the central banks. Under this theory, the vaccine passport would merely be a gateway device for what would eventually become a central bank digital currency system that would both monitor vaccine compliance and institute a social credit score to make sure you behave as "a good citizen." To memorialize this thesis, I began a Twitter thread on May 3, 2020, which predicted what might unfold over the next two years.

How I became Involved in the Fight for Freedom and The Malone Doctrine

In September of 2021, I became very distraught while watching my predictions unfold before my eyes. All over the world, societies were falling into a dystopian nightmare that most of the people around seemed to be fine with, as they gave their freedoms away without so much as a single thought or suspicion. There were a number of residents on Maui (Hawaii) who did consider the draconian mandatory vaccines and passports an assault on our freedoms. I attended the multiple local rallies and protests that ensued. I met many different kinds of people who all had one thing in common, a belief in fundamental bodily sovereignty. There was no red team-blue team dynamic, but rather a wide range of different races and creeds and belief systems were represented. We began to call ourselves team humanity. After I presented Dr. Malone and his wife, Jill, with the Malone Doctrine, they asked me to be on the board of the Malone Institute (maloneinstitute.org). Over the next few months, I informed Dr. Malone that I would be monitoring the life insurance companies and funeral home results to confirm what we had suspected, namely, that excess mortality and disability were being caused by the vaccine.

Data Fraud, Life Insurance, CDC Excess Mortality & US Disability

In the first week of January 2022, OneAmerica CEO Scott Davison made comments at an Indiana Chamber of Commerce meeting that were picked up by a reporter for The Center Square Margret Menge:

> "We are seeing, right now, the highest death rates we have seen in the history of this business – not just at OneAmerica," the

company's CEO Scott Davison said during an online news conference this week. "The data is consistent across every player in that business."

Davison said the increase in deaths represents "huge, huge numbers," and that's [*sic*] it's not elderly people who are dying, but "primarily working-age people 18 to 64" who are the employees of companies that have group life insurance plans through OneAmerica.

"And what we saw just in third quarter, we're seeing it continue into fourth quarter, is that death rates are up 40% over what they were pre-pandemic," he said.

"Just to give you an idea of how bad that is, a three-sigma or a one-in-200-year catastrophe would be 10% increase over pre-pandemic," he said. "So 40% is just unheard of."

Most of the claims for deaths being filed are not classified as COVID-19 deaths, Davison said.

He said at the same time, the company is seeing an "uptick" in disability claims, saying at first it was short-term disability claims, and now the increase is in long-term disability claims.

When I appeared on Bannon's War Room to say that the vaccine program was based on fraudulent data, I said I would be monitoring the results of insurance companies and funeral homes for the fruits of their fraud. Unlike financial frauds, the damage here was not monetary, but human lives and long-term health. Two individuals came forward to assist my effort.

The first was Brook Jackson, the Ventavia Research Group whistleblower who witnessed and reported data corruption in the Pfizer 28-day clinical vaccine trial. She oversaw 1,000 of the 44,000 patients enrolled in the clinical trial, and the most egregious thing she witnessed was the unblinding of the patients. As a direct violation of Pfizer's own protocols, the data should have been thrown out. Instead, after reporting the irregularities to the FDA, Brook was fired, and the data made their way into the clinical trial. The number of COVID patients during the 28-day trial was so small that results from Brook's clinical research site alone could have mathematically altered the results (95% effectiveness) such that the vaccine appeared completely ineffective. Real-world experience has proven that at a bare minimum the vaccine does not stop infection or transmission. This was the original sin and genesis of the fraud.

Brook's sites were awarded by Pfizer for doing a great job. It can be easily inferred that *other* sites also engaged in unblinding of the patient data. The impact of this unblinding is essential to understand. Unblinding introduces bias, potentially leading a doctor to assume that an ill subject doesn't have COVID because the doctor sees that the patient had the vaccine, which the doctor assumes is effective. In this way, unblinding the study results in underreported disease and overreported efficacy. Brook is currently engaged in a lawsuit with Pfizer; she recently learned that one of Pfizer's defenses is that the fraud was OK because the US government was *aware* of it. The fact that this is not appearing as a mainstream media story is a testament to the corruption of the press by government money.

The second person who came forward was Josh Stirling. Josh was a former #1-ranked institutional investor Wall Street insurance analyst who worked for Sanford C. Bernstein Research. He had us focus on the loss ratio of the group life and disability divisions of the life-insurance companies. We did that because that is a stable and very profitable sector for insurance companies. These are the typical death benefit and disability policies offered to midlevel employees when they join a corporation. The death benefit is expected to be rarely collected, as (statistically speaking) healthy working-age people with good jobs rarely die.

What we found was stunning and confirmed what the OneAmerica CEO saw in January. Some of the major insurers saw increases in their fourth-quarter loss ratios ranging from 25% to 45% over 2019 base line levels, and there was a continued rise from the third quarter of 2021. Many CEOs blamed this huge increase on COVID, developing a strange new concept they termed "indirect COVID." Disability also saw a marked increase and continues to climb today. Josh and I suspected that the reason these insurance companies didn't observe these types of losses and excess mortality in the early part of the year was that corporate vaccine mandates instituted by the Biden administration began in the fall of 2021, and that coincided with the huge uptick in the deaths and losses the insurance companies experienced in the third and fourth quarters of 2021. Remember these are working-age people who, as a group, were not affected by COVID in 2020 before the vaccines were deployed. Suddenly at the end of 2021, however, they experienced a huge uptick in excess all-cause mortality and disability. The only thing that changed from 2020 to 2021 was the vaccination program and the mandates.

Next Josh looked at the CDC excess mortality data. The data as presented on the CDC website were not very helpful, as all ages were lumped together. It was, in itself, damning, however, as it showed two spikes of excess mortality. The first spike was during the fall-winter of 2020; then there was a subsequent spike in the fall of 2021, which was almost but not quite as high. That alone would suggest gross incompetence by our health officials, given the introduction of the supposedly miraculous vaccines. However, Josh was able to grab the data from the website and break it down by age. He developed baseline mortality analyses from 2015 to 2019 (before COVID) and then developed excess mortality charts over time for each age group. What he found was stunning. His analysis effectively confirmed the results we had seen from the life insurers in their financial reports. Millennials saw an acceleration of excess mortality into the second half of 2021 to new all-time highs, a stunning 84% above baseline. The rate of change during the fall vaccine mandates was particularly striking. We called this the smoking gun chart. The virus wasn't suddenly killing younger people in the fall of 2021. Suicides didn't magically increase in that three-month period, nor did overdoses or missed cancer screenings. The only thing that changed was that genetic vaccine products were pushed upon the millennial generation via government and corporate mandates. We summarized this stunning finding in a series of graphs (see following pages).

The second most important discovery that Josh found was the shift in the proportion of excess mortality from old to young that occurred from 2020 to 2021. In 2020 there were 592,000 excess deaths with 126,000 under the age of 65 (approximately 21%). In the second year of the outbreak, there were 512,000 excess deaths with 181,000 under the age of 65 (approximately 35%). The millennials saw the greatest percent increase in mortality of 45% from 42,000 to 61,000. This shift in mortality to younger groups cannot be due to COVID, because the virus was already mutating and becoming less virulent, and we had already determined that the virus killed mostly older people with comorbidities. It's important to note that 45,000 more people under the age of 65 died in year two than in year one. Did the virus suddenly preferentially target younger folks? The only thing that changed in

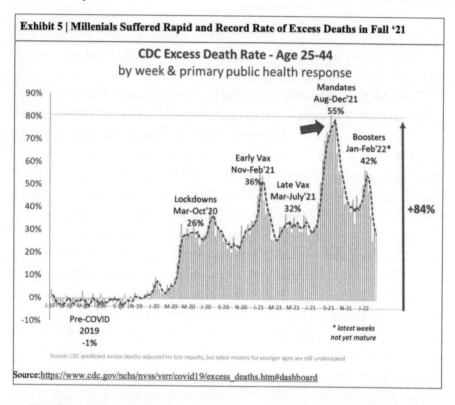

Exhibit 5 | Millenials Suffered Rapid and Record Rate of Excess Deaths in Fall '21

CDC Excess Death Rate - Age 25-44
by week & primary public health response

Source: https://www.cdc.gov/nchs/nvss/vsrr/covid19/excess_deaths.htm#dashboard

The Millennial generation suffered its worst-ever excess mortality last fall, and these deaths occurred the same time as vaccine mandates were announced, and boosters approved. This young population is not particularly at risk to COVID, and the size and timing of this spike in fall of 2021 raises clear questions about potential contributions from the vaccines and boosters. As you know, mortality reporting for younger-age people is also typically much slower (due to slower reporting on nonhospital deaths), so the recently elevated levels for the age group persisting into early 2022 will most likely develop further and may signal for continuing elevated mortality among working age in 2022.

year two was the introduction of the vaccine and the subsequent mandates, making them the obvious culprit. The authorities and the corporate media refuse to acknowledge, much less comment on, these data. If the data were acknowledged, the obvious question would arise—are the vaccines really safe and effective?

In the first week of June 2022, Josh and I discovered another database collected by the US Bureau of Labor Statistics, the department responsible for the monthly household survey that delivers the employment report every month. The survey routinely asks a number of questions related to disability.

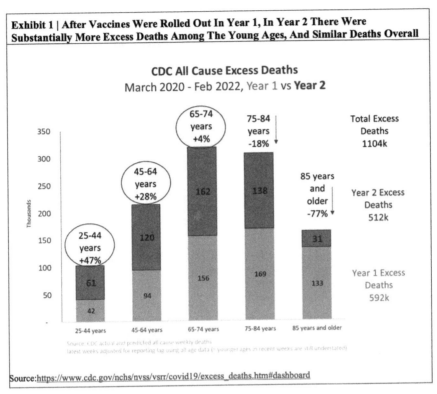

Exhibit 1 | After Vaccines Were Rolled Out In Year 1, In Year 2 There Were Substantially More Excess Deaths Among The Young Ages, And Similar Deaths Overall

Mix of excess deaths by age shift from year 1 to year 2 shows deaths among elderly declined, while deaths among the young increased substantially. As younger ages are less exposed to COVID and most other health issues, this suggests other health policies likely were substantial contributors. It is also worth noting that the substantial declines in excess deaths among the elderly, is likely substantially due to "pull forward" of deaths among elderly into 2020, rather than evidence of actual underlying mortality improvement in 2021.

The surveyed respondents indicate whether they are disabled or someone else in their home identifies as disabled. This number is not derived from claims or doctors' notes, but rather is based on self-identification. This is important to note because it gives a very good real-time snapshot of disability trends in the US. Prior to the vaccines, the run rate was about 29 million, give or take, for the last five years. As of June 2022, the number of Americans self-identifying as disabled has increased by 13.7% over the prior run rate. This represents a numerical increase of about 4 million Americans. The graph below shows the steep rate of change that continues today.

In my opinion, these data reflect an ongoing national disaster. I suspect the labor shortages we are seeing are heavily influenced by this number and can also explain much of the inflation in wages we are seeing. Again, the simplest explanation for this increase is that the disability is caused by the vaccines. When pressed, the establishment claims that this is due to long COVID [50]. However, most officials and corporate press outlets are ignoring and obfuscating this national tragedy [50].

I also took a look at funeral home results. Consistent with our findings from the other databases, business has been quite good for publicly traded funeral homes, with the number of funeral contracts in the second half 2021 accelerating into the end of the year and continuing into 2022. The commentary from funeral home executives during the first quarter of 2022 was

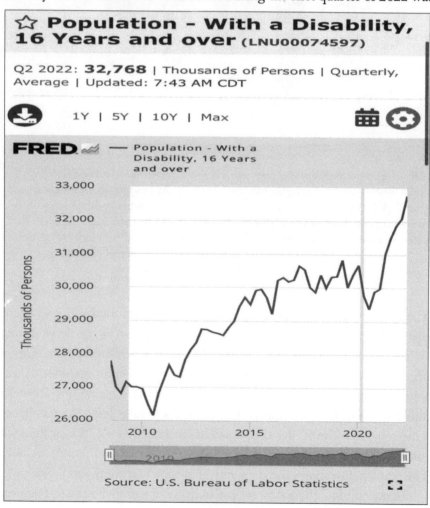

interesting. They were mostly surprised by their own results, and one executive even said the deaths his company was seeing could not all be explained by COVID. Service Corporation International hit all-time highs and has been outperforming the S&P 500 return for well over a year and a half. One would expect that with the introduction of "miracle" vaccines and the pandemic itself being over, funeral homes would be seeing business return to normal trends. Unfortunately, that assumption is wrong, and funeral homes are currently growth stocks.

Central Bank Exploitation of the COVID crisis

In the late nineties, corporate fraud took over, and we had a 50% stock market correction. The Federal Reserve responded by turning on the money spigot: they lowered interest rates, and the money found its way into the real estate market, which turned into an unsustainable bubble.

Real estate was being hypothecated through collateral debt obligations and mortgage-backed security. Wall Street levered up 20-to-1, 30-to-1 on their balance sheets to make money and thought the party would go on forever. But inevitably, the Fed started to raise interest rates, and the whole thing collapsed. The problem with this bank fraud was that it was systemic in nature. The central banks had to step in and buy this fraudulent debt.

So, this fraud still remains today, on the Federal Reserve's balance sheet, and on the balance sheets of countless other banks. In other words, the fraud didn't go away. It was just baked in and hidden. As a consequence, financial collapse is a mathematical certainty. Then, because the economy collapsed globally, governments started spending like drunken sailors. The last twelve years have been a ballooning of the central bank-government bubble, the sovereign debt bond bubble. Who's going to save that bubble? Who's going to be the buyer of all that debt when this bubble finally blows up? Answer: No one. Many who are aware of the situation are just surprised the system has lasted this long.

It looked like it was ready to burst in 2019, and then, conveniently, COVID-19 showed up, which granted emergency powers to all central banks. Governments went on another spending spree, printing money, and this allowed them to kick the proverbial can down the road for another two years. Here we are in 2022, and the financial system is unraveling again. And the reason why COVID was important is that the Federal Reserve was able to plug the hole in what was beginning to become a liquidity debt crisis. The Fed printed 65% more money. The money stock went up 65% year over year in 2020, and that was able to paper it over. Then, when the economy was shut down, it was an external shock, not an internal shock, so when they reopened with all the money in the system, we had a recovery for a year and a half. Stock markets went crazy, credit markets went crazy, and we went back up again.

But here we are two years later, with inflation caused by the bad policies of the Biden administration, the EU, the money growth. COVID also broke a lot of supply chains. Basically, we hadn't had inflation in goods and services for the last twelve years. We had inflation in assets, stocks, and bonds. But what's going on now is the real economy is feeling the effects of the inflation, the bad policies. We're starting to see the US dollar go up, and the dollar is a reserve currency of the world. Over the last twenty-two years, there's been a tremendous growth in what's called dollar-denominated debt. We have about $15 trillion in dollar-denominated debt. So, when you see the dollar going up, that's indicative of a debt crisis because money's becoming tight. There are fewer dollars out there. People are scrambling for dollars. And the reason why I think we're imminently going to collapse is we've never seen a commodity inflation cycle with the dollar going up at the same time.

You can make the case that it's intentional because the policies are so bad that they're shutting down energy production. Before the Ukraine War, Biden's first executive order on Day 1 of his administration was to shut down the Keystone pipeline. So, here we are. I think we're at the end.

COVID provided cover for the central banks and the governments, but it also allowed for a control system. If everything's going to collapse, wouldn't it be nice to have a control system where travel is restricted, you can blame it on a virus, you create vaccine passports, which then get linked to digital IDs, and then central bank digital currency. So, I think COVID was a convenient excuse. As we roll through time, I'm starting to think this was a plan. I don't have evidence, but the fact that we're not stopping what's going on suggests to me that it's a conspiracy of interests, and they don't want to stop the rollout of these vaccines. And the longer this goes on, the more convinced I become that COVID may have been a plan. I used to say it was a convenient excuse, but the longer this goes on, the more ridiculous this becomes. This has the appearance of ill intent.

If stock markets become seriously unhinged and we start getting declines of more than 40% in the indices, the Federal Reserve may start buying stocks outright, which will result in a neofeudalism system that will only magnify already-existing discrepancies between the haves and the have-nots. At this point, there is no market mechanism to punish anybody for making bad decisions. Bad decisions by large investors and banks are bailed out by the central banks. The moral hazard is so high that if you just are a C-suite executive at a major Fortune 500 company, you're going to become phenomenally wealthy, and you do not have to really be particularly skilled. You're going to be one of the lords, and the workers and everybody else are going to be struggling to make ends meet. That's what's been going on for the last twelve years. The economy for the most part has been an economy of the big and those close to the printing machine. If you're trying to actually create a small business, if you're a worker at one of these corporations and you don't get a lot of stock options, you're not getting ahead.

Final Thoughts and Implications

My successful career in stock picking was predicated on pattern recognition, developing a thesis with limited information, taking an early initial position, and being proved right or wrong over time. Essentially, I had learned to become a stock-picking "conspiracy theorist." Those who want to call me that now are welcome to do so, but I believe that my thesis on the link between global debt, central banks, and COVID will be borne out over time and has already gained more legitimacy since my initial "crazy predictions" in May of 2020. With regard to vaccine data, every week that rolls by produces more evidence of malfeasance by Pfizer and Moderna, unearthed by Dr. Naomi Wolf's dedicated volunteers who peruse the clinical trial data that the FDA wanted to hide for seventy-five years. The evidence of excess death and disability continues to pile up. I have never been more convinced that the vaccines not only don't work, but that these are the deadliest vaccines ever introduced into the human population. The US government is guilty of democide with their forced mandates, and countless corporations and government agencies are also liable for forcing employees to accept injections of experimental vaccines that employ novel gene therapy-based technology.

The corporate media and large tech companies are also complicit due to their censorship of critical vaccine information and, in my opinion, are accessories to wrongful death. Once we open up the Overton window (the range of allowable public, political, or scientific discourse) on this topic, and the majority of the population learns what has happened, we will see a tremendous loss of trust in our institutions. At a bare minimum the NIH, CDC, FDA, and Health and Human Services structures need to be razed and built up again anew. In addition, the politicians, doctors, university administrators, and media-tech complex that pushed this vaccine program will have much to answer for in the coming years. They, too, will have lost the public's trust and will need to rethink their institutions and governance.

PART TWO
DIAGNOSIS – LIES AND THE DAMAGE DONE

Ask yourself, has the US government and the WHO earned our trust? Do they have any right to set and police global health policies? US President Joe Biden and (now former) NIAID Director Anthony Fauci apparently think so, based on the modifications to the international health regulations that they submitted for consideration to the World Health Organization during late January 2022.

On January 18th, 2022, the United States Department of Health and Human Services proposed amendments to the International Health Regulations (IHR). These IHR amendments also cede control to WHO "regional directors," who are given the authority to declare a Public Health Emergency of Regional Concern (PHERC). These proposed IHR amendments advocate for "an adaptable incentive regime, [including] sanctions such as public reprimands, economic sanctions, or denial of benefits." Properly understood, the proposed IHR amendments are directed toward establishing a globalist architecture of worldwide health surveillance, reporting, and management. Consistent with a top-down view of governance, the public will not have opportunities to provide input or criticism concerning the amendments. The anticipated impacts include increased global and national surveillance, a forty-eight-hour deadline for national governments to respond to WHO determinations and mandates, secret WHO Intelligence operations, weakened national sovereignty, and an abbreviated six-month amendment review timeline. Voting on these amendments was scheduled to occur during President Biden's trip to the UN's 75th World Health Assembly in Geneva, Switzerland, but was postponed until a following meeting currently scheduled for November 2022. The postponement was largely a consequence of objections from African member states who were concerned with the implied loss of national sovereignty.

The following is a brief summary of US government and WHO COVIDcrisis lies we have heard over the last couple of years and builds upon the list initially developed by Dr. Scott Atlas, who served as a COVID advisor to the Trump administration:

Lies the US government has told all of us

- SARS-CoV-2 coronavirus has a far higher fatality rate than influenza virus by several orders of magnitude.
- Everyone has a significant risk of death from COVID-19.
- No one has immunity, because this virus is new ("novel"), and so expedited vaccine development and deployment is essential.
- Everyone is dangerous and spreads the infection.
- Asymptomatic people are major drivers of the spread of disease.
- Locking down—closing schools and businesses, confining people to their homes, stopping non-COVID medical care, and eliminating travel—will stop/eliminate the virus.
- Masks will protect everyone and stop the spread.
- Immune protection can only be obtained with a vaccine.
- Natural immunity conferred by infection and recovery is short-lived and inferior to vaccine-induced immunity.

Who was responsible for these lies?

- Deborah Birx (who was trained by Anthony Fauci)
- She wrote virtually all official White House guidance to state Governors.
- This usurped constitutional authority of states to set public health policies
- Anthony Fauci
- Francis Collins

What were their policy decisions?

- "Flatten the Curve"… Then "Stop all cases."
- No masks. Then all masked.
- Lockdowns: School closures, business shutdowns, limits on medical care, a host of restrictions, mandates, and quarantines.
- Perverse financial incentives for hospitals to overdiagnose COVID-19, overuse Remdesivir and ventilation, and cause a massive wave of iatrogenic (drug/doctor caused) excess death.
- Stop early treatment and block repurposed drug use.
- "Come back to the hospital when your lips are blue."

What was the effect of their policy decisions?

- Virus? >1,000,000 American deaths attributed to the virus. One of the highest mortality rates per capita in the world.
- Lockdowns? Caused massive deaths and severely harmed millions of families and children, especially working class and poor.

A better alternative was known by March 2020, known as "targeted protection," and was described in the following media:

STAT: John P. Ionnidis, "A fiasco in the making? As the coronavirus pandemic takes hold, we are making decisions without reliable data," March 17, 2020.

New York Times – David L. Katz, "Is Our Fight Against Coronavirus Worse Than the Disease?" March 20, 2020.

Washington Times – Scott W. Atlas, "Widespread isolation and stopping all human interaction will not contain the COVID-19 pandemic," March 26, 2020.

CNN – Martin Kulldorff, (in Spanish—he could not get it published in English), "Abrir o no abrir las escuelas: la experiencia sueca," ["To open or not to open schools: the Swedish experience"] August 20, 2020.

What were the alternative policies proposed?

- Increase the protection of the high-risk groups with an unprecedented focus.
- Reopen society, including medical care, schools, businesses, and hospitals.
- Carefully monitor hospital capacity and supplement when needed.

This set of policy recommendations was codified on October 4, 2020, as the Great Barrington Declaration.

According to Wikipedia:

> The Great Barrington Declaration was an open letter published in October 2020 in response to the COVID-19 pandemic and lockdowns. It claimed harmful COVID-19 lockdowns could be avoided via the fringe notion of "focused protection," by which those most at risk could purportedly be kept safe while society otherwise continued functioning normally. The envisaged result was herd immunity in three months as SARS-CoV-2 swept through. Authored by Sunetra Gupta of the University of Oxford, Jay Bhattacharya of Stanford University, and Martin Kulldorff of Harvard University, it was drafted at the American Institute for Economic Research in Great Barrington, Massachusetts, signed there on 4 October 2020, and published on 5 October. The document presumes without evidence that the disease burden of mass infection can be tolerated, that any infection confers long term sterilizing immunity, and makes no mention of physical distancing, masks, contact tracing, or long COVID, which has left patients suffering from debilitating symptoms months after the initial infection.

Why did the public believe the lockdown advocates?

- Culture of trust (of the credentialed class)

- Fear (actively weaponized against the public by the government, WHO, and corporate media)
- Demonization of opposing views (globally coordinated propaganda and censorship campaign)
- Legacy media, social media, and political campaigns

Key messaging deployed to support the lies

- If you are against lockdowns, you are selfish and choosing the economy over lives.
- If you are against lockdowns, you are for allowing the infection to spread without mitigation and therefore in favor of unnecessary and preventable deaths.
- Active destruction and denial of fundamental public health ethics
- "If a school is implementing a testing strategy, testing should be offered on a voluntary basis. It is unethical and illegal to test someone who does not want to be tested, including students whose parents or guardians do not want them to be tested." CDC, October 13, 2020.

Mandating vaccines for children

- "But we're never going to learn about how safe this vaccine is unless we start giving it. That's just the way it goes." Eric Rubin, MD, Editor in Chief, *New England Journal of Medicine*, October 26, 2021 (FDA Advisory meeting on vaccine approval in children)

How to restore trust in science?

- Admit errors in public forums.
- Change Leadership.
- Strengthen conflict of interest rules and add term limits on government agency leadership positions.
- Clarify definition of "public health emergency" with strict time limits, adding legislative action requirement to extend.

- Restore appropriate roles of health agencies to advise, rather than set rules.
- Fact-check the media.
- Decentralize research funding.
- Introduce new transparency and accountability.
- De-anonymize reviews of papers and grants.
- Increase independent oversight to government agencies and committees.
- Evaluate universities regarding ethics, free debate.
- New training programs, including logic and ethics for journalists, doctors, and scientists.

SECTION 1:
MEDICINE, SCIENCE, PHILOSOPHY, AND PSYCHOLOGY

CHAPTER 9
Science versus Scientism

Eighteen months into the COVIDcrisis, many people suddenly realized that Dr. Anthony Fauci, longstanding director of the National Institutes of Allergy and Infectious Diseases (NIAID), was not the benign, selfless, fatherly protector of public health that corporate media had made him out to be. I had known for decades of his failure to follow the clinical research standards that should apply to scientists. I had lived through the consequences of his aggressive moves to gather power and money at the expense of other scientists and federal agencies. My decades of professional experience in dealing with the NIAID in the context of grant and contract peer review, combined with Jill's PhD research project concerning the NIH peer review system, had left me with little respect for Dr. Fauci's professional integrity.

It was June 9, 2021; Robert Kennedy's shocking book *The Real Anthony Fauci* was yet to be printed. At that point in time, I thought that Tony could do nothing that would shock me. And then he gave the infamous Chuck Todd interview wherein he equated himself to science, and it suddenly became clear that Dr. Fauci had lost all perspective and was suffering from what can best be described as megalomania. He appeared to be channeling the Sun King, Louis XIV, who made the infamous statement, "The State is me." The interview transcript speaks for itself, but to fully appreciate the interpersonal interaction between interviewer and interviewee, it is helpful to view the video recording.

> It's very dangerous, Chuck, because a lot of what you're seeing as *attacks on me*, quite frankly, *are attacks on science . . . .* So, *if you are trying to get at me as a public health official and scientist, you're really attacking not only Dr. Anthony Fauci, you are attacking science* [51].

The impetus for Dr. Fauci's breathtakingly arrogant statement was a video from Sen. Marsha Blackburn (R-TN), who floated a theory that the Director of the NIAID was colluding with Facebook CEO Mark Zuckerberg to develop an approved narrative about COVID-19, presumably in order to cover up NIAID and Fauci complicity in coronavirus gain-of-function research performed at the Wuhan Institute of Virology, which many were coming to believe offered the best explanation of the origin of the SARS-CoV-2 virus. Chuck Todd called the Senator's accusation a "really wild, fantastical conspiracy."

According to Senator Blackburn, "Dr. Fauci was emailing with Mark Zuckerberg from Facebook, trying to create that narrative." He was "cherry-picking information so that you would only know what they wanted you to know, and there would be a narrative that would fit with this cherrypicked information" [52].

Due to the filing of a couple of pending lawsuits [53] and Freedom of Information Act requests that disclosed Fauci and Zuckerberg's email correspondence, we now know that Senator Blackburn's accusations had significant merit.

When Dr. Fauci's arrogant elitism was revealed on camera to the world, a turning point was reached in the coordinated efforts to deify the most powerful scientific bureaucrat/politician in modern history. In a prescient opinion piece published in the *Washington Times* in April of that year, Everett Piper predicted Dr. Fauci's fall from grace [54]:

> More than a year ago, Americans welcomed Anthony Fauci into their homes as a sober scientist who was helping them make sense of a deadly new virus. But he has worn out that welcome.
>
> William F. Buckley's heirs are absolutely right, and here's why. Anthony Fauci is no longer viewed as our nation's sober "scientist" because he's not one. Instead, he has shown himself to be a political opportunist and our country's new high priest of "scientism."

Piper quoted G.K. Chesterton, who said, "I never said a word against eminent men of science. What I complain of is a vague, popular philosophy which supposes itself to be scientific when it is really nothing but a sort of new religion and an uncommonly nasty one." "Predicting the rise of what he and others labeled 'scientism,'" said Piper, *"Lewis warned of a dystopia where public policy and even moral and religious beliefs would be dictated by oligarchs only too eager to assume the role of our new cultural high priests."* C.S. Lewis further said [55]:

> [T]he new oligarchy must more and more base its claim to plan us on its claim to knowledge. If we are to be mothered, mother must know best. This means they must increasingly rely on the advice of scientists, till in the end the politicians proper become merely the scientists' puppets.

So, what is scientism, and how is the concept important for understanding the COVIDcrisis? Merriam-Webster defines it as "an exaggerated trust in the efficacy of the methods of natural science applied to all areas of investigation (as in philosophy, the social sciences, and the humanities)." The term was popularized by F.A. Hayek, who defined it as the "slavish imitation of the method and language of Science." And Karl Popper defined it as "the aping of what is widely mistaken for the method of science" [56].

In the Chuck Todd interview that (appropriately) prompted almost universal derision around the world, Todd confronted Fauci with Republican politicians' accusations of collusion with Facebook leadership to establish an approved narrative that would protect his status and support his various unilateral authoritarian policies:

> I don't even know where to begin, but it's a sitting United States Senator. It's the most, what I would call the most extreme version of what I have heard. You've got Kevin McCarthy doing his own version of this. Marco Rubio, you're aware of the critiques. You've been debunking this. How do you debunk that? She's got it in her own head. Again, a sitting US Senator that represents the State of Tennessee? What do you say to that?

Note the gaslighting of a female US Senator from a rural state. Fauci replied, "You know Chuck, I don't have a clue what she just said, I don't have a clue what she is talking about," and shrugged his shoulders.

"Neither do we," Todd interjected, without defining who "we" is, another classic propaganda strategy.

Fauci continued, "I mean so, welcome to the club. I have no idea what she is talking about."

"And I am sorry, I do not want to be pejorative about a United States Senator, but I have no idea what she is talking about.

"And, and you know, Chuck, if you go through each and every one of the points, which are so ridiculous, as, as, as, as, you know, just painfully ridiculous, but nonetheless, if you go through each and every one of them, you can explain and debunk it immediately."

Notice how smoothly Fauci avoids answering Todd's question. This all may have been a classic distraction feint by a smooth and experienced DC bureaucrat that got out of hand. There are, in fact, multiple lines of evidence demonstrating collusion between Facebook and the US government as well as the World Health Organization. For example, Facebook has publicly stated it is assisting efforts of the White House, the CDC, and the WHO to censor unwanted speech about vaccines. In fact, this government-Big Tech collusion began before the COVIDcrisis, when Representative Adam Schiff (D-CA) wrote to Facebook and Google leadership directly, to urge censorship of vaccine "misinformation" [57].

Children's Health Defense responded to the Schiff letter with a nine-page March 4, 2019, open letter to Facebook CEO Mark Zuckerberg and subsequently filed a lawsuit [58]. The statement detailed the government request to suppress and purge Internet content critical of its vaccine policies. This succinct summary of the dangers of the US government's scientism regarding vaccination policies highlights the "use of so-called 'independent fact-checkers,' which, in truth, are neither independent nor fact-based." Consequent to a lawsuit brought by journalist John Stossel, we learned from Facebook's legal team that the supposed battle against "misinformation" has been a farce. Facebook admitted that the "fact-checks" social media use to police what Americans read and watch are just "opinion" [59].

The CHD letter to Zuckerberg provides multiple clear and compelling data—and/or logic-based examples of the consequences of US government-promoted scientism in action [60]. For example:

For your company to take on the role suggested by Mr. Schiff, you would essentially be engaging in the practice of censoring information about vaccines on behalf of the government. There is no other way to logically interpret his letter, in which he expresses his expectation that your company will take measures to stop Facebook users from seeing what he calls "antivaccine" information, a term he treats synonymously with "medically inaccurate information about vaccines." Mr. Schiff expresses his concern that certain information might discourage parents from vaccinating their children, and he describes any such information as "a direct threat to public health."

Hence, Mr. Schiff's true criterion for determining what information constitutes a "threat" is not whether it is truthful and accurate, but whether or not it accords with the goal of achieving high vaccination rates. In a truly Orwellian fashion, he then defines any information that could undermine that goal as "medically inaccurate." He is, in short, employing the logical fallacy of begging the question. When he says that certain information threatens public health, what he really means is that it threatens current public health policy.

Mr. Schiff's false statements are indicative of the problem of how the government systematically misinforms the public about vaccine safety and effectiveness. The CDC itself is a leading purveyor of misinformation about vaccines. For example, a literature review by the prestigious Cochrane Collaboration [author's note: 2009] on the safety and effectiveness of the influenza vaccine concluded that the fundamental assumptions underlying the CDC's universal flu shot recommendation are unsupported by the scientific evidence and, furthermore, that the CDC has deliberately misrepresented the science in order to support its policy.

In a foreshadowing of the censorship, propaganda, defamation, and coordinated (old and new) media policies that have directly contributed to the catastrophic global mismanagement of the COVIDcrisis, the WHO (specifically, Director-General Dr. Tedros Adhanom Ghebreyesus, who is an Ethiopian microbiologist, malaria researcher, and politician—with no medical training) issued a September 4, 2019, press release stating [61],

The World Health Organization welcomes the commitment by Facebook to ensure that users find facts about vaccines across Instagram, Facebook Search, Groups, Pages, and forums where people seek out information and advice.

Facebook will direct millions of its users to WHO's accurate and reliable vaccine information in several languages, to ensure that vital health messages reach people who need them the most.

The World Health Organization and Facebook have been in discussions for several months to ensure people can access authoritative information on vaccines and reduce the spread of inaccuracies.

Vaccine misinformation is a major threat to global health that could reverse decades of progress made in tackling preventable diseases. . . .

Major digital organizations have a responsibility to their users—to ensure that they can access facts about vaccines and health. It would be great to see social and search platforms come together to leverage their combined reach.

We want digital actors doing more to make it known around the world that #VaccinesWork. . . .

These online efforts must be matched by tangible steps by governments and the health sector to promote trust in vaccination and respond to the needs and concerns of parents.

The US government, Fauci, and specifically the CDC have illegally conspired to restrict freedom of speech. In March 2020, Zuckerberg communicated by email with White House Chief Medical Advisor Anthony Fauci, proposing a collaboration between Facebook and the government on COVID-related information; Fauci agreed to this collaboration, and Zuckerberg made an offer of some kind (so far undisclosed) connected to that collaboration. One month later, in April 2020, Facebook began affirmatively directing users to the CDC's information on COVID, and in May the company announced a new, more stringent policy against COVID "misinformation" [62].

In January 2021 the White House stated that its "direct engagement" with Facebook would cause the organization to "clamp down" on so-called vaccine misinformation.

In May 2021, Dr. Fauci reversed the government's previous denunciations of the lab-leak hypothesis of COVID's origins. That was followed almost immediately by Facebook's removal of its ban on content suggesting that COVID was "manmade or manufactured." The close proximity of these paired events supports an inference that Facebook works jointly with—and willingly takes direction from—the federal government about what COVID-related speech to censor and what not to censor [63, 64].

Furthermore, Zuckerberg has contributed $35 million to the CDC (through the vehicle of the CDC Foundation), and Facebook has donated millions of dollars in free advertising to the CDC [65]. In 2021, a Facebook whistleblower revealed that Facebook censors vaccine-related content based on a secret "vaccine hesitancy" algorithm, which determines whether and to what extent the content (even if completely accurate) could induce vaccine hesitancy in viewers. Facebook banned "vaccine misinformation" and implemented the "vaccine hesitancy" algorithm pursuant to an understanding, agreement, or "meeting of the minds" with its federal "partner," the CDC. Facebook says openly that it defers to the CDC and WHO for "authoritative information" [66]. Moreover, Facebook openly states that it blocks content "which public health experts *have advised us* could lead to COVID-19 vaccine rejection" or "[other] negative outcomes."

In light of all these facts, it is an eminently reasonable inference that the "public health experts" who "advise" and give direction to Facebook on which content to censor and suppress include federal health officials, and that Facebook's deference was reflective of and pursuant to an agreement or understanding between Facebook and the CDC. The US government and Facebook have sought to evade scrutiny by keeping the details of their collaboration largely secret.

In February 2020, Facebook "opened its Menlo Park, Calif., headquarters to the WHO for a meeting with tech companies (including Alphabet Inc.'s Google and Twitter Inc.), where a WHO official discussed the companies' role in spreading 'lifesaving health information'" [67]. Moreover, in September 2019, the WHO publicly stated that it had "discussions for several months" with Facebook about removing "inaccuracies" from its pages [61]. While the WHO is ordinarily not a federal actor, it appears that the WHO engaged in joint action with the CDC and acted as CDC's agent-in-fact in the effort to stamp out so-called COVID "misinformation," making the WHO a federal actor in this context.

In sum, Anthony Fauci—appropriately designated "America's high priest of scientism" by Everett Piper—has been dishonest with the American people throughout the COVIDcrisis and has repeatedly substituted opinion for science-based factual information, directly contributing to one of the greatest losses of life, freedom, and livelihood in the history of mankind. This is an embodiment of the true essence and nature of scientism.

But Dr. Fauci is not the only one who has crossed the line between science and scientism. The lockdowns, masking, and social distancing policies were all based not on science, but on the opinions of the people at the top of the administration—policies not to be questioned by scientists or laypeople.

I have spent my whole professional life dealing with the new priesthood of scientism, and it has always infuriated me. Scientism has nothing to do with the scientific method that I was so rigorously trained in. In my experience, those who ascribe to this substitute religion are typically second- or third-rate intellects who exploit a broken system of public funding of the "scientific" enterprise to build personal status and power, typically coupled to a cult of personality.

So how does scientism differ from the "science," which Dr. Fauci claims to embody?

As before, let's turn to examining both the meaning and practice of what is science, at least that version of science that I have been taught and practiced for over forty years. When defining science, I personally prefer the point of view nicely summarized by Steve Savage [68]:

> Science is a verb.
>
> In an allusion to the John Mayer song "Love Is A Verb," Dr. Cami Ryan noted that as with the word "Love," "Science" is a legitimate noun. But in both cases, it is the action, the process, and the effort—the verb—that really matters.
>
> Science is a verb in the sense that it is a method (activity) involving the making of hypotheses, the design of experiments, and the analysis of data. *But a critical part of the scientific process is the conversation phase after the experimentation is done.* Scientists share their findings with the broader community through

publications or presentations at meetings. What happens next is a back-and-forth discussion including a critique of methods or interpretation, and a comparison with previous findings.

If there are flaws in the experimental design or interpretation, other scientists will point that out. *To participate in the conversation, scientists need to be willing to hear and respond to feedback.* If there are conflicting results, it may require additional hypothesis making and experimentation. *Only when the conversation runs its course do the conclusions become a part of accepted scientific understanding.*

A bit of personal background would probably be helpful here. The mentor who really taught me the process of "doing science" (ergo, the verb) was basically a scientific ascetic, in that he was quite austere in manner, habits, and practice. Both an MD and a PhD, he was a practicing board-certified pathologist focused on breast cancer research, with training at the Armed Forces Institute of Pathology back when that really meant something. I was just a college junior hoping to get into medical school. I thought if I didn't get in becoming a bench scientist in the areas of virology and molecular biology might be a pretty good fallback plan. I do not know why he took me in, but he did, and I worked at the bench in his laboratory every minute I could spare for two solid years.

Talk about a harsh taskmaster. Every week, during the group lab meeting, it was stand and deliver. What is the positive control? What is the negative control? What is the hypothesis? What are the findings, alternative findings, and the limitations? Week after week, surrounded by mature scientists, physicians, graduate students, and lab technicians who were all much more experienced than I was. He still lives in my brain, and when I click into analytical scientist mode, I have to restrain my inner asshole. I have to be particularly careful when reviewing manuscripts, grants, or contracts, lest I end up a scientific nihilist—nothing is ever good enough. But from him I learned how to do rigorous scientific investigations, how to think about experimental design, how to interpret data, and how to find the holes in almost any research paper. That is my origin story as a scientist.

My mentor was particularly attuned to the nuances of scientific bias and how it can so easily compromise scientific research and interpretation.

For all my remaining years I will recall his admonition to avoid building hypotheses on sand rather than firm rock, just as I will never forget his looking me in the eye and telling me that he had no time for false modesty.

There were two key papers at the core of his teaching concerning the scientific method. The first is titled "Strong Inference: Certain Systematic Methods of Scientific Thinking May Produce Much More Rapid Progress Than Others" [69]. Strong inference is possible when results from an experimental paradigm are not merely consistent with a hypothesis, but they provide decisive evidence for one particular hypothesis compared to competing hypotheses.

The second is titled "The Method of Multiple Working Hypotheses" and describes a method for avoiding bias associated with a single hypothesis to which one may become overly attached. This method requires devising as many competing alternative hypotheses as possible and then designing experiments that differentiate between the alternatives. First published in 1890 in the journal *Science* [70], it was later republished in the same magazine in 1965 with the subtitle "With This Method the Dangers of Parental Affection for a Favorite Theory Can Be Circumvented" [71]. All of the above is grounded in a profound respect for the inherent complexity of biological systems. Humility in the face of our ignorance, not hubris concerning all that scientists may know (or think they know). This leads to a step-by-step process for "doing" robust science:

State the problem (or hypothesis) that you wish to clarify or resolve. This can be the most difficult part—in my experience, once you can clearly articulate the problem, it gets a lot easier to solve.

Come up with as many alternative hypotheses to explain the phenomenon (or problem) as you can.

Discuss the alternative hypotheses with others (particularly those who are not invested in the favored explanation) and try to get them to help you to see additional alternative explanations. Getting input from "outsiders" can be particularly useful at this point. More on that later.

Design experiments which can eliminate the various alternative explanations. Sometimes this will require multiple experiments.

Perform the experiments and record the data. Always include rigorous positive and negative control experiments performed at the same time. Optimally, a "strong inference" study will provide a definitive result, demonstrating that one of the hypotheses is clearly correct, and the others can be rejected. However, a well-designed and interpreted experiment can often raise more questions than it resolves.

Repeat step 5 to confirm and perhaps reconsider step 4 prior to repeating step 5.

Continue until a definitive result concerning the original question is obtained.

You can see, even without detailed knowledge of clinical trial design, good clinical practices, or knowledge of regulatory affairs, that the process used to determine COVID-19 vaccine policies was not scientifically sound research. The determination of mask policy was even less so. These were examples of scientism in practice—and applied to the global population.

I like to divide the world up into three domains: the known, the knowable unknown, and the unknowable. I believe that there is an objective approximation of "truth" within the realms of the known and the knowable unknown. In my belief system, it is the job of the scientist to master knowledge of as much of the known as possible, and then to venture into the knowable unknown for the purpose of capturing and bringing fragments of that world into the domain of the known. Good scientists are (by nature, training, and practice) like pioneers or traders who move between the realms of known and knowable unknown. Upon bringing back some fragment of what they believe to be truth to the realm of the known, they then subject one another to a form of "intellectual torture by criticism" when seeking to correctly interpret that fragment.

In contrast to those who practice science, I believe it is the job of philosophers and those who focus on the spiritual realm to provide some structure to the unknowable, to help us grapple with the mysteries of what happens after death or the existence of a higher power or purpose, which resist measurement and quantitation. As far as I am concerned, the answers to these eternal questions are matters of faith, not of science. Personally, I am convinced that there is something deeply mysterious and wonderful about sentient beings, including ourselves—an emergent property that defies rational explanation and cannot be quantified on some utilitarian or economists' spreadsheet. As a scientist, my sense is that this is not something that can be reduced to the domain of the known, as it defies measurement—at least at this point in time. In my internal model of the world, this emergent property of sentient beings, the basis for this luminous transcendent wonder that we often call the soul, resides in the realm of the unknowable unknown. It seems to live in the realm of the unconscious rather than the analytical conscious mind. And since it cannot be measured or quantitated, it defies utilitarian optimization.

To my mind, this is a key reason why the suggestion from the World Economic Forum that by 2030 we will own nothing and be happy rings so hollow. Such statements emerge from the profound hubris of those who believe that they can engineer happiness on a global scale. Happiness is something that emerges from the individual soul, and not something that can be algorithmically optimized. Historically, every time it has been attempted, the result has been destruction of mind, initiative, and soul on a massive scale. In my opinion, the key human parameter that the philosophical systems of Utilitarianism and Marxism miss is often referred to as "agency," whereby people act as individual members of the society. Individualism represents a cultural opposite to collectivism.

The process of science—science as a verb—is intrinsically incremental and fundamentally conservative (in the classic sense of the word, not the modern political sense). Knowledge progresses in small steps, much like biological evolution. A modification is made, a theory or hypothesis is designed, and tests are developed, data are collected, and results are interpreted, discussed, and challenged, and then the whole process begins again. Often experiments raise new questions, resulting in a seemingly endless loop of test, analyze, interpret, retest. Step by step, true knowledge builds—on rock, not on sand. This is the opposite of the surety and hubris of those who practice scientism.

Human beings perform this process of science, and they all share inherent cognitive flaws that will introduce bias at all stages of the process. The structural outline of these flaws is revealed by the study of human thought, of the process of cognition. The core problem is that our conscious mind does not perceive reality directly. We receive sensory input from the world, but we filter that information based on very personal internal models of reality, which we have built up since birth. These models are the products of our personal experiences as well as the external models that we have assimilated through interactions with others (parents, teachers, mentors, etc.). Having built these internal models of reality through both personal trial and error as well as external interactions, humans generally use a form of abstract mental tokens that we call "words" and "language" to process and integrate these models—a process that we call conscious thought (as opposed to unconscious information processing).

In sum, this process of human conscious thought gives rise to three key problems. The first problem is that words and language, as abstract representational tokens, carry intrinsic bias. Words are an internal approximation of some intrinsic meaning, and at a deep cognitive level they have no objective meaning; they are representations that always require reference to other (imperfect) words to yield some subjective sense of "truth." In other words, words and the internal meaning that we assign to them bias our ability to think—to discern an accurate interpretation and meaning of the raw data received from our senses. This is why the manipulation of the meaning of words for propaganda purposes is so insidious. This practice incrementally destroys our ability to accurately comprehend, to think, to make meaning of external reality, resulting in what Dr. Joost Meerloo refers to as *menticide*, the rape of the mind [72].

The second problem is even more profound. Cognitive psychology studies, particularly involving the process of hypnotic suggestion (which lies at the heart of the mass formation process), clearly demonstrate that the human mind will reject sensory data that are inconsistent with its internal model of reality. In other words, if our internal models (which can be considered as "paradigms") are inconsistent with some external sensory reality, we will typically reject the true objective reality and force the incoming sensory data to fit our internal cognitive models.

The third problem with the internal models of reality, which we all hold and use to interpret the raw data we receive from our senses, is that membership in communities and organizations often requires sharing models (or *paradigms*) with others in the community, regardless of their validity. Since the time of Aristotle, the field of science (both the noun and the verb) has expanded to encompass such a vast scope of knowledge and inquiry that it has become necessary to parse all of this into subsets of increasingly finer divisions, which are often referred to as scientific disciplines. This is also true for other fields of human thought, practice (for example, the trades), and philosophy. Each of these have then developed their own paradigms and models of reality, acceptance of which (or reaction to which) basically defines the community of practitioners of that discipline. Those who fail to meet a minimal level of acceptance of the paradigms that define a scientific discipline are typically rejected by other members; they are heretics.

Paradigms can be viewed as models that provide benefit or utility for solving the problems that practitioners of a scientific discipline encounter. But all paradigms are limited, because they do not represent reality itself, but learned models used to help make sense of the perceived stream of sensory data, which is the true reality. Because of the various forms of cognitive bias discussed above, as a scientific discipline approaches the edges of the accuracy or usefulness of a model to interpret reality, practitioners face an increasingly difficult task when trying to solve problems at those boundaries. As a consequence, these "mature" scientific disciplines face two choices: They either must modify the model or jettison it for a new model, or they must force the reality to conform to the model and avoid those problems that do not fit within the model.

Practitioners who seek to modify or jettison a model that defines a scientific discipline in favor of a new model that better fits the data are typically labeled heretics and rejected by the "tribe" of "true" practitioners. One common example of this type of reasoning is the logic error known as "no true Scotsman," which we have seen quite a lot of during the COVIDcrisis. For example, the statement is made that "All vaccinologists agree that the COVID genetic vaccines are safe and effective." When countered with the true statement "I am a vaccinologist and I do not think that these genetic vaccines are safe and effective," the response is, "Therefore, you are not a true vaccinologist." When scientific disciplines' models are reaching their limits of usefulness for problem solving, a new scientific priesthood typically arises. These high priests of scientism use this same faulty logic to defend the models or paradigms that have come to define the discipline's "truth."

Just like biological evolution, sometimes there are bursts of innovation, insight, and eventually knowledge. Thomas Kuhn was one of the first to look rigorously at science, discovery, innovation, and knowledge development and detailed his findings in his classic work *The Structure of Scientific Revolutions* [73]. He largely introduced the concepts and terms *paradigm* and *paradigm shift* into the epistemology of science, and his scholarship and teaching form the last main pole in the tent of my understanding of the practice of science (as a verb).

Kuhn was perplexed by the abrupt shifts that often occur in scientific knowledge. What were the conditions and causes for the changes in scientific thought that triggered these bursts of innovation, insight, and knowledge? Examples of such include the insight that the Earth is round and revolves around the sun (which allowed much more effective problem solving in the critical domain of navigation, among other things) and, of course, the discovery of the double-helix structure of DNA (which allowed rapid resolution of many problems in biology and genetics in particular). These breakthroughs in thought, which then led to explosions of both scientific insight as well as greatly enhanced problem-solving capabilities, were each associated with heretical models inconsistent with the established "truths" jealously held by scientists of the day. Interestingly, Kuhn realized that these breakthroughs often come from outsiders or newcomers rather than practitioners of a scientific discipline. In large part, this can be explained by the need to learn and accept the dominant models that define a scientific discipline in order to "join the guild." The very act of assimilating the model constrains the ability to see the limitations of the model and increases the risk that practitioners will "force the data to fit the model."

This is why (in my opinion), if we wish to better perceive "truth" and reality, and to solve the difficult problems that lie outside the limits of our current knowledge and belief systems, outsiders and heretics are essential. Thus, in my own laboratory work, I always seek out and actively listen to the newcomers. They may not know the language of insiders, but they are often the only ones who can see scientific realities that insiders cannot perceive. These rare insights are often not the product of "logical" thought, but seem to arise from somewhere outside of the conscious mind and, strangely, often arise independently in multiple places at about the same time. As far as I am concerned, that is one of the great mysteries and wonders of our shared humanity and a key argument in favor of personal agency—of freedom.

In cultures dominated by the concepts of utilitarian collectivism and Marxism, there seems to be little room for heresy, insight, innovation, and the paradigm shifts that are the hallmark of true science—as opposed to scientism.

CHAPTER 10
Repurposed Drugs

There are many paradoxes in the COVID-19 data from the western nations concerning disease and death attributed to SARS-CoV-2 infection. One of the most problematic is the result of widespread systemic reporting bias, in which disease and deaths WITH evidence of infection are grossly over-reported as disease and deaths FROM infection by SARS-CoV-2. In the case of the United States, the truth is that we may never be able to resolve this, to get to the bottom of what really went on, due to perverse political and financial incentives to overreport COVID-19 deaths (while also minimizing toxicity of the vaccines). But there is no question that if you are admitted to a Western hospital located in an economically developed country with a COVID-19 diagnosis, your risk of death during that hospitalization has been amazingly high. In contrast, countries with low COVID-19 mortality are often economically underdeveloped, with Haiti and many African nations providing notable examples.

In my opinion, many of those hospital deaths were avoidable—many were iatrogenic (due to medical error). Iatrogenic disease is the result of diagnostic and therapeutic procedures undertaken on a patient. Again and again, I hear academics, physicians, hospitalists, and relatives of patients speak of the horrors of hospital-based treatment of COVID-19, of the unnecessary isolation of the patients, of the horrible and inhumane treatment that patients are receiving, of the toxicity of the FDA-approved and the Anthony Fauci-promoted drug Remdesivir (globally nicknamed by nurses and orderlies "run, death is near"), and of the contribution of bad intubation and ventilation practices to those outcomes.

But they never, ever acknowledge that their mismanagement of these hospitalized patients has contributed to the death toll. The hospitalists have often slavishly followed the limited inpatient guidance protocols of the NIH (which has never before been in the business of setting national treatment standards before COVID), while failing to even be willing to try the alternative inpatient and outpatient treatments that many independent physicians have developed and successfully implemented while saving many thousands of patients' lives. Clearly, what is needed is a way to keep patients from ever getting to the hospital and receiving these dysfunctional treatments associated with high levels of iatrogenic disease and death.

There is no question in my mind that early COVID-19 treatment saves lives, and many different repurposed drug treatment protocols for treating this disease have become popular despite withering criticism and gaslighting from FDA, NIH, corporate media, and hospitalist physicians. For examples of successful early treatment protocols, see those developed by FLCCC, the late Dr. Vladimir Zelenko, Drs. George Fareed and Brian Tyson, and the European doctors who practice under the banner of Ippocrateorg.org. In just one example, while in the USA Ivermectin has been vilified by both FDA and the press, worldwide adoption of Ivermectin for treatment of COVID-19 disease is now at 45%, with many of these nations that permit or encourage Ivermectin use having remarkably low hospitalization and mortality rates!

> Ivermectin is currently used for about 27% of the world's population. Countries where COVID-19 mortality is close to zero may not have incentive to adopt treatments. When excluding these countries, Ivermectin adoption is about 45%. We excluded countries where the cumulative mortality over the preceding month was less than 1 in 1 million, according to the data at https://ourworldindata.org/. For the estimated population coverage, isolated use, some regions, mixed usage, and many regions use a factor of 0.05, 0.25, 0.5, and 0.75, respectively. For the source reference, please see [74].

While many of these alternative early treatment and hospital treatment protocols rely on drug combinations that typically include Hydroxychloroquine plus Azithromycin—the combination championed by Dr. Didier Raoult [75]—or Ivermectin, there are many other drugs and combinations that have shown substantial efficacy in both outpatient and inpatient treatment environments.

For example, those who have followed my work over the last two years may be familiar with the data supporting the use of Famotidine [76, 77] with or without Celecoxib [76, 78–80]. Unfortunately, despite passing peer review, publication of much of this work and associated findings was actively blocked by various academic journals [81] and ridiculed by lay press including the *Washington Post* [82, 83], despite having been demonstrated to have benefits in clinical trials including a randomized Phase 2 clinical trial [84–86]. How and why journalists with no medical training working for the *Washington Post* became arbiters of medical truth continues to elude me. Who knows how many lives could have been saved if the corporate press had just focused on doing solid reporting rather than trying to influence clinical treatment practices while attacking physicians who were just trying to do their jobs.

Although it may seem like both the government and the corporate press in the United States have been particularly hostile to early treatment protocols employing cheap generic drugs for COVID-19, things have been even more difficult for Italian physicians providing early treatment. This makes the following studies even more remarkable!

An Italian team working in a traditional hospital setting has published two peer reviewed studies, one in the *Lancet*-affiliated journal eClinical Medicine [87] and the other in Frontiers in Medicine [87, 88]. The clinical treatment protocol tested in the clinical trial associated with these publications is built around COX-2 inhibitors, which are a type of nonsteroidal anti-inflammatory drug (NSAID) that specifically blocks COX-2 enzymes (Nimesulide, which is available in EU but not USA or Celecoxib, which is available in both EU and USA). In the case of either of these agents being contraindicated due to patient preexisting conditions, the combined COX-1 and COX-2 inhibitor Aspirin (careful, this is also used to treat horses...) was substituted. The corresponding clinical trial is called "A Simple Approach to Prevent Hospitalization for COVID-19 Patients," and the title of this clinical trial pretty much sums up how to keep people out of the hospital [89]:

"A Simple Approach to Prevent Hospitalization for COVID-19 Patients"

Here is the resulting recommended outpatient clinical treatment protocol [89]:

I. Non-steroidal anti-inflammatory drugs (NSAIDs)

Relatively selective COX-2 inhibitors §# (for myalgias and/or arthralgias or other painful symptoms)

§ based on the ratio of concentrations of the various NSAIDs required to inhibit the activity of COX-1 and COX-2 by 50 percent (IC50) in whole blood assays

#unless contraindicated

Nimesulide *

100 mg b.i.d p.o, after a meal, for a maximum of 12 days.

Or

Celecoxib *

Initial oral dose of 400 mg, followed by a second dose of 200 mg on the first day of therapy. In the following days, up to a maximum of 400 mg (200 mg twice a day) should be given as needed for a maximum of 12 days

* Should the patient have a fever (≥37.3°C) or develop laboratory signs of hepatotoxicity associated with nimesulide, or if there are contraindications to celecoxib, these drugs should be substituted with aspirin (a COX-1 and COX-2 inhibitor) (500 mg twice a day p.o.— after a meal). Patients receiving these treatments should also be given a proton pump inhibitor (e.g., lansoprazole--30 mg/day; or omeprazole—20 mg/day; or pantoprazole--20 mg/day).

Approximately 3 days after the onset of symptoms (or longer if the physician is seeing the patient for the first time), a series of hematochemical tests should be performed (blood cell count, D-dimer, CRP, creatinine, fasting blood glucose, ALT). Nimesulide/celecoxib (or aspirin) treatment can continue if inflammatory indexes (CRP, neutrophil count), ALT, and D-dimer are in the normal range.

II. Corticosteroids*

Dexamethasone (for persistent fever or musculoskeletal pain or if hematochemical tests are repeated a few days later and there is even a mild increase in the inflammatory indexes—CRP, neutrophil count

–, *or if the patient has a cough and oxygen saturation (SpO2)<94– 92% occur).*

8 mg p.o. for 3 days, then tapered to 4 mg for a further 3 days, and then to 2 mg for 3 days. This makes a total of 42 mg dexamethasone over 9 days.

*The duration of corticosteroid treatment also depends on the clinical evolution of the disease.

III. Anticoagulants

Low–molecular weight(LMW) heparin* (when the hematochemical tests show even a mild increase in D–dimer, or for thromboembolism prophylaxis for bedridden patients)

Enoxaparin, at the prophylactic daily dose of 4,000 U.I subcutaneously—i.e., 40 mg enoxaparin. Treatment recommended for at least 7–14 days, independently of the patient recovering mobility.

*unless contraindicated (e.g., ongoing bleeding or platelet count<25 × 109/L)

IV. Oxygen therapy

Gentle oxygen supply in the early phase of the disease, possibly before pulmonary symptoms manifest, in the presence of progressively decreasing oxygen saturation—as indicated by an oximeter—or following a first episode of dyspnoea or wheezing.

Conventional oxygen therapy is suggested when the respiratory rate is >14/min and oxygen saturation (SpO2) < *94–92%*, but it is required with SpO2 <90% at room air. With liquid oxygen, start with 8–10 liter/min and monitor SpO2 every 3–4 h. Titrate oxygen flow rate to reach target SpO2 *>94%*. Then the rate of oxygen administration can be reduced to 4–5 liter/min (but continue SpO2 monitoring every 3–4 h). With gaseous O2, start with 2.5–3.0 liter/min, but monitor SpO2 more frequently than with liquid oxygen, and titrate flow rates to reach target SpO2 *>94%*. Hospitalization could be considered, if feasible, when oxygen saturation (SpO2) ≤ 90% at room air, despite conventional oxygen therapy.

V. Antibiotics

Azithromycin* (with bacterial pneumonia or suspected secondary

bacterial upper respiratory tract infections, or when hematochemi-
cal inflammatory indexes (CRP, neutrophil count) are markedly
*altered)*500 mg/day p.o. for 6–10 days depending on the clinical
judgement

   * *Should the patient be at risk of or have a history of cardiac*
arrhythmia or present other contraindications, cefixime (400 mg/day
p.o for 6–10 days) or amoxicillin/clavulanic acid (1 gr three times a
day for 6–10 days) can be considered as alternatives to azithromycin.

In a separate study completed by the IppocrateOrg Association Working
Group for the Early Outpatient Treatment of COVID-19 (which is pre-
dominantly an association of Italian physicians and scientists), an alterna-
tive protocol also demonstrated effectiveness in outpatient treatment of
COVID-19. That protocol was published in a preprint server and is titled:
"Early Outpatient Treatment of COVID-9: A Retrospective Analysis of 392
Cases in Italy" [90].

   These researchers conclude:

This is the first study describing attitudes and behaviors of physicians caring for COVID-19 outpatients, and the effectiveness and
safety of COVID-19 early treatment in the real world. COVID-
19 lethality in our cohort was 0,2%, while the overall COVID-19
lethality in Italy in the same period was between 3% and 3,8%.
The use of individual drugs and drug combinations described in
this study appears therefore effective and safe, as indicated by the
few and mild ADR reported. Present evidence should be carefully
considered by physicians caring for COVID-19 patients as well as
by political decision makers managing the current global crisis.
The protocol used for this study is more typical of the protocols
used in the United States. As is often the case in the United States,
the general treatment protocol developed by the Ippocrateorg team
is staged by disease severity and can be found at this webpage:
https://ippocrateorg.org/en/2020/12/15/how-to-treat-covid-19/.

A summary table of the treatment received for the 392 summarized Italian
cases can also be found on that website, and the treatments administered

include the use of aspirin (which is well known to have anticoagulant properties due to its activity on platelets).

Irrespective of the excess death and disease associated with the mandated genetic vaccines, there is no doubt in my mind that the concerted and coordinated propaganda and information control efforts of the United States Government Department of Health and Human Services, acting in alignment and as sponsors of Big Tech and Corporate Media censorship, have cost large amounts of unnecessary death and disease due to both iatrogenic causes during hospitalization as well as by suppression of life saving early treatment protocols. The data supporting this conclusion increase almost daily. The unresolved issue remains: will anyone be held accountable for this avoidable tragedy?

CHAPTER 11
mRNA Vaccines. The Largest Human Experiment Ever

More than 5.41 billion people worldwide have received a dose of some type of COVID-19 vaccine, equal to about 70.5 percent of the world population [91]. In the United States as of October 17, 2022, 265.59 million US residents have received at least one dose, and 226.59 million have completed the initial vaccination protocol [92], out of a total population of 335.49 million (67.5%). Of the 613.25 million mRNA vaccine doses administered, 375.64 million of these doses were manufactured by Pfizer/BioNTech, and 237.61 doses by Moderna. Between the US and the EU, a total of nearly 1.5 billion doses have been administered. This accomplishment involved a novel technology, a large-scale manufacturing process, and a product that was created, passed nonclinical and clinical development, manufactured, distributed, and globally deployed in less than three years. In terms of logistics alone, this is undeniably a major achievement.

At a meeting of the Special Committee of the European Union Parliament held on 11 October 2022 to discuss the findings regarding the COVID-19 pandemic and recommendations for the future, a Pfizer executive confirmed that their vaccine had never been tested for its ability to prevent the transmission of SARS-CoV-2 virus before being put on the market. Data emerging since the introduction of the vaccine indicate that it is in fact unable to do so, thereby negating the claim that COVID-19 passports are necessary to protect others [93]. In other words, governments throughout the world employed a wide range of propaganda and censorship methods to promote these products as both safe and effective at stopping the spread of SARS-CoV-2 infection, despite the fact that there were no studies that even tested how well they would prevent the spread of COVID-19. It is not an exaggeration to state that this massive deployment has been the largest clinical experiment performed on human beings in the history of the world.

All of the mRNA vaccine doses administered in the United States (to both citizens and military personnel) have been provided under "Emergency Use Authorization" (EUA). Although the FDA has licensed the Pfizer/ BioNTech and Moderna vaccines for some age cohorts, the manufacturers have elected not to distribute or market the licensed products in the United States. The reasons for this are not clear but appear to include liability issues as well as the fact that additional clinical studies, safety monitoring (pharmacovigilance), and product disclosures are required by the FDA once the licensed products are marketed. From the standpoint of the vaccine manufacturers, EUA is the preferred pathway for marketing their products. A single purchaser (the US government) provides complete liability indemnification, a guaranteed market with very little oversight, and manages both the distribution and marketing. Manufacturers are prohibited from marketing unlicensed products, but the US government has been doing it for them and has coordinated with corporate media, social media, and large technology firms to suppress any discussion of the risks or limitations of the products. From the standpoint of the manufacturers, this is all profit and no risk—the perfect business model. In the United States, all of this was paradoxically primarily overseen by the Department of Homeland Security and National Security apparatus rather than the Department of Health and Human Services [63]. Why would the vaccine manufacturers ever consider taking up the burden of producing and marketing the licensed version of these products when they had both DHS and HHS doing their bidding for them?

Emergency Use Authorization is a process defined by US federal law for the introduction of "a drug, device, or biological product intended for use in an actual or potential emergency." Continued use of these unapproved vaccines requires a determination "that there is a domestic emergency, or a significant potential for a domestic emergency." Once the government declares that the emergency has passed, "A declaration under this subsection shall terminate." In other words, when the emergency is over, the EUA expires, and the vaccines (that are currently being distributed) would revert to their status as not approved, licensed, or cleared for commercial distribution. These products remain experimental and are intended to be used for a limited time during an ongoing emergency.

What if the largest experiment on human beings in history is a failure?

As Ed Dowd described in Chapter 8, life insurance companies have been reporting alarming increases in all-cause mortality and disability in working-age people. We may be experiencing both a huge human tragedy as well as a profound failure of the US government to serve and protect its citizens. We may be forced to conclude that the genetic vaccines that were so aggressively promoted have failed and the federal campaign to prevent early treatment with lifesaving drugs has contributed to a massive, avoidable loss of life. In addition, the federal workplace vaccine mandates may have caused a massive loss of life in workers who were forced to accept a toxic vaccine at a higher frequency than the general population.

Furthermore, we have also been living through the most massive, globally coordinated propaganda and censorship campaign in the history of the human race. All major mass media and the social media technology companies have coordinated with governments and the vaccine manufacturers to stifle and suppress any discussion of the risks of the genetic vaccines and/or alternative early treatments.

There must be accountability. We are not just talking about grinding the First Amendment into the mud; we are talking about an avoidable mass casualty event caused by a mandated experimental medical procedure—for which all opportunities for the victims to inform themselves about the potential risks have been methodically erased from both the Internet and public awareness by an international corrupt cabal operating under the flag of the "Trusted News Initiative" while systematically silencing Physicians who raised concerns via a globally coordinated censorship and defamation campaign [64]. George Orwell must be spinning in his grave.

I hope I am wrong. I fear I am right.

When is mRNA not really mRNA?

There are those who believe that I bear personal responsibility for the morbidity and mortality associated with the COVID-19 mRNA vaccines because of my pioneering work in the use of mRNA as a transient "gene therapy" method, with the entry level application for vaccine purposes. (These findings and insights were documented in the nine original patents covering DNA and mRNA gene therapy and DNA and mRNA vaccination and a tenth patent covering mucosal DNA and mRNA vaccination [1–8, 94].) This accusation has been echoed by many angry social media detractors looking for someone to blame for the adverse events that have been associated with these vaccines. Therefore, I thought it would be worthwhile to examine some of the differences between what I originally envisioned and the current molecules being injected into our bodies.

Gene therapy and the origins of mRNA as a drug or vaccine

The idea of permanent "gene therapy" was originally envisioned by Richard Roblin, PhD, and academic pediatrician Dr. Theodore Friedman in 1972 [95]. An article in the January 2015 UC San Diego News nicely summarizes the underlying logic of "Gene Therapy" as envisioned by Friedman and Roblin.

> The idea of gene therapy, which quickly captured the public imagination, was fueled by its appealingly straightforward approach and what Friedmann has described as "obvious correctness": Disarm a potentially pathogenic virus to make it benign. Stuff these viral particles with normal DNA. Then inject them into patients carrying abnormal genes, where they will deliver their therapeutic cargoes inside the defective target cells. In theory, the good DNA replaces or corrects the abnormal function of the defective genes, rendering previously impaired cells whole, normal, and healthy. End of disease [96].

The core idea captured in the original nine patents that stem from my work between 1987 and 1989 was that there are multiple key problems with this concept. But the work continued to progress, until 1990, when the first patient was treated by gene therapy, a four-year-old girl with a congenital illness called adnoside deaminase deficiency. The experiment failed, and the child was not cured.

> Nonetheless, media attention and hype about gene therapy continued to be rampant, fueled in part by overenthusiastic opinions by some scientists. Things crashed in 1999, when an 18-year-old

patient named Jesse Gelsinger, who suffered from a genetic disease of the liver, died during a clinical trial at the University of Pennsylvania. Gelsinger's death was the first directly attributed to gene therapy. Subsequent investigations revealed numerous problems in the experimental design.

What is wrong with the original "gene therapy" concept? There are multiple issues, and here are just a few:

1. Can you efficiently get genetic material ("polynucleotides") into the nucleus of the majority of cells in the human body so that any genetic defects can be fixed? In short, no. Human cells (and the immune system) have evolved many, many different mechanisms to resist modification by external polynucleotides. Otherwise, we would already be overrun by various forms of parasitic DNA and RNA—viral and otherwise. This remains a major technical barrier, one that "transhumanists" overlook in their enthusiastic but naive rush to play God with the human species. What are polynucleotides? Basically, the long chain polymers composed of four nucleotide bases (adenine, thymine, guanine, and cytosine in DNA, with the thymine replaced by uracil in RNA) that carry all genetic information (that we know of) across time.

2. What about the immune system? Well, this was one of my breakthroughs way back in the late 1980s. What Ted Friedman originally envisioned was the simple idea that if a child had a genetic birth defect causing the body to produce a defective molecule or not produce a critical protein (such as Lesch-Nyhan syndrome or adenosine deaminase deficiency), this could be simply corrected by providing the "good gene" to complement the defect. What was not appreciated was that the immune systems of these children were educated during development to either recognize the "bad protein" as normal/self, or to not recognize the absent protein as normal/self. So, introduction of the "good gene" into a person's body would cause production of what was essentially a foreign protein, resulting in immunologic attack and killing of the cells that now have the "good gene."

3. What happens when things go wrong and the "good gene" protein is toxic? Unfortunately, that is the case with the spike protein that is central to the COVID mRNA vaccines. I get asked all the time, "What can I do

to eliminate the RNA vaccines from my body?" To which I have to answer: "Nothing." There is no technology that I know of that can eliminate these synthetic mRNA-like molecules from your body. You just have to hope that your immune system (T cells) will attack and "clear" the cells that have taken up the foreign genetic information and break down the offending large molecule that causes them to manufacture the toxic protein. Since virtually all current "gene therapy" methods are inefficient and deliver the genetic material randomly to a small subset of cells, there is no practical way to surgically remove the scattered, relatively rare transgenic cells.

4. What happens if the "good gene" lands in a "bad place" in your genome? It turns out that the structure of our genome is highly evolved; despite having sequenced the human genome, we are still relative neophytes in our current understanding of it. Sequencing the human genome is akin to knowing all of the letters in a large book, which is a long way from being able to read the book and understand what you have read. Insertional mutagenesis—genetic alteration of chromosomes by inserting DNA (possibly from viruses)—is used to generate new insights into genetics, from fruit flies to frogs to fish to mice. But when new DNA is inserted into chromosomes, it can cause many unexpected things to happen—like development of cancers, for example. This is why there is so much concern about the potential for the mRNA-like particles used in the vaccines to travel into cells' nuclei and insert or recombine with the cells' DNA after reverse transcription [97]. Normally, with DNA-based gene therapy technologies, the FDA requires genotoxicity studies for this reason, but the FDA did not treat the mRNA vaccine technology as a gene therapy product [98] and so did not require that the vaccine developer/manufacturers do this research.

The original idea behind using mRNA as a drug (for treatment or vaccine purposes) was that mRNA is typically degraded quite rapidly once released into a cell. The stability of an individual mRNA molecule is regulated by a number of genetic elements but typically ranges from thirty minutes to a couple of hours. Therefore, if natural (or synthetic that mimics natural) mRNA is introduced into the body, it should last for a very short time. And when asked, "How long does the injected mRNA last after injection?" Pfizer, BioNTech, and Moderna have all replied that it lasts for only a few hours.

But we now know the "mRNA" in the vaccines incorporates a synthetic nucleotide called "pseudouridine" and can persist in lymph nodes for at least sixty days after injection [9]. This is not natural, and this is not really mRNA. These molecules have genetic elements similar to those of natural mRNA, but they are far more resistant to the enzymes that normally degrade mRNA. In addition, they seem capable of producing large amounts of protein for extended periods and evading normal immunologic mechanisms for eliminating cells that produce foreign proteins.

Regarding Pseudouridine and mRNA

Uridine   Pseudouridine

What is pseudouridine? Natural mRNA is composed of the same bases as RNA (adenine, guanine, and cytosine), with the exception of uracil, which is replaced with uridine. Pseudouridine is a modification of uridine that occurs in natural human mRNAs in a highly regulated manner. This is in sharp contrast to the random incorporation of synthetic pseudouridine that occurs in the manufacturing of the Moderna and Pfizer/BioNTech COVID-19 vaccines.

A 2022 paper in the journal *Molecular Cell* has shed light on some of the mechanisms of action associated with natural pseudouridine modification [99]. It appears that various highly regulated cellular enzymes modify the normal uridine nucleotide subunit to form pseudouridine in specific mRNAs and specific locations within those mRNAs while they are being made in the cell. "Pre"-mRNA is produced as a single long chain polymer of A, U, G, and C, then "spliced" to remove some sections (which are typically degraded) and leave others in the final "mature" mRNA product, which is then used to guide protein production. Pseudouridine modifications occur at locations associated with alternatively spliced RNA regions, are enriched near splice sites, and overlap with hundreds of binding sites for RNA-binding proteins [99].

Relevant to the mRNA vaccines, Erin Borchardt et al. suggested in 2020 that pseudouridine is one factor that controls how long an mRNA stays around in your body. While the Borchadt review highlights just how much "remains to be understood" with respect to the consequences of "mRNA pseudouridylation in cells," it explores several known effects it has on the immune system, potentially explaining the immunosuppression (increasingly being referred to as an acquired immunodeficiency syndrome, i.e., AIDS) that is sometimes observed after multiple mRNA vaccine doses:

> . . . incorporating RNA modifications, including pseudouridine, in foreign RNA allows for escape from innate immune detection. This makes RNA modification a powerful tool in the field of RNA therapeutics where RNAs must make it into cells without triggering an immune response and remain stable long enough to achieve therapeutic goals. In addition, the presence of modified nucleosides in viral genomic RNA could contribute to immune evasion during infection.

The authors further describe several other ways such modified RNA can suppress immune response, concluding that

"Pseudouridine likely affects multiple facets of mRNA function, including reduced immune stimulation by several mechanisms, prolonged half-life of pseudouridine-containing RNA, as well as potentially deleterious effects of [pseudouridine] on translation fidelity and efficiency."

Summary

Based on this information, it appears to me that the extensive random incorporation of pseudouridine into the synthetic mRNA-like molecules used for the Pfizer/BioNTech and Moderna vaccines may well account for much or all of the observed immunosuppression, DNA virus reactivation, and remarkable persistence of the synthetic "mRNA" molecules observed in lymph node biopsy tissues [9]. Many of these adverse effects were reported by Kariko (a vice president at BioNTech) et al. in a 2008 paper and could have been anticipated by regulatory and toxicology professionals if they had bothered to consider these findings prior to allowing emergency use authorization and global deployment of an immature and previously untested technology. Therefore, neither the agencies nor the manufacturers can claim ignorance; rather, what we have seen is more appropriately classified as "willful ignorance."

Finally, based on a review of the scientific data, including the articles cited above, it is my opinion that the random and uncontrolled insertion of pseudouridine into the manufactured "mRNA"-like molecules creates a population of polymers that may resemble natural mRNA, but that have a variety of properties that are clinically relevant. These characteristics and activities may account for many of the unusual effects, unusual stability, and striking adverse events associated with this new class of vaccines. These molecules are not natural mRNA and do not behave like natural mRNA.

The question that most troubles and perplexes me at this point is why the biological consequences of these modifications and associated clinical adverse effects were not thoroughly investigated before widespread administration of random pseudouridine-incorporating "mRNA"-like molecules to a global population. Biology, and particularly molecular biology, is highly complex and interrelated. Change one thing over here, and it is really hard to predict what might happen over there. That is why one must do rigorously controlled nonclinical and clinical research. Once again, it appears to me that the hubris of "elite" high-status scientists, physicians and governmental "public health" bureaucrats has overcome common sense; well-established regulatory norms have been disregarded; and patients have unnecessarily suffered as a consequence. These products do not use natural mRNA, and referring to them as mRNA vaccines is misleading. I recommend that these products, which employ a synthetic unnatural polymer, should be designated using a different term, such as Ψ [for pseudouridine]-mRNA genetic medicines.

The Spike Protein and Cytotoxicity

The Ψ-mRNA vaccines are associated with a wide range of adverse events. We covered some associated with the random insertion of pseudouridine. There are other types of adverse events that appear to be associated with the chemicals (including polyethylene glycol) used to coat the Ψ-mRNA. But the spike protein itself, which the Ψ-mRNA causes your cells to manufacture, can also be toxic. In fact, current data suggest that this may be one of the most toxic biological molecules ever used for vaccination purposes.

Let's review the science on spike protein toxicity. First, we need to understand a little bit about the SARS-CoV-2 spike protein.

The only differences in the actual protein sequence between the spike protein of the original "Wuhan" strain of the virus and that coded for by the vaccines are two amino acids in the S2 region of the protein. These were not introduced to make the vaccine version less toxic (as some "fact-checkers" have asserted), but rather to make it better able to stimulate an antibody response. Whether in the vaccine or the virus, the S1 subunit (which includes the receptor binding domain to which the majority of "neutralizing" antibodies are directed) gets cut free to circulate in the blood, bind ACE2 receptors, interact with platelets and neurons, open up vascular endothelial tight junctions, etc. There is no difference between the S1 subunit released from the vaccine spike protein and the S1 subunit released from the virus spike protein. They are the same thing!

Now, how much protein does this free S1 subunit produce and for how long compared to natural infection?

One might expect that the answers were well understood and characterized by Pfizer *before* the vaccines were widely deployed. Surely the FDA required that these studies be performed? Unfortunately, no such studies were done until an academic group published a paper at the end of January 2022 [9]. Without going into too much detail, we now know the vaccine mRNA does not break down rapidly and can be found via fine needle biopsy in human lymph nodes for sixty days (this was the end point of the study, so the amount of time is actually unknown). The amount of spike protein found in plasma was higher in the recently vaccinated than in recovering COVID-19 patients. So, the vaccine produces far more spike S1 subunit for far longer than the natural infection does.

But is the S1 subunit actually a toxin?

First question, does the spike S1 subunit cross the blood–brain barrier and get into the brain? Why yes, thank you for asking, it does! This was verified in a 2021 paper published in *Nature Neuroscience* titled "The S1 Protein of SARS-CoV-2 Crosses the Blood–Brain Barrier in Mice." [100].

Next question, does S1 damage the brain when it hits nerve cells (neurons)? Yes, it looks like it does [101]! And, yes, there are neurological consequences to these brain-related pathogenic mechanisms involving the spike protein [102]07198, Palma, Spain.University of the Balearic Islands (UIB, including cerebrovascular, sensitive, motor, cognitive, and diffuse brain disorders favoring blood–brain barrier disruption, inflammation, hypoxia, and secondary infections. It is important to recognize that there is no significant difference between the symptoms of long COVID (PASC) and postvaccination syndrome [103]: Harvard Medical School, Boston, MA, United States.</auth-address><titles><title>Long COVID or Post-acute Sequelae of COVID-19 (PASC).

There is also evidence of brain endothelial attack [104]. On the basis of this alone, the spike protein can be fairly classified as a toxin, but there are several other important papers on the various toxic effects of the spike protein. These include that the SARS-CoV-2 spike protein directly damages lungs, triggers both large and small persistent blood clots, and triggers inflammation [105–107].

What did various "fact-checking" organizations have to say regarding the SARS-CoV-2 spike protein (and the analogous vaccine-encoded protein)?

Catalina Jaramillo, who was trained at the Columbia School of Journalism, said in a post on July 1, 2021, at factcheck.org that "COVID-19 Vaccine-Generated Spike Protein is Safe, Contrary to Viral Claims." Tom Kertscher, a contributing writer for PolitiFact who has no training in medicine or biology, wrote, "No sign that the COVID-19 vaccines' spike protein is toxic or cytotoxic." Another doozy came from Beatrice Dupuy of the Associated Press on June 9, 2021: "Spike Protein Produced by vaccine not toxic."

So, I ask you, who was correct? The actual scientists or the fact-checkers? Is the spike S1 subunit a toxin? A toxin is a harmful substance produced by living cells, so, yes indeed, the S1 subunit is a toxin. How many people developed brain damage or lost their life or that of a loved one because they accepted a vaccine based on the falsehoods propagated by these grossly unqualified "fact-checkers"? Do they have criminal liability for their falsehoods and propaganda?

The "New Updated Vaccine"

In September 2022 the United States government began rollout of a modified "bivalent" mRNA vaccine for COVID-19. In this case, "bivalent" refers to two different mRNAs encoding two different spike proteins. Per Karine Jean-Pierre, who currently serves as the chief spokesperson for the executive branch of the United States government, these products are "new updated vaccines" and not boosters (September 8, 2022, press conference). Inconveniently, the FDA defines these EUA products as "updated boosters" [108]. According to the FDA, the products were only tested on mice prior to authorization for human use, and the FDA and CDC hope that these products will provide increased protection against the Omicron variant BA.5 (which is currently being displaced by newer Omicron variants). What "increased protection" means is not clear; it could be from any or all of infection, replication, spread, disease, and death. While the anticipated benefits are undefined and unknown, the FDA expects the risks and side effects of the modified product to be similar to those of the current monovalent products.

Emergency Use Authorization for these new products was granted as amendments to the previous EUA for the monovalent products—without any human clinical data or review and advice by the semiindependent Vaccine Related Biologics Review Committee (VRBAC). Neither the FDA nor the CDC considered or discussed the risk that these products could further exacerbate the development of immune imprinting, also known as "original antigenic sin," in our highly inoculated population.

Immune Imprinting

From the standpoint of the approved narrative, one of the major unresolved COVIDcrisis mysteries has been why so many who are "fully vaccinated" (whatever that means) against SARS-CoV-2 still develop infection and COVID disease. A big advocate of vaccine mandates, Canadian Prime Minister Justin Trudeau was infected (despite apparently being fully vaccinated) in January of 2022 and was reinfected (despite receiving three doses of the mRNA inoculum) in June, just four and a half months later. Despite having received four doses of the mRNA inoculum, Dr. Anthony Fauci himself was infected and developed COVID disease in mid-June, 2022. Another advocate of vaccine mandates, California Governor Gavin Newsom, was also infected and developed COVID [109] just ten days after his fourth injection. And more recently, the fully inoculated Director of the CDC, Dr. Rochelle Walensky, was also infected with SARS-CoV-2 and developed COVID-19 disease (and her infection recurred despite a Paxlovid treatment course). Notice a pattern?

In many countries, data during the first two quarters of 2022 indicate that the majority of individuals who were hospitalized due to COVID were "fully vaccinated." Of course, in most Western countries, the majority of the population is vaccinated, so this general finding requires some qualification and correction for the resulting sampling bias. The Canadian COVID Care Alliance distributed a video in February of 2022 that "busted" the myth that COVID-19 is a pandemic of the unvaccinated by using public health data from Ontario, Canada [110]. As discussed in the video, many of these data point toward "negative effectiveness" of the vaccines—meaning that those who are "fully vaccinated" are *more* likely to get COVID disease than those who are "unvaccinated." A word of caution about this conclusion: The term "unvaccinated" is increasingly misleading, as, over time, a larger and larger fraction of the total population has become infected and so has not only been previously infected (and immunologically primed) by infection with one or more of the seasonal "cold" betacoronaviruses (which share large numbers of both B and T cell epitope antigens with SARS-CoV-2), but have also been "boosted" by natural infection with one of the variants of SARS-CoV-2. This problem (or artifact) has become increasingly true since the onset of the Omicron variant. For example, a (non-peer-reviewed) summary of Canadian COVID-19 cases, hospitalized cases, and deaths between May 1, 2022, and June 5, 2022 [111], found that fully vaccinated patients accounted for nine in every ten COVID-19 deaths in Canada over the month, four in every five of which had received three inoculations. Most people know from their own personal experience that whether or not their associates were previously infected or vaccinated (or both), they were highly likely to also get infected by the Omicron variant of SARS-CoV-2.

Just to be crystal clear on this point, when I stood on the steps of the Lincoln Memorial on January 23, in front of 40,000 people at the Defeat the Mandates Rally, I said,

> Regarding the genetic COVID vaccines, the science is settled. They are not working, and they are not completely safe.
>
> Now we have Omicron. These vaccines were designed for the original Wuhan strain, a different virus. Whether they made sense for protecting our elderly and frail from the original virus

is irrelevant. So let's stop arguing about that. We must look forward.

These vaccines do not prevent Omicron infection, viral replication, or spread to others. In our daily lives, with our friends, with our families, we all know that this is true.

These genetic vaccines are leaky, have poor durability, and even if every man, woman, and child in the United States were vaccinated, these products cannot achieve herd immunity and stop COVID. They are not completely safe, and the full nature of the risks remain unknown.

The *Washington Post* called me a liar and spreader of discredited misinformation at the time for making this statement, but since then it has become a widely accepted scientific truth—one that is self-evident to anyone who is not caught up in the mass formation psychosis. The sin that triggered the defamation was apparently stating an inconvenient truth before it became widely accepted. Now, there are many, many scientific papers showing that my statement on the steps of the Lincoln Memorial that cold winter day was completely accurate and if anything was too conservative [112–117].

I think we can now safely say that it was the *Washington Post* who was peddling mis- or dis-information (or just plain old-fashioned propaganda) in their defamatory article.

Previously, any potential for negative effectiveness of the vaccines was actively denied by researchers involved in the design and creation of the genetic vaccines. Unfortunately, however, data and the passage of time have proven their assertions of safety to have been premature [118]; so much for highly confident "experts," their predictive powers of inference, and their personal hubris.

There are multiple working hypotheses for why we sometimes see negative effectiveness of these genetic inoculations for preventing COVID-19 disease. Examples include:

1. Antigenic or immune imprinting, otherwise known as "original antigenic sin" [119].
2. Antibody-dependent enhancement (ADE) [120].

3. Other forms of vaccine-enhanced disease (VAED) [121].
4. Vaccine-induced acquired immunodeficiency (VAIDS) of one type or another [122].
5. In the context of widely deployed "leaky" vaccines, evolutionary selection of SARS-CoV-2 variants that can escape the pressure of vaccine-induced immune responses.

I am sure there are many more hypotheses, and it is always important to recognize that more than one thing can be happening at the same time. However, the preponderance of data increasingly points to immune imprinting as the leading explanation for the observed public health data suggesting negative effectiveness of the mRNA inoculations—which are currently marketed as vaccines but are functionally being deployed more as immuno-therapeutics.

(Of course, the corporate media—either completely unaware of or in denial about their profound incompetence as objective mediators of scientific discussions of complicated immunologic topics—have once again interjected their aggressively provaccination, deny-any-problems agenda into the discussion. And then, once again, the even more scientifically unqualified "fact-checkers" have followed suit. But, by now, that is to be expected.)

So, what is immune imprinting or "original antigenic sin"? A group of influenza virus researchers describes it well:

> We define immune imprinting as a lifelong bias in immune memory of, and protection against, the strains encountered in childhood. Such biases most likely become entrenched as subsequent exposures back-boost existing memory responses, rather than stimulating de novo responses. By providing particularly robust protection against certain antigenic subtypes, or clades, imprinting can provide immunological benefits, but perhaps at the cost of equally strong protection against variants encountered later in life [123].

The authors address the use and limitations of the two terms "immune imprinting" and "original antigenic sin" and find the former term a generally better fit to the actual data [123]. They also provide a nice summary of the issues at hand, which are directly applicable to coronavirus vaccines and evolved SARS-CoV-2 variants:

> Antibody responses are essential for protection against influenza virus infection. Humans are exposed to a multitude of influenza viruses throughout their lifetime and it is clear that immune history influences the magnitude and quality of the antibody response. The 'original antigenic sin' concept refers to the impact of the first influenza virus variant encounter on lifelong immunity. Although this model has been challenged since its discovery, past exposure, and likely one's first exposure, clearly affects the epitopes targeted in subsequent responses. Understanding how previous exposure to influenza virus shapes antibody responses to vaccination and infection is critical, especially with the prospect of future pandemics and for the effective development of a universal influenza vaccine.

Now that the data concerning the "how" and "why" of this odd and unfortunate relationship between the SARS-CoV-2 genetic vaccines, viral evolution of immune escape mutants, and negative effectiveness are really starting to come into focus, it does appear that immune imprinting is playing a big role. It also appears that it is not the general vaccinated population, but rather the subset of people who develop chronic infections that is driving development of the antibody escape mutant viruses. And it is certainly not the general *unvaccinated* population, to the extent that there even are any individuals (other than newborn infants) who are immunologically naive to SARS-CoV-2. And speaking of newborns, the data concerning immune imprinting demonstrate that vaccinating very young children with a genetic vaccine that expresses a spike antigen from a virus that has not been circulating for a very long time (with the consequent immune imprinting) is either malevolence or madness—or both.

It can take a long time and often require interactions between many different people, together with quite a bit of trial and error, to get to the bottom of complicated problems. This is generally true in science as well as in life, which makes the rapid progress toward understanding the immunologic and virologic processes driving negative effectiveness of the vaccines, as well as SARS-CoV-2's immune escape from vaccine-induced protection, all the more remarkable.

See these references for numerous examples of relevant scientific literature [9, 124–129]. When you look them over, make sure you note the progression from smaller, more fringe journals to mainstream medical journals.

After that introduction to immune imprinting, you should be ready for an extended discussion of an important peer-reviewed paper that was published in *Science* in June of 2022. I particularly appreciate the authors' (Catherine Reynolds et al.) recognition that protection against SARS-CoV-2 involves both an antibody/B cell as well as a T cell component and their detailed discussion on the topic [130].

The paper's abstract meets the criteria for "bombshell," in my opinion:

We investigated T and B cell immunity against [Omicron] in triple mRNA vaccinated healthcare workers (HCW) with different SARS-CoV-2 infection histories. B and T cell immunity against previous variants of concern was enhanced in triple vaccinated individuals, but magnitude of T and B cell responses against [Omicron] spike protein was reduced. Immune imprinting by infection with the earlier [Alpha] variant resulted in less durable binding antibody against [Omicron]. Previously infection-naive HCW who became infected during the [Omicron] wave showed enhanced immunity against earlier variants but reduced [neutralizing antibody] potency and T cell responses against [Omicron] itself. Previous Wuhan Hu-1 infection abrogated T cell recognition and any enhanced cross-reactive neutralizing immunity on infection with [Omicron].

In short, I think what the authors are saying is that Omicron is evolving to not only escape prior neutralizing antibodies generated from either vaccination (with spike protein derived from Wuhan Hu-1) or infection with Wuhan Hu-1 OR Alpha, but also infection with Omicron is reducing both T cell and B cell (antibody) responses to itself. This is not good news.

The authors state that "Across several studies, 2 or 3-dose vaccination is protective against severe disease and hospitalization, albeit with poor protection against transmission." But the cited references demonstrate that vaccine effectiveness against severe disease and hospitalization varies in a time-dependent manner and is in the 40–70% range, which means that 30–60% of fully vaccinated persons were *not* protected against hospitalization. Based on this, my conclusion is that the genetic vaccines are not even working well to prevent hospitalization—and Omicron causes *mild* COVID disease. ("Working well" and "not working" are subjective judgments, so I ask any "fact-checkers" reading this to please bugger off.)

The authors continue with a discussion of why the rate of breakthrough infections is so high: "A rationale for this . . . comes from mapping of virus neutralization . . . showing this to be the most antibody immune-evasive [variant], with titers generally reduced by 20-40-fold." The references cited to support this statement use words like "considerable escape" and "striking antibody evasion" to describe the reduced protection against Omicron, which "extensively but incompletely escapes" neutralization with vaccine-induced antibodies [131–134].

I think that you can see the pattern. And suffice to say, it does not support the position taken by the *Washington Post.*

The authors attribute the apparent relative attenuation of severe symptoms in the vaccinated "to the partial protection conferred by the residual neutralizing [antibodies] and the activation of primed B cell and T cell memory." They have shifted the focus from B cells (antibody-based immune responses) to T cells (cytotoxic or killer cells). Note the cautious wording? T cells would likely be activated in those were previously infected, as well. Are the "unvaccinated" groups in question immunologically naive? Is there a comparison to naturally infected patients? Let's look at some of the references cited. For example, the manuscript titled SARS-CoV-2 vaccination induces immunological T cell memory able to cross-recognize variants from Alpha to Omicron [135], in which the experimental control group in this study was previously infected individuals who had mild disease.

One paper, "Ancestral SARS-CoV-2-specific T cells cross-recognize the Omicron variant" [136], does include the previously infected and shows a relatively lower drop in preservation of T cell cross-recognition in vaccinated than the previously infected but does not appear to compare those numbers statistically. Another, "Vaccines elicit highly conserved cellular immunity to SARS-CoV-2 Omicron," by Liu et al., does not compare the vaccinated to the naturally infected, and although it asserts that the vaccines provide substantial protection from Omicron, it does not demonstrate this in any way other than showing persistence of T cell responses.

Interestingly, "T cell reactivity to the SARS-CoV-2 Omicron variant is preserved in most but not all individuals" [137] examined both natural infection and vaccinated individuals and demonstrated that, in most patients *from either category*, T cell responses are relatively preserved against Omicron. But there is a caveat—a subset of patients lose T-cell responses to Omicron in addition to losing B-cell (antibody) responses. This may explain why some patients report repeated or chronic Omicron infection, and these patients may be driving further development of escape mutant viruses.

Another study [138] focused on memory B cells (which is not easy to do) over time in vaccinated individuals with no assessment of naturally infected patients. Their findings varied from some of the other studies, but they had some interesting observations:

> Omicron-binding memory B cells were efficiently reactivated by a 3rd dose of wild-type vaccine and correlated with the corresponding increase in neutralizing antibody titers. In contrast, pre-3rd dose antibody titers inversely correlated with the fold-change of antibody boosting, suggesting that high levels of circulating antibodies may limit the added protection afforded by repeat short interval boosting.

We can conclude from this study that booster timing is important. Too late, and you become as susceptible to infection and disease as—or even more than—the naturally infected. Too early, and the antibodies still circulating from the prior inoculation will interfere with the boost.

Reynolds et al. make several important conclusions. Regarding B cell immunity after three vaccine doses, the researchers noted that:

> Healthcare workers (HCW) were identified with mild and asymptomatic SARS-CoV-2 infection by ancestral Wuhan Hu-1, [Alpha], [Delta] and then [Omicron variants] during successive waves of infection and after first, second and third

mRNA (BioNTech BNT162b2) vaccine doses. By three vaccine doses antibody responses had plateaued, regardless of infection history. We found differences in immune imprinting indicating that those who were infected during the ancestral Wuhan Hu-1 wave showed a significantly reduced anti-RBD (receptor binding domain) titer against [Beta], [Gamma] and [Omicron] compared to infection-naive HCW.

To simplify, if you were first infected with Wuhan Hu-1, then vaccinated, then infected with Omicron, your antibody levels against the important part of spike (the receptor binding domain) would be *lower* than those who have never been infected.

The authors noted that memory B cell frequency against the S1 protein of several variants (the part that binds to cells) "was boosted 2-3 weeks after the third vaccine dose compared to 20-21 weeks after the second vaccine dose." But they also found that, regardless of whether the patient was previously infected or not, memory B cell frequency was similar against the original Wuhan variant and the Delta variant but was *lowered against Omicron* 2-3 weeks after the third dose and 20-21 weeks after the second dose. Memory B cell frequency is an indirect indicator of long-term protection. This is more evidence of immunologic escape of Omicron from B-cell mediated (antibody) control.

Next, they compared T cell responses after three vaccine doses. For Omicron they found that more than half of the subjects (27 of 50; 54%) had *no T cell response at all*, irrespective of infection history, while only 8% had no T cell response against the ancestral strain. They got a similar result when they compared "peptide pools" for the original Wuhan strain and for the Omicron variants: "42% (21/50) of [healthcare workers] make no T cell response at all against the [Omicron] mutant pool." This is problematic. Nearly half of these triple-vaccinated healthcare workers failed to generate *any* T cells against Omicron. T cells are major contributors to immune protection, and generating both B and T cell responses is the whole reason to use an mRNA vaccine.

So now we have evidence suggesting both poor B and poor T responses to Omicron, implying that Omicron has evolved to escape both B and T cell adaptive immunity.

But why and how did this happen?

Essentially, Reynolds et al. "showed that priming with one pool resulted in impaired responses to the other." Omicron infection boosted immunity to prior strains in triple-vaccinated healthcare workers, but not so much against itself. For the triple-vaccinated healthcare workers who had *not* been infected by prior viral strains, including Omicron, the neutralizing antibodies against Omicron were rapidly lost after the third vaccination.

If the healthcare workers were first infected by the original Wuhan strain (as I was), the researchers demonstrated that they do not make an increased neutralizing antibody response after Omicron infection (compared to those who are triple-vaccinated but not previously infected by the original Wuhan strain):

[Omicron] infection can boost binding and [neutralizing antibody] responses against itself and other [variants] in triple-vaccinated previously uninfected infection naive [healthcare workers], but not in the context of immune imprinting following prior Wuhan Hu-1 infection. Immune imprinting by prior Wuhan Hu-1 infection completely abrogated any enhanced [neutralizing antibody] responses against [Omicron] and other [variants].

So, prior infection by the Wuhan variant seems to block production of neutralizing antibody responses during and after Omicron infection. In summary, Omicron infection resulted in enhanced, cross-reactive antibody responses against all variants tested in the three-dose vaccinated infection-naive healthcare workers, but not those with previous Wuhan Hu-1 infection, and less so against Omicron itself. And that is pretty much proof of the immune imprinting effect.

So, we have evidence of immune imprinting with B cell (antibody) responses, but what about T cell responses?

The authors found a rapid loss of any detectable T cell immunity against S1 (the main antigen in the vaccines) shortly after the third dose of vaccine:

Fourteen weeks after the third dose (9/10, 90%) of triple-vaccinated, previously infection-naive [healthcare workers] showed no cross-reactive T cell immunity against [Omicron] S1 protein.

This is not good. In addition, fully vaccinated healthcare workers who were infected by Omicron had significantly reduced T cell responses to Omicron itself, which the authors said was "basically a set up for either chronic Omicron infection or rapid Omicron re-infection."

Furthermore:

> Importantly, none (0/6) of [the healthcare workers] with a previous history of SARS-CoV-2 infection during the Wuhan Hu-1 wave responded to [Omicron] S1 protein . . . . This suggests that, in this context, [Omicron] infection was unable to boost T cell immunity against [Omicron] itself; immune imprinting from prior Wuhan Hu-1 infection resulted in absence of a T cell response against [Omicron] S1 protein.

The findings consistently show that people initially infected by Wuhan Hu-1 in the first wave and then reinfected during the Omicron wave do not evidence either T cell or B cell immunity against Omicron. This is very bad news and yet more evidence for initial immune imprinting, reinforced by repeated vaccination with the original Wuhan virus-derived spike mRNA vaccine, causing an inability to respond to Omicron. It really sounds like these patients may become the breeding ground for the next wave of Omicron variants.

The researchers investigated how prior infection differentially imprinted Omicron T and B cell immunity:

> To investigate in more detail the impact of prior SARS-CoV-2 infection on immune imprinting, we further explored responses in our longitudinal [healthcare workers] cohort. We looked initially at the S1 RBD (ancestral Wuhan Hu-1 and Omicron VOC) antibody binding responses across the longitudinal cohort at key vaccination and SARS-CoV-2 infection timepoints, exploring how different exposure imprinted differential cross-reactive immunity and durability. This revealed that at 16-18 weeks after Wuhan Hu-1 infection or [Alpha] infection, unvaccinated [healthcare workers] showed no detectable cross-reactive S1 RBD binding antibodies against [Omicron].

In other words, Omicron has evolved to completely evade any antibodies generated from natural infection by either the original Wuhan or the Alpha strain.

Hybrid immunity (the combination of prior infection and a single vaccine dose) significantly increased the S1 RBD binding antibodies against [Omicron] (p < 0.0001) compared to responses of infection-naive [healthcare workers], which were undetectable after a single vaccine dose. This increase was significantly greater for prior Wuhan Hu-1 than [Alpha] infected [healthcare workers].

Good news. Prior infection with either the original Wuhan or Alpha strain, followed by a single mRNA dose, resulted in detectable antibodies to Omicron, although this worked better if you were first infected with the original Wuhan rather than the Alpha strain.

However, 20–21 weeks after the second vaccine dose, differential [Omicron] RBD [antibody] waning was noted with almost all (19/21) of the [healthcare workers] infected during the second [Alpha] wave no longer showing detectable cross-reactive antibody against [Omicron] RBD.

So one dose of the mRNA vaccine after natural infection is good; two doses is not. This indicates a profound differential impact of immune imprinting on Omicron-specific immune antibody waning between healthcare workers who were previously infected by Wuhan Hu-1 and those who were infected by Alpha. Fourteen weeks after the third vaccine dose previously infection-naive healthcare workers who were infected during the Omicron wave showed increased Omicron binding responses, but those with prior Wuhan Hu-1 infections did not, indicating that individuals previously infected with Wuhan Hu-1 were immune imprinted to not boost antibody binding responses against Omicron despite having been infected by Omicron itself.

Three doses of mRNA vaccine in people who were never infected with virus shows antibody production against Omicron spike protein, but not if the healthcare workers were previously infected with the original Wuhan strain first.

In fact, infection during the [Omicron] wave imprinted a consistent relative hierarchy of cross-neutralization immunity against [variants of concern] across different individuals with potent cross-reactive [neutralizing antibody] responses against [Alpha]), [Beta], and [Delta] (Fig. 6, D and E). Comparative analysis of [neutralizing antibody] potency for cross-neutralization of [variants of concern] emphasized the impact of immune imprinting, which effectively abrogates the [neutralizing antibody] responses in those vaccinated [healthcare workers] infected during the first wave and then reinfected during the [Omicron] wave.

If the healthcare workers were first infected by the Wuhan strain, then vaccinated, then reinfected with Omicron, the immune imprinting associated with being first infected and then vaccinated pretty much destroyed their ability to respond effectively to Omicron. In other words, forcing healthcare workers who were previously infected with the Wuhan strain to take three doses of the mRNA vaccine pretty much destroyed their ability to mount an effective immune response against Omicron.

This may explain the recent observation of widespread COVID-19 illness in the (mandated) highly vaccinated healthcare worker population of Houston Methodist and many other hospitals across the United States as reported by Emily Miller in the *Epoch Times*, August 2022 [139].

> The first U.S. hospital system to enforce a COVID-19 vaccine mandate for all employees could now be facing a staffing shortage because of a rise in infections.
>
> Houston Methodist now has hundreds of employees out of work because they tested positive for the virus that causes COVID-19. At the same hospital system in 2021, 153 staff members who refused to get vaccinated quit or were fired. Now, Houston Methodist's leadership is trying to avert a crisis.
>
> "What is worrisome is the climbing number of our employees who cannot work because they are home sick with COVID-19. Almost 400 employees tested positive last week," Dr. Robert Phillips, Houston Methodist's executive vice president and chief physician executive, wrote in an internal email on July 12, 2022.
>
> "While most of these employees are getting COVID-19 from the community, it is vital that we don't face a situation where too many employees are out sick, and we find ourselves with a staffing shortage."
>
> Houston Methodist, with a workforce of about 28,000, was the first hospital system in the country to mandate the COVID-19 vaccine for all of its employees. It also was the first system

in the nation to mandate the vaccine for its private healthcare providers who are credentialed members of its medical staff. The hospital later required all its employees to get a vaccine booster by March 1.

While most employees got vaccinated and stayed, the system is having trouble with staffing as the vaccines prove increasingly worse at protecting against infection as new variants of SARS-CoV-2, the virus that causes COVID-19, emerge.

"The spike in cases is happening all over the country and is likely attributed to the highly contagious and more vaccine-resistant Omicron subvariant," Phillips wrote. "BA.5 is now the most infectious variant so far and is thought to be four times more vaccine evasive than the last dominant variant."

I spoke directly to current and former Houston Methodist employees to verify this account and was told that the hospital was under severe stress as it sought to find staff, and that it had to reduce qualification requirements for many positions in order to remain in operation. For example, one staff member reported, "You should see the emergency room at Houston Methodist, it is like a war zone, with patients in the ER lobby and ER hallways" due to lack of staffing." Another stated, "I know it's a revolving door and turnover is high."

Unresolved is whether those fully vaccinated people who land in the hospitals and/or die from Omicron were first infected with another strain prior to becoming fully vaccinated. What do Reynolds et al. conclude about all of this?

Molecular characterization of the precise mechanism underpinning repertoire shaping from a combination of Wuhan Hu-1 or B.1.1.7 (Alpha) infection and triple-vaccination using ancestral Wuhan Hu-1 sequence, impacting immune responses to subsequent VOCs, will require detailed analysis of differential immune repertoires and their structural consequences. The impact of differential imprinting was seen just as profoundly in T cell recognition of B.1.1.529 (Omicron) S1, which was not recognized by T cells from any triple-vaccinated HCW who were initially infected during the Wuhan Hu-1 wave and then re-infected during the B.1.1.529 (Omicron) wave. Importantly, while B1.1.529 (Omicron) infection in triple-vaccinated previously uninfected individuals could indeed boost antibody, T cell and MBC responses against other VOC, responses to itself were reduced. *This relatively poor immunogenicity against itself may help to explain why frequent B.1.1.529 (Omicron) reinfections with short time intervals between infections are proving a novel feature in this wave.* It also concurs with observations that mRNA vaccination carrying the B.1.1.529 (Omicron) spike sequence (Omicron third-dose after ancestral sequence prime/boosting) offers no protective advantage.

In summary, these studies have shown that the high global prevalence of Omicron infections and reinfections likely reflects considerable subversion of immune recognition at both the B, T cell, antibody binding, and neutralizing antibody levels, although with considerable differential modulation through immune imprinting. Some imprinted combinations, such as infection during both the Wuhan Hu-1 and Omicron waves, confer particularly impaired responses. Yet the government continues to advise that all of our children, most of whom have already been infected and have cleared the virus with very little problem, should get vaccinated.

These data may help explain the negative effectiveness being observed with "full" mRNA vaccination in those who have been infected by Omicron.

Safe, Effective, Ethical?

In August 2022, the first risk-benefit assessment of the safety, efficacy, and ethics of SARS-CoV-2 boosters for young, previously uninfected adults under forty years old was reported by a team of highly qualified public health professionals from leading academic institutions. Titled "Covid-19 Vaccine Boosters for Young Adults: A Risk-Benefit Assessment and Five Ethical Arguments against Mandates at Universities" [140], the analysis relies on currently available data (prior to the bivalent boosters) and is focused on assessing risks and benefits of mRNA booster vaccination for students at North American universities who risk disenrollment due to third-dose Covid-19 vaccine mandates.

In their analysis, the authors note that "Proportionality is a key principle in public health ethics. To be proportionate, a policy must be expected to produce public health benefits that outweigh relevant harms, including harms related to coercion, undue pressure, and other forms of liberty restriction." Using CDC and sponsor-reported adverse event data, the study demonstrates that booster mandates may cause a net expected harm: For each COVID-19 hospitalization prevented in previously uninfected young adults, the data indicate that 18 to 98 serious adverse events are likely to occur, including 1.7 to 3.0 booster-associated myocarditis cases in males, and 1,373 to 3,234 cases of grade ≥3 reactogenicity, which interferes with daily activities. Based on this analysis, it is clear that the risks of these mRNA booster inoculations greatly outweigh the benefits in this age group. Furthermore, given the high prevalence of postinfection immunity, the authors note that this risk-benefit profile is even less favorable in those who have been previously infected.

Regarding the ethics of mandated mRNA booster vaccination in this cohort, the authors conclude the following:

> University booster mandates are unethical because: 1) no formal risk-benefit assessment exists for this age group; 2) vaccine mandates may result in a net expected harm to individual young people; 3) mandates are not proportionate: expected harms are not outweighed by public health benefits given the modest and transient effectiveness of vaccines against transmission; 4) US mandates violate the reciprocity principle because rare serious vaccine-related harms will not be reliably compensated due to gaps in current vaccine injury schemes; and 5) mandates create wider social harms. We consider counter-arguments such as a desire for socialization and safety and show that such arguments lack scientific and/or ethical support.

CHAPTER 12
Preventable Deaths and Vitamin D$_3$

We had an inexpensive lifesaving solution both before and during the pandemic . . .

The inconvenient truth is that even at the beginning of the COVID-19 crisis, a very simple, inexpensive, and effective treatment was available that could have saved the majority of lives lost [141–143]. All that the WHO and national public health bureaucracies (including the US HHS) had to do was to recommend and support people taking sufficient Vitamin D3. This failure to act traces back to the unscientific bias and provaccine obsession of Dr. Anthony Fauci. And once again the legacy corporate media, while being paid by the US government and the pharmaceutical industry to promote vaccination, acted by censoring, defaming, and suppressing the ability of physicians to inform people of scientific truth. The disease you suffered, the loss of life among your family and friends, could have been greatly reduced by simply getting enough Vitamin D3. This is another example of what happens when unelected bureaucrats are allowed to control free speech and the practice of medicine. Crimes against humanity.

The effectiveness of Vitamin D3 as an immune system-boosting prophylactic treatment for influenza and other respiratory RNA viruses was first discovered in 2006 [144, 145]. Despite that fact that this treatment is amazingly effective for preventing death (by strengthening your immune system), it has never been investigated by the NIH or promoted by the CDC or by the US government for the prevention or treatment of influenza. One major issue has been that of uncontrolled variables of dosing, timing of dosing, and disease status have resulted in inconsistent clinical trial results (much as we have seen with the Ivermectin and Hydroxychloroquine COVID trials). However, when Vitamin D3 is given prophylactically at sufficient doses, there is clear and compelling evidence that Vitamin D blood levels of around 50 ng/ml or above will substantially reduce symptomatic infection, severe disease, and mortality.

Long-standing worldwide public health policy is that Vitamin D should be taken at sufficient levels (typically supplemented in milk products) to prevent the bone disease called rickets. But this is just a minimal level to prevent a very obvious debilitating disease. The recommended Vitamin D levels in our milk are not sufficient for the subtler immune system-boosting effects of this critical vitamin/hormone. Our bodies' way of normally producing Vitamin D requires a lot of sunlight, but life in the modern world and northern latitudes make this difficult, particularly in winter months, which is often when the respiratory viruses cause the most disease and death. In a sense, disease and death from influenza and other respiratory RNA viruses are a lifestyle disease. Just the way things are. Significantly avoidable unnecessary death.

As I write the above, I am reminded that I recently spoke with a scientist and physician who was on a team at the Department of Defense (DoD) in 2006 that had discovered a surprising finding while analyzing data from warfighters. He and his team had been looking for things that could help explain why some soldiers got bad disease from circulating influenza viruses, while others did not. I hear a lot of stories, but this one was a first for me.

In any given year, soldiers pretty much all get exposed to the same influenza virus variants, so why the differences in medical outcomes? Important to keep in mind that lots of data suggest that the 1918 "Spanish Flu" that swept the world at the close of WW I and caused so many deaths in relatively young people may well have come from young US midwestern recruits exposed to pig influenza viruses. This version of the 1918 influenza origin story goes along the lines that these young farmer recruits brought a human-adapted pig influenza virus from US to the European battle theater, where it incubated in the infectious disease petri dish of the horrible conditions of trench warfare and then was spread worldwide to civilians by returning soldiers. The "Spanish Flu" label that the US mainstream media of the time applied to the disease was yet another case of propaganda designed to deflect responsibility for a lethal infectious disease outbreak (from the US Government). In any case, you can understand why the DoD and the

Walter Reed Army Institute of Research in particular has a long history of influenza virus research—starting long before the CDC, NIH, or NIAID ever existed.

This DoD research scientist and his team had conducted a retrospective study that tied higher baseline vitamin D levels to lowered respiratory virus infection and disease (influenza), using a military database to correlate vitamin D levels to flu levels and death. The DoD believed that if he presented his research to Dr. Fauci, then Director of NIAID (National Institutes of Allergy and Infectious Diseases), the US government might change direction by investing in this line of research and developing corresponding treatment guidelines. The DoD saw the potential of reducing influenza disease and death with this safe prophylactic and directed him to contact Dr. Fauci to discuss this finding.

This scientist told me that he scheduled the meeting as assigned and presented his rock-solid data to Dr. Fauci. He was then informed by Dr. Fauci that US policy is to control influenza in the USA with vaccines, not therapeutics. End of story. No funding or support available for future work. Therefore, NIAID had no interest in pursuing Vitamin D3 as a prophylactic for respiratory diseases, such as influenza, and the DoD dropped the follow-up. That means that over fifteen years ago, Dr. Fauci had already set the policies that informed the US government's present response to the COVIDcrisis. Because that policy extends well beyond flu, it is the response that the US Government falls back on for all infectious disease outbreaks, including those that emerge due to a pandemic or viral biothreat. The official policy, set by Dr. Fauci, is that the US government wants vaccines for respiratory viruses above all else, and no other prophylactic solutions are to be promoted. With that background in mind, why would anyone expect anything else other than an exclusive USG obsession with a vaccine solution for an infectious respiratory disease such as COVID-19, even if there are excellent, cheap alternatives already available?

The data for the use of Vitamin D3 are extremely strong; there are now even randomized clinical trials supporting its use for the treatment of COVID [146], as well as many retrospective clinical trials showing its efficacy. The title of a major meta-analysis study published in October 2021 is "COVID-19 Mortality Risk Correlates Inversely with Vitamin D3 Status, and a Mortality Rate Close to Zero Could Theoretically Be Achieved at 50 ng/mL 25(OH)D3: Results of a Systematic Review and Meta-Analysis," and that title pretty much says it all [147]. Yet the NIH treatment guidelines found on their website in May 2022, state that:

> Recommendation: There is insufficient evidence to recommend either for or against the use of Vitamin D for the prevention or treatment of COVID-19.

The CDC's website says nothing about the link between Vitamin D3 levels and decreased severe disease and death in respiratory virus diseases, including COVID. The NIH guidelines cite a single study in which Vitamin D was given to COVID patients in the intensive care unit (late-stage COVID) in Brazil as the sole criterion for their evaluation of Vitamin D. They even mention that this paper is flawed, writing that:

> It should be noted that this study had a small sample size and enrolled participants with a variety of comorbidities and concomitant medications. The time between symptom onset and randomization was relatively long.

Yet this admittedly flawed work is the cited study from which the NIH determined that there is no link between Vitamin D levels and reduced incidence and disease due to SARS-CoV-2, while ignoring all other data including superior studies. Clear documentation of the scientific bias that has resulted in so many poor public health management decisions throughout the current outbreak.

There is nothing in the CDC guidelines about the meta-analysis studies, retrospective studies, and even randomized clinical trials concerning preventative use of Vitamin D3—just an oblique reference to clinicaltrials. gov if one wanted more information. Can this be explained by anything other than regulatory capture by the US government institutes within the department of Health and Human Services, including CDC, NIH, and FDA?

With an emerging infectious disease, it is drugs and therapeutics that are often the first line of defense. Physicians use deductive reasoning together with the currently available pharmacopeia when confronted with a new infectious disease or even any unknown disease. This is how they are taught to respond to a newly identified disease of any kind, because it is a very effective way to treat when faced with an unknown or even unclear diagnosis when there is no proven treatment plan [148]. Begin by treating the symptoms until you can figure out the underlying pathophysiology.

With COVID, it became clear early on that the front-line physicians were able to develop effective therapies using this strategy. There were many drugs, and many treatments (including prophylactic Vitamin D3) that worked. These physicians made deductions and treated the symptoms. The numbers of lives saved using this method are astounding, but the government literally said that physicians should not use these treatments. Instead, the government instructed patients to go home and wait until their oxygen levels were so low that their lips were turning blue. That was criminal on the part of the HHS and US government. Truly a crime against humanity.

There are doctors who ignored these "official" guidelines and behaved like doctors should act—when they are committed to the Hippocratic oath. They saved lives. They formed quiet communities with other doctors to find viable treatments. Dr. George Fareed and Dr. Brian Tyson are two such doctors who have saved thousands and thousands of lives, as documented in their book titled *Overcoming the COVID-19 Darkness: How Two Doctors Successfully Treated 7000 Patients* [149]. Compare the case studies and protocols in this book and the many complementary case histories of physicians working on the front lines (in the USA Drs. Peter McCullough, Pierre Kory, Paul Marik, Vladimir [Zev] Zelenko, and Richard Urso; and Didier Raoult and his colleagues in France, to name just a few examples) to what happened when the US government became involved in dictating medical treatments for COVID.

Unfortunately, the US government did not support any of this frontline physician work and in fact worked hard to undermine early multidrug treatment using licensed drugs. Precisely as Dr. Fauci did fifteen years ago when his learned of the role of vitamin D3 for the reduction of disease and death in respiratory diseases.

To further illustrate the enormous tragedy of this historic bias, just think of all the elderly who could have had a few more good years, whose grandchildren could have benefited from their wisdom, but instead died of the flu just because no one ever told them to keep their Vitamin D3 levels up. Because Dr. Fauci believes that vaccines should always be the first line of defense.

This also relates back to the faulty logic of vaccine-induced herd immunity. A logical fallacy that through the use of vaccines we could control influenza to a significant extent in the US population. This is flawed because 1) influenza is constantly mutating to escape existing vaccines, 2) there is a large seasonal unvaccinated world population, and travelers are constantly bringing new strains to the USA, 3) the vaccines are at best 40% (and often much less) effective at preventing influenza disease (sound familiar?), and 4) there are enormous animal reservoirs that harbor and constantly develop new influenza virus strains. But due to the world's success in eradicating smallpox, "official" public health (and Mr. Bill Gates) cannot seem to understand that not all viruses are a DNA virus (like smallpox) that mutates extremely slowly and is only found in humans. Comparing smallpox to a rapidly mutating RNA respiratory virus with a large animal reservoir is both illogical and naive.

But let's take a step back in time, a decade back. Let's imagine that Dr. Fauci had authorized the DoD or some other research entity to do a well-designed randomized clinical trial concerning the benefits of adequate D3 levels in preventing respiratory virus disease. If such a trial had been funded, results would have shown that higher vitamin D3 supplementation to achieve blood levels greater than 50 ng/ml helped prevent disease and death caused by influenza virus. Let's imagine that five years later (at the latest), a CDC guideline for D3 levels was put in place (particularly for the elderly). For sake of discussion, let's even throw out a number. A conservative number, based on what we know now. That 50% of the people who have died from influenza could have been saved if they had sufficiently high vitamin D3 blood levels. Per a CDC website, on average 35.7 thousand people die per year of influenza. In other words, about 357,000 people have died of influenza over the last decade. Which means if 50% were saved by providing Vitamin D3 supplements, then 178,500 people could have been saved over the last decade in the USA by simply having the CDC advocate nationally for prophylactic administration of Vitamin D3. Think about that. A simple, pennies-per-day treatment that never happened. Why? Because Dr. Fauci believes that the USA uses vaccines to treat flu, and that vaccine-induced herd immunity is key—a fallacy that he has never revisited in his own mind.

Now let's fast-forward to COVID-19. How many people could have been saved from just having their levels of vitamin D3 brought up to 50 ng/ml (or higher!)? We knew about vitamin D3 and its benefits in helping patients resist disease and death from influenza viruses. It really didn't take a randomized clinical trial to understand the link between D3 and RNA respiratory virus morbidity and mortality. The USA alone could have saved hundreds of thousands of lives. Let alone all of the possible lives that could have been saved in the rest of the world. That these lives were unnecessarily lost is not acceptable in any way, shape, or form.

Many people (and physicians) rely on the CDC and NIH to guide them in healthcare and wellness decisions. It is way past time that these organizations step up to the plate and do their job and stop relying on the unscientific biases of highly influential bureaucrats. That job being to protect the health of the public. Not advancing the interests of the pharmaceutical industry and its shareholders.

CHAPTER 13
Scientific Fraud at the Centers for Disease Control and Prevention (CDC)

The CDC has withheld critical data on boosters, hospitalizations…

> "The C.D.C. Isn't Publishing Large Portions of the
> COVID Data It Collects"
> *New York Times*, February 21, 2022

Two full years into the pandemic, the agency leading the country's response to the public health emergency has published only a tiny fraction of the data it has collected, several people familiar with the data said. Much of the withheld information could help state and local health officials better target their efforts to bring the virus under control…

"*The C.D.C. is a political organization as much as it is a public health organization,*" said Samuel Scarpino, managing director of pathogen surveillance at the Rockefeller Foundation's Pandemic Prevention Institute. "*The steps that it takes to get something like this released are often well outside of the control of many of the scientists that work at the CDC.*"

Let me translate that quote for you. Basically, a nongovernmental spokesperson for the "official" public health scientific community is throwing Rochelle Walinsky, Director of the CDC, under the bus and saying that the politicians forced us (the CDC) to commit scientific fraud by withholding key data.

The medical practitioners who have stood up to the lies and tyranny— who have been harassed, jobs lost, medical licenses lost, smeared and libeled—are right. The data are being withheld. The mainstream media owes a whole lot of professionals—scientist and physicians—a huge apology. The corporate media has to stop being the mouthpiece for the government. This is not communist China! Furthermore, the federal government owes the American people a huge apology. People in the government who have lied to the American people need to be charged and must be held legally accountable. *We the people must demand to see ALL of the data from the CDC and the FDA.*

Let's talk about these data. The CDC is using cumulative data from the beginning of the vaccine rollout in early 2021 to prop up the lie that these vaccines are effective against Omicron. The CDC is clearly hiding the data about safety. The (thoroughly biased) NYT piece above writes further on this.

> Pfizer's data supported the safety of the vaccine, but researchers said the effectiveness wasn't there with two shots.
>
> "It was effective in the younger kids so those six months to two years but in the two to four-year-old age group it didn't quite meet the levels of antibody response they expected to see," said Dr. Christina Canody, BayCare Pediatric Service Line Medical Director.
>
> Now instead of just having an EUA meeting about two doses, Pfizer is continuing their trial for three doses and will present that data once they have it...
>
> Concern about the misinterpretation of hospitalization data broken down by vaccination status is not unique to the C.D.C. On Thursday, public health officials in Scotland said they would stop releasing data on Covid hospitalizations and deaths by vaccination status because of similar fears that the figures would be misrepresented by anti-vaccine groups [150]

Precisely what we have been saying.

Why is this important?

If the CDC released the age-stratified data for COVID, it would be clear that a vaccine is not necessary for most if not all Americans. If the vaccine risk ratio of those vaccinated and hospitalized were published for Omicron—it would be clear that the vaccine benefit is not observed.

The FDA has not revealed what the efficacy of the boosters for children is. They have not released the safety data. They have withheld the safety data on the vaccines for children and adults. *This must stop. We are deep into outright scientific fraud territory.*

Let's remember where this started... We have been manipulated from the VERY start of this pandemic. The government has been deciding what has been written, removed, censored by media and the big tech giants. This is propaganda.

I am citing these historic references from the beginning of 2020 to show that our government has been involved in scientific fraud from the beginning. Do not forget—this goes back to 2020:

1. "World Health Organization holds secretive talks with tech giants Google, Facebook, and Amazon to tackle the spread of misinformation on coronavirus." *Daily Mail.* February 17, 2020:

> "Google, Facebook, Amazon, and other tech giants spent a day in secretive talks with the World Health Organization to tackle the spread of coronavirus misinformation. *Social media companies including Twitter and YouTube have already been working to remove posts about the virus that are proved to be fake.*
>
> The World Health Organization (WHO) has offered to work directly with the companies on fact-checking in a bid to speed up the process.
>
> Posts on the virus that needed to be removed have ranged from those calling it a fad disease or *created by the government* to claims it can be treated with oregano oil."
>
> Companies at the meeting agreed to work with WHO on collaborative tools, better content, and a call center for people to call for advice, CNBC reported.

2. Bloomberg. "Amazon, Alphabet among tech firms meeting with White House on coronavirus response." *LA Times.* March 11, 2020:

> White House officials discussed combating online misinformation about the coronavirus and other measures during a teleconference Wednesday with tech companies including Alphabet Inc.'s Google, Facebook Inc., and Twitter Inc.

U.S. Chief Technology Officer Michael Kratsios led the call, which also included representatives from Amazon.com Inc., Apple Inc., Microsoft Corp., IBM Corp. and other companies and tech trade groups.

The discussion focused on information-sharing with the federal government, coordination regarding telehealth and online education and the creation of new tools to help researchers review scholarship, according to a statement from the White House's Office of Science and Technology Policy.

"Cutting edge technology companies and major online platforms will play a critical role in this all-hands-on-deck effort," Kratsios said in a statement. He said his office would unveil a database of research on the virus in coming days."

3. "White House asks Silicon Valley for help to combat coronavirus, track its spread and stop misinformation." *Washington Post*. March 11, 2020:

The White House on Wednesday sought help from Amazon, Google, and other tech giants in the fight against the coronavirus, hoping that Silicon Valley might augment the government's efforts to track the outbreak, disseminate accurate information...

The requests came during a roughly two-hour-long meeting between top Trump administration aides, leading federal health authorities and representatives from companies including Cisco, Facebook, IBM, Microsoft, and Twitter, as Washington sought to leverage the tech industry's powerful tools to connect workers and analyze data to combat an outbreak that has already infected more than 1,000 in the United States.

Three participants described the phone-and-video conversation on the condition of anonymity because the session was private. Most tech companies in attendance either did not respond or declined to comment.

The evidence above makes it crystal clear that the government has been manipulating data from the start. Now that Omicron is here, the vaccines are clearly not working. Now that we have data from other countries that there are safety issues with these products, we must demand transparency and a stop to the manipulation of the American people. Free speech is free speech. *Scientists and physicians must be allowed to discuss data on the Internet. We ALL must be allowed to discuss data. It is time to stop the madness.*

How this all ties into the globalists is becoming more and more clear, as summarized in an article titled "The Next Step for the World Economic Forum," by Roger Koops and published by the Brownstone Institute on February 20, 2022:

> It has been obvious since early 2020 that there has been an organized cult outreach that has permeated the world as a whole. It's possible that this formed out of a gigantic error, rooted in a sudden ignorance of cell biology and long experience of public health. It is also possible that a seasonal respiratory virus was deployed by some people as an opportunity to seize power for some other purpose.
>
> Follow the money and influence trails, and the latter conclusion is hard to dismiss.
>
> The clues were there early. Even before the WHO declared a pandemic in March 2020 (at least several months behind the actual fact of a pandemic) and before any lockdowns, there were media blitzes talking about the "New Normal" and talk of the "Great Reset" (which was rebranded as "Build Back Better").
>
> Pharmaceutical companies such as Pfizer, Johnson & Johnson, Moderna, and Astra-Zeneca were actively lobbying governments to buy their vaccines as early as February 2020, supposedly less than a month after the genetic sequence (or partial sequence) was made available by China.
>
> As a person who spent his whole professional career in pharmaceutical and vaccine development, I found the whole concept of going from scratch to a ready-to-use vaccine in a few months simply preposterous.
>
> Something did not add up.

For more readings on this, I highly recommend the online journal Brownstone Institute.

Natural Immunity

Just as with most other viral infections, it has been clear throughout the COVIDcrisis that while protection from the disease caused by SARS-CoV-2 is afforded by recovery from actual infection ("natural immunity") and is robust and long-lasting, the CDC and US HHS repeatedly denied that this was the case until the "reorganization" of August 2022, in which CDC policy positions were modified to reflect that "The risk for medically significant illness increases with age, disability status, and underlying medical conditions but is considerably reduced by immunity derived from vaccination, *previous infection*, or both, as well as timely access to effective biomedical prevention measures and treatments." The CDC itself had previously published a report titled "COVID-19 Cases and Hospitalizations by COVID-19 Vaccination Status and Previous COVID-19 Diagnosis—California and New York, May–November 2021" [151] that demonstrated that this was the case during February 2022. The report analyzed COVID-19 cases in California and New York in 2021 from May 30 to November 20 and compared the risk of new SARS-CoV-2 infection among four groups of people: those who were unvaccinated without a prior case of COVID-19, those vaccinated without prior COVID-19, those unvaccinated with prior COVID-19, and those vaccinated with prior COVID-19. During the delta wave of COVID-19, the incidence of SARS-CoV-2 infection among those with "enhanced" immunity due to both vaccination and prior infection was 32.5-fold lower in California and 19.8-fold lower in New York, whereas rates among those vaccinated alone (without prior COVID-19) were only 6.2-fold lower in California and 4.5-fold lower in New York. The rates among those with natural immunity were 29.0-fold lower in California and 14.7-fold lower in New York. The authors note that hospitalization rates followed a similar pattern. The report finally acknowledged what many have known for a long time, that recovery from SARS-CoV-2 natural infection provides excellent protection from repeat infection as well as from hospitalization and death for the delta variant of COVID-19. Subsequent data from around the world have since demonstrated that this is even more true in the case of the Omicron variants.

The Vaccine Adverse Events Reporting System (VAERS)

When Congress passed the National Childhood Vaccine Injury Act of 1986, the law included the first national vaccine reporting system in the United

States. The Vaccine Adverse Events Reporting System (VAERS) began operating in 1990 and is jointly operated by the FDA and CDC.

This law requires doctors and other vaccine providers to report serious vaccine injuries, hospitalizations, deaths, and other serious health problems following vaccination to VAERS. The VAERS system also has a voluntary component through which people other than physicians and vaccine providers can report cases. This is a "passive" reporting system, and no mandatory reporting is required. Prior to Covid-19, there were over 500,000 reports of adverse events in the VAERS systems [152]. There are no penalties to physicians or healthcare providers for not reporting adverse events. It is estimated that between less than 1 to 10 percent of all vaccine-related health problems are actually reported to VAERS. There is no verification system within VAERS for follow-up, and the government does not verify reports. Although VAERS is known to be inaccurate, the CDC and FDA have not corrected these deficiencies. Even though these issues have plagued the system since the inception of the program.

The paper "Safety of mRNA vaccines administered during the initial 6 months of the US COVID-19 vaccination program: an observational study of reports to the Vaccine Adverse Event Reporting System and V-safe" was published in the *Lancet* on March 7, 2022 [153]. The authors are CDC employees, although no conflicts of interest as such were reported to the *Lancet*. The legacy media immediately promoted the study as documenting that the vaccines are safe and effective. With severe side effects being of short duration and rare.

I began reading this paper with my usual wary eye, and what jumped out at me was that the conclusions reached by the legacy media did not match what I, as a trained physician and scientist, found important. This is because they are journalists, not scientists, and do not have training in scientific methodologies. Please remind me, why are we relying on journalists and the media to interpret science when they are not trained for this? In any case, here are some of the headlines from the mainstream media:

- "THE *LANCET* INFECTIOUS DISEASES: Large U.S. study confirms most mRNA COVID-19 vaccine side effects are mild and temporary" (Medical Express and many others)

- "Huge study finds most COVID-19 vaccine side effects were mild for Pfizer-BioNTech and Moderna" (*USA Today*)
- "Side Effects of COVID mRNA Vaccines Are Mild and Short, Large Study Confirms" (Medscape)
- "No link between Covid vaccine and deaths, says major US study: Just 4,500 people died out of the 298 million vaccines considered in the study" (*Evening Standard*)

Wait, let's back up a bit here and do our own due diligence and thinking! The *Lancet* paper documents the percentage of severe adverse events (6.6%), compared to nonsevere adverse events (92.1%). By the way, death was a separate category determined to be around 1.3% of all adverse events. So, what does this mean? A severe event ratio (including death) of 7.9% of all reported adverse events is high—very high! That means that about 1 in 13 people has a severe adverse event out of all adverse events reported, as defined by the VAERS system (quote from the *Lancet* paper below):

> VAERS reports were classified as serious if any of the following outcomes were documented: inpatient hospitalization, prolongation of hospitalization, permanent disability, life-threatening illness, congenital anomaly or birth defect, or death.

One out of every eight reported adverse events were classified as serious! But "somehow" what the Medscape Headline concludes is that the side effects are "mild and short." This is just not accurate.

But let's dig deeper. One has to look at the actual numbers of people affected by adverse events. Not just at the percentage points of the various adverse events. So, let's take a look under the hood and figure out what this all means.

First, there are many caveats to this paper. This data are only for the first six months after the vaccine rollout, so no children and almost no teens were vaccinated during this period (the 15–18-year-old age range began to get vaccinated around May 2021, but the data analysis started January and ends June 2021). Why the cutoff at six months? There were data that extend for fourteen months—which the paper could have easily included—and those data included information involving children.

The paper relied on a literature search to make many of its claims. However, there is a significant issue with the literature search as presented. By using too many search words, highly technical and long phrases, the search that the authors used did not yield many papers that discus the health impact they were searching for (for example: "BNT162b2" OR "mRNA-1273" OR "mRNA COVID-19 vaccine" AND "reactogenicity" OR "side-effects" OR "adverse effects" OR "health impact"). A more simplified search done in January 2022 by my team found many more peer reviewed papers that discussed the health impacts of these vaccines that the authors evidently did not discover, based on their following statement:

> Among 429 results, few publications described health impacts following vaccination by BNT162b2 (Pfizer-BioNTech) or mRNA-1273 (Moderna). Available literature included reports of manufacturer-sponsored phase 1–3 clinical trials, observational and cross-sectional studies among specific groups (e.g., transplant recipients or employees of a specific health-care system), and reviews or society recommendations that discussed reactogenicity and adverse events following mRNA vaccination.

Then, there are issues regarding conflict of interest of the authors. The authors are CDC employees. As we have recently been warned from the *New York Times*, the CDC is now a political organization that has been hiding data from physicians, public health officers, and the public. They have been supporting what the executive branch wants to hear, by publishing that which they feel fits that narrative that vaccines are "safe and effective." You know, not publishing data—so as to avoid "vaccine hesitancy." As such, each and every author on the publication has a significant conflict of interest. This is a big red flag.

Next, we have very good documentation that the VAERS system, which is the vaccine injuries national system for tracking injuries, traditionally undercounts the actual adverse events by a wide range, depending on vaccine type and/or adverse event. This is because the VAERS system is not a mandatory reporting system. I found one study of various vaccine adverse events using the VAERS system that showed a rate of about 50% of vaccines adverse events are underreported, with a large variability range. Other studies report a much higher underreporting rate, but going with 50% is probably a good, conservative number. This means that whatever data are presented by the VAERS system most likely represents an undercount of at least half of the cases.

Or... at least that is what would have been my estimate until the Cell paper titled "Immune imprinting, breadth of variant recognition, and germinal center response in human SARS-CoV-2 infection and vaccination" *came out showing that the synthetic mRNA hangs around in the lymph node germinal ce*nters for at least sixty days—and continues to produce spike protein as well as spike protein antigen, for at least that duration of time. Physicians and medical professionals have been informed by the CDC and the FDA that the side effects of the vaccine occur within a short time frame after vaccination. The FDA has stated that the mRNA degrades rapidly. So, adverse events (such as myocarditis) outside of the time limits imposed by the VAERS reporting system do not get recorded. There is a good likelihood that the adverse events and deaths reported to the VAERS system grossly underestimate these events, as event reporting is time-limited. More studies will have to be conducted, but clearly the VAERS system only works if adverse events are reported. If vaccine-related events are happening two months out, as the data from the Cell paper suggest may well be the case, we really don't have any idea of what the adverse event rate is.

So, here are some of the highlights from the *Lancet* paper, using the VAERS data for the first six months of vaccine administration:

Frequency of reports of death are 1 in 66,666 for EACH dose administered. That is, 15 deaths per one million doses administered. For two doses, the risk is much higher—as risk actually increases with each dose. The risk would be at least doubled, in my opinion. By three doses, the risk would be much higher. At least tripled, in my opinion.

Frequency of adverse events: 1 in 953 for EACH dose administered. For two doses, the risk is much higher—as risk actually increases with each dose. The risk would be at least doubled, IMO. By three doses, the risk would be much higher. At least tripled, in my opinion.

Frequency of severe adverse events ("inpatient hospitalization, prolongation of hospitalization, permanent disability, life-threatening illness, congenital anomaly or birth defect, or death") is 1 in 11,056 for EACH dose administered. For two doses, the risk is much higher—as risk actually increases with each dose. The risk would be at least doubled, in my opinion. By three doses, the risk would be much higher. At least tripled, in my opinion.

Because this vaccine is being administered to hundreds of millions of people, this is an unacceptable risk for the young and healthy, as this is a disease of those with comorbidities and elderly. The USA is still discussing mandating the vaccine to school-aged children; please stop and think about these adverse event numbers.

By the way—for brevity, I am skipping data from many of the tables, data that show percentage and types of adverse events. Please go to the paper and read it. The adverse event list is quite varied.

These data for the first six months of the vaccine rollout are skewed, as this manuscript doesn't report all age cohorts, and the adverse events reported in the VAERS are grossly underreported, as discussed above.

Next, the paper sought to use the V-safe survey system to determine quality of life issues after vaccination. The V-safe survey system revealed that 26% were unable to do normal activities and 16% were unable to work after vaccination.

Then came the new variant called Omicron in early December 2021. This variant, although more infectious, was also much less pathogenic. In my opinion, vaccinating for a mild cold in the healthy, young person versus loss of significant quality of life issues, even in the short term, is unacceptable. By midspring, 2021, 99.5% of the cases in the USA were the Omicron variant, per the CDC. We know that for most healthy people, Omicron is nothing more than a cold and for the young is usually a very mild cold and often asymptomatic. To use a gene-therapy-technology-based vaccine with a high-risk profile and uncharacterized long-term effects against a mild variant is the height of scientific ignorance and arrogance. It is time to stop.

Finally, the discussion at the end of the paper is misleading at best. The authors state that there is no pattern to heart-related deaths after analysis by the authors. The methodology or data from that analysis, if there was actually such an analysis, is not presented in the paper. There is no analysis presented. This analysis does not include children or adolescents. The risk of myocarditis to young men is much higher—we know this. The Hong Kong data show 1 in 2,700 in boys.

Frankly, the CDC is again obfuscating the data to suit their own political agenda. And the *Lancet* is letting the CDC get away with yet more propaganda cloaked as semiscience. This is unacceptable.

An article titled "COVID-19: Is the US compensation scheme for vaccine injuries fit for purpose?" in the *British Medical Journal (BMJ)* documents that the national system for compensating the COVID-19 vaccine-injured has not paid out a single claim [154]. The Countermeasures Injury Compensation Program (CICP) was set up to address vaccine injuries associated with vaccines and other countermeasures during a pandemic or biothreat event. Due to specific federal legislation, a person cannot sue a manufacturer for an injury caused by a vaccine or other product listed as a countermeasure; they can only seek compensation from CICP by filing a claim. Shockingly, after 1.5 years after the rollout of gene therapy vaccines, the US government through CICP has only approved one claim and has yet to pay out a single dollar to anyone vaccine-injured or for death benefits to those who have died.

The table below is from the VAERS Summary for COVID-19 Vaccines through 4/8/2022 [155]. It shows the extensive vaccine injuries and deaths

All charts and tables below reflect the data release on 4/15/2022 from the VAERS website, which includes U.S. and foreign data, and is updated through: **4/8/2022**

High-Level Summary	COVID19 vaccines (Dec'2020 – present)	All other vaccines 1990-present	US Data Only COVID19 vaccines (Dec'2020 – present)	US Data Only All other vaccines 1990-present
Number of Adverse Reactions	1,226,314	878,073	805,921	763,579
Number of Life-Threatening Events	30,292	14,498	12,474	9,979
Number of Hospitalizations	149,527	83,881	60,386	39,191
Number of Deaths	26,976*	9,633*	12,471	5,346
# of Permanent Disabilities after vaccination	50,100	21,021	13,626	12,997
Number of Office Visits	187,892	50,861	159,407	48,604
# of Emergency Room/Department Visits	127,373	213,173	99,299	203,562
# of Birth Defects after vaccination	1,037	196	502	107

*Note that the total number of deaths associated with the COVID-19 vaccines is more than double the number of deaths associated with <u>all other vaccines combined</u> since the year 1990.

reported. The government is quick to point out that these are reported injuries and deaths to the US government, which will not be fully investigated by the CDC and so therefore can't be verified. If this isn't a Catch-22, I don't know what it.

When a public health emergency was declared in 2020, the 2005 Public Readiness and Emergency Preparedness Act went into effect. That meant any injuries or deaths arising from the vaccines would have to be filed with the Countermeasures Injury Compensation Program (CICP), as opposed to the usual route with the US's national Vaccine Injury Compensation Program (VICP).

The *BMJ* article reports that since then, thousands of people have filed claims of injuries and deaths from the vaccines, but not a single person has collected any compensation. Whereas, under the national vaccine program (VICP), compensation has been awarded in 36% of the 24,909 claims filed with around $4.7bn paid out since 1988 [155].

The CICP payouts are limited to only the most serious injuries and death. The claims have to be made within a year after vaccination, and the program has a much higher burden of proof than the VICP. Loss of income under the CICP is limited to $50,000 a year, and no compensation is included for pain or emotional distress (or for attorney fees). Under the traditional vaccine injury program, payouts for lost wages are not capped, and compensation for pain and suffering is much higher.

Of concern is that the filing of a case must be completed within a year, but there is at least one person who has documented the electronic filing of her case, only to find on follow-up that the CICP had no record of her case. Concerns arise that such dropped cases will then be unable to be refiled, due to the time limits for filing. The backlog of cases now appears so large, the processes so opaque, that the CICP system seems irrevocably broken.

The CICP is a "horrible program," says Peter Meyers, emeritus professor at George Washington University Law School in Washington, DC. "You basically submit your application for compensation, it's then dealt with secretly, and you don't have a right to have a lawyer paid for by the program. You don't have a right to a hearing. We have no idea how these cases are being processed. . . . There is such a lack of transparency in this program that it's frightening" [150].

Furthermore, the CICP program resolves claims through an

administrative process, not a judicial one (unlike the VICP). In order for a claim to be won through the CICP program, the legal burden of proof has to be beyond a reasonable doubt. That is a virtually unattainable demand. Particularly for an experimental vaccine for which the adverse events are not completely known and for which the government has stymied research efforts to determine just what those adverse events are. The CDC has also hidden the large portions of the data it is collecting for these vaccines. This means that the administrators adjudicating the injury claims would also not have the information that the CDC knows on the adverse events from these vaccines, making it virtually impossible to win many of the CICP vaccine injury cases.

Currently, a small group of senators including Senators Ron Johnson, Mike Lee, Mike Braun, and Cindy Hyde-Smith have introduced the Countermeasure Injury Compensation Amendment Bill to reform the CICP to make its processes and payouts comparable to the VICP program. The bill also proposes the creation of a commission to identify injuries caused by COVID vaccines, and it would also allow claims to be resubmitted.

This harks back to the issues of the mRNA, the lipid nanoparticles, as well as the spike protein issues with these mRNA vaccines.

We now know that the "mRNA" from the Pfizer/BioNTech and Moderna vaccines that incorporates the synthetic nucleotide pseudouridine can persist in lymph nodes for at least sixty days after injection [9]. This is not natural, and this is not really mRNA. These molecules have genetic elements similar to those of natural mRNA, but they are clearly far more resistant to the enzymes that normally degrade natural mRNA, seem to be capable of producing high levels of protein for extended periods, and seem to evade normal immunologic mechanisms for eliminating cells that produce foreign proteins not normally observed in the body.

We also don't know the full effects of the nanolipid particles used, although we know that they aggregate in various organs, including ovaries and brain. We also know that they are very inflammatory. We know that the spike protein is cytotoxic. So, adverse events are going to persist for months after vaccination. That includes myocarditis.

So these long-term and unusual adverse events, most of which haven't even been investigated to the full extent needed or even recognized, will not be included in the Countermeasures Injury Compensation Program.

Then there is the government's "Vaccine Adverse Event Reporting System" (VAERS), which does consider vaccine injuries past a certain time. These adverse events, which may not show up for weeks or even months after vaccination, are not getting entered into the VAERS system. Further distorting what is known and knowable about this global "mRNA vaccine" experiment.

Isn't it time to take a good, hard look at what is happening?

In order to fight corruption, we must first expose it. But when our government is determined to hide embarrassing data, obfuscate facts, and deny culpability, what chance do we have?

The government in the USA has agreed to provide liability for the vaccine-injured in this country, relieving the pharmaceutical industry of this burden. It is time they did their job and lived up to their obligations.

CHAPTER 14
Bioweapons, the Future Is Here

Would the Russian invasion of Ukraine be justified if it were for biodefense?

Even before I was deplatformed by Twitter (according to Twitter lawyers for posting the famously accurate Canadian COVID Care Alliance video concerning the many fraudulent aspects of the original Pfizer mRNA vaccine clinical trials [156]), before I was deplatformed by LinkedIn with no explanation at all immediately before the infamous Joe Rogan hit #1757 [157], where I said the three little words "mass formation psychosis" that caused the Silicon Valley overlords to lose bladder control, many feared that I was "controlled opposition." Detractors still make that claim, presumably because of my long-standing interactions with the "biodefense" sector of the US military-pharmaceutical-industrial complex.

I have spent most of my career deeply involved in the US Biodefense enterprise. I have worked closely with biodefense research teams at USAMRIID, DTRA, and MIT Lincoln Lab. I was once a business partner with a retired CIA officer who was deeply involved in the DoD biodefense enterprise, and I have copublished with another. I once worked for the Dynport Vaccine Company, which had the DoD contract for "advanced development" (basically, clinical testing) of virtually all biodefense medical countermeasures for the US Department of Defense. My father worked as a federal defense contractor all his life, as did my father-in-law. In my father's case, it was mostly in high-energy systems, including developing technologies for protection against the electromagnetic surge effects of "the bomb" as well as exploding foil—the technology used to trigger "the bomb." My father-in-law ran the Raytheon special projects division—basically a CIA gadget and technology shop. Think "Q" from the James Bond series. This is a byzantine world that I have deep understanding of, and direct experience with, for virtually my entire life. I lived by the mantra that all DC bureaucrats know—keep your head down, because if they cannot see you, they cannot shoot you.

But I never really allowed myself to confront the possibility that we might not be the good guys, the white hats. Until I experienced what we have all been through over the last two years. A government (or really multiple governments and transnational organizations) that clearly believes that it is justified in disregarding fundamental principles of bioethics and the common rule. And like many others, once I saw that, it was like having backed into a light switch and suddenly the entire room was lit up, and I could never unsee what was revealed. Are we always the good guys? Or is this just more interchangeable Spy vs Spy gaming, where ethics and roles are fungible and "situational." A world in which there are no good guys, no white hats. Just a matter of media spin, perspective, and realpolitik. The world as envisioned by Henry Kissinger and Klaus Schwab.

And by the way, "biodefense" is big business. Yet more weapons of war.

Most of us who are not deep into the mass formation process at this point can see the coordinated pivot from legacy media pushing the COVID fear-porn to the same outlets who have pushed the Ukraine/Russia Conflict as "Putin crazy bad man—Zelenskyy good man" theme. But almost as soon as the shooting war started, a more nuanced and complex counternarrative cropped up.

That counternarrative involves the deep ties between children of key Democratic party leaders and Ukrainian petroleum industry interests. Then there are USA-sponsored bioweapon research facilities located throughout Ukraine, including along the Russian border. Which, by the way, I have an active-duty Lieutenant Colonel inform me on April 10, 2022, that "we" blew up those same USA-sponsored bioweapon research facilities. That it was not Russia who destroyed these facilities. Who to believe? An active-duty officer or the mainstream corporate media? Then there are the legitimate Russian concerns about NATO efforts to geopolitically encircle Russia. Other issues include whether Zelenskyy is really just a western puppet, rather than being the populist leader that has been pitched to us. Not to mention the surreptitious hand of World Economic Forum meddling in all of this. As all of this alternative information began to build, things started looking a lot more complicated than just "Putin crazy bad man—Zelensky good man."

I love to illustrate key points with stories based on personal experience. I have been told by people who would know (including Major General [ret.] Philip K. Russell, MD) that over many decades, the total expenditures of the US Government in developing biowarfare agents exceeded the money spent on thermonuclear weapons. A case can be made that modern understanding and technology relating molecular biology, microbiology, and virology is fundamentally a "civilian" byproduct of a massive investment in biowarfare tech by US, USSR, and other governments.

The latest evolution of these technologies is that we appear to have the CCP of the People's Republic of China, which seems to recognize no ethical boundaries, pushing the limits of the brave new world of "transhumanism." Which in turn becomes a justification for Western nations basically making the argument that "since they are doing it, we have to do it." An increasingly sophisticated next generation of biological warfare. Where people become the weapons of choice for governments that give us no choice.

Transhumanism can be thought of as a subset of human augmentation—augmenting humans with genetic engineering, information technology, cybernetics, bioengineering, artificial intelligence, and molecular nanotechnology. This will result in an augmented version of mankind. Human augmentation is considered one of the next "horizons" in modern warfare. However, at the core, these new engineered beings will still be fundamentally human. At the core of transhumanism is the belief advances in genetic, wearable, and implantable technologies will overtake the rate of biological evolution and it will be useful, appropriate, and ethical to artificially expedite the natural evolutionary process. Akin to breeding animals for different specialized purposes using advanced technologies. That transhumanism will be used for good.

Back to my story illustrating just what we are up against here, as told by those I used to hang with (who may be inflating their self-importance). The history that was relayed to me is that the real event that catalyzed the fall of the former Soviet Union was actually the development breakthrough of a binary (two-part) bioweapon that could be delivered via an airborne route. So lethal was this weapon that it could basically stop tank commanders and their crews in their tracks. According to this version of reality, the major military tension and strategic concern between the former USSR and Western Europe involved Russian Tank battalions that were poised to be able to blitzkrieg all the way to the English Channel. A threat that the western European states were acutely aware of, consequent to Hitler having successfully deployed the same strategy. The west basically had no way to mitigate this threat, or so the story goes, so it was always hanging over any geopolitical tensions that would arise from time to time between the European NATO states and the USSR. Apparently, the potent binary biologic would kill or incapacitate the tank crews so quickly that it negated the risk of blitzkrieg. Of course, this is just one story told me by friends in high places—so it remains unverified.

The point is that biological warfare agents are potent, cheap, easy to manufacture (particularly compared to thermonuclear devices), readily deployed, and have changed the tide of history on many occasions. Including all the way back to the "Indian" wars of American history, where smallpox was basically weaponized from time to time against indigenous peoples in North America. And probably all the way back through recorded history [158].

So now we have the emerging rich documentation of US-sponsored bio-labs scattered across what had increasingly become the US client state called Ukraine. If you want further to dip your toe into that topic, dive down that rabbit hole, please see the following:

- "EXCLUSIVE: Deleted Web Pages Show Obama Led an Effort To Build a Ukraine-Based BioLab Handling 'Especially Dangerous Pathogens.'"
- Recovered by the National Pulse, the article raises serious questions about US government activity in Ukraine, stretching back almost two decades [159].
- "BREAKING: Biden official says US working with Ukraine to prevent bio research facilities from falling into Russian hands.
- "Ukraine has biological research facilities, which in fact we are now quite concerned Russian troops, Russian forces may be seeking to gain control of" [160].
- "US Embassy Quietly Deletes All Ukraine Bioweapons Lab Documents Online—Media Blackout" [161].
- "China urges US to release details of bio-labs in Ukraine" [162].

- "China urges US to reveal details of US-backed biological labs in Ukraine—including types of viruses stored" [163].
- "Russia Negotiator Charges It Now Has Evidence of 'Biological Weapons Components' in Ukraine That Show 'Good Reason' for Invasion" [164].
- "What have Fauci's friends been up to in Ukraine?" [165].

Here's the point. Once upon a time, the US engaged in thermonuclear war brinksmanship with the USSR because of Russian missiles being placed on Cuban soil. The weapons of war have evolved. Bioweapons technologies have matured. What would the USA do if Russia was transforming Mexico into a client state and had placed biowarfare research laboratories along our southern border? Would we invade? I strongly suspect so. So, let's be honest with ourselves… Are "we" the good guys or the bad guys here? At a minimum, one has to conclude that this is a complicated question to answer.

The United Nations (UN) Bioweapons Convention
CONVENTION ON THE PROHIBITION OF THE DEVELOPMENT, PRODUCTON AND STOCKPILING OF BACTERIAL (BIOLOGICAL) AND TOXIN WEAPONS AND ON THEIR DESTRUCTION
ARTICLE I

Each State Party to this Convention undertakes never in any circumstances to develop, produce, stockpile or otherwise acquire or retain:

1. microbial or other biological agents, or toxins whatever their origin or method of production, of types and in quantities that have no justification for prophylactic, protective, or other peaceful purposes;
2. weapons, equipment or means of delivery designed to use such agents as toxins for hostile purposes or in armed conflict.

Deborah G. Rosenbaum, the US Assistant Secretary of Defense for Nuclear, Chemical, and Biological defense programs (ASD, NCB) [166], testified to the House subcommittee on Intelligence and Special Operations on April 1, 2022:

I can say to you unequivocally there are no offensive biologic weapons in the Ukraine laboratories that the United States has been involved with [167, 168].

With this testimony, the US Department of Defense has made a clear statement that there were no *offensive* biological weapons that the US was involved with. Did you catch that sleight of hand? No *offensive* biological weapons. Why would the US admit to such a thing? Wouldn't that set off alarm bells in the international community? The answer is that developing and even stockpiling biological weapons is allowed under Article I of the convention on the prohibition of the Development, Production and Stockpiling of Bacteriological (Biological) and Toxin Weapons and on their Destruction (BWC). This international convention (treaty) allows that the development, production, and stockpiling of *"defensive"* biological weapons are perfectly legal.

In order to understand this, we have to carefully parse what the treaty actually says. To do that, one must remove the word salad from Article I above and rewrite it to say what they are saying, without saying it. So, let's examine Article 1 by breaking it down into parts IA and IB.

ARTICLE 1A: Each State Party to this Convention undertakes never in any circumstances to develop, produce, stockpile, or otherwise acquire or retain:

1. Microbial or other biological agents, or toxins whatever their origin or method of procurement.
2. Weapons, equipment or means of delivery designed to use such agents or toxins for hostile purposes in armed conflict.

ARTICLE 1B: Each State Party to this Convention can develop, produce, stockpile, or otherwise acquire or retain:

1. Microbial or other biological agents, or toxins whatever their origin or method of procurement for the justification for prophylactic, protective, or other peaceful purposes.

Any government could drive a train through this loophole. As long as a signatory of this convention is developing, stockpiling, acquiring, or retaining biological or toxin weapons for PROTECTION (undefined what constitutes defensive versus offensive bioweapons), they are not breaking the convention. And by the way, USA close ally Israel is not a signatory to the convention. *Wow*. As someone who has spent much of my (post-9/11) professional life in this sector of biodefense, I had never examined or really thought about the actual wording of the treaty. And I have no clear idea of what a "defensive" bioweapon would be. The term seems to be a non sequitur. If a bioweapon exists, to my mind it is intrinsically capable for offensive use. But apparently, if a bioweapon is for "prophylactic, protective, or other peaceful purposes," then it is defensive in nature. This appears to be another case of Orwellian twisting of the meaning of words by our government, but is entirely consistent with prior USG and "deep state" communications about the Ukrainian biolabs that acknowledge the existence of these laboratories, as well as the US DoD/DTRA role in funding them [169]. Of course. the implication embedded in this careful word parsing by ASD(NCB) Rosenbaum is that the United States Department of Defense has been developing "defensive" bioweapons in the Ukraine biolabs. And based on her résumé, it appears to me that she would likely have firsthand knowledge, and appears to have had a hand in supervising some aspect of this activity [166]. This is consistent with the official statement by the US Ukrainian embassy that "The Biological Threat Reduction Program's priorities in Ukraine are to consolidate and secure pathogens and toxins of security concern and to continue to ensure Ukraine can detect and report outbreaks caused by dangerous pathogens before they pose security or stability threats" [170].

As an aside, an unconfirmed confidential source (active Lt. Colonel, US Army) has told me that the bombing of these sites, which apparently occurred soon after the initial Russian invasion of Ukraine, was not performed by Russia, but rather that the bombing was "by our side." This is consistent with the initially widely reported (but now largely Internet-scrubbed) "Russian" attacks of these "biolabs" during the initial wave of attack [171]. Giving the benefit of the doubt, such action would pretty much be what one would expect, from a tactical and strategic standpoint, even if all activities at those "biolabs" were only "to consolidate and secure pathogens and toxins of security concern," lest any materials, documents, or computer files become at risk for falling into

Russian hands, where they might be weaponized for political advantage. At this point, this is secondhand information and needs to be verified further.

Of additional relevance is that senior DoD/DTRA colleagues have repeatedly told me that it is the position of that agency that "nonlethal" or "incapacitating" biowarfare agent development and deployment is *not prohibited* by the BWC. Good to know, but not consistent with how I read the actual language of the treaty. Except for the justification for prophylactic, protective, or other peaceful purposes, as discussed above.

The following provisions, as described by the United Nations, do not include development of biological weapons or toxins that are for prophylactic, protective, or other peaceful purposes [172]. Not including such agents in the United Nations provisions of this treaty is a glaring omission and can only have occurred by intent. So, as long as a country is involved in the following activities for prophylactic, protective, or other peaceful purposes, they are not violating the convention.

Key Provisions of the Convention

Article	Provision
Article I	Undertaking never under any circumstances to develop, produce, stockpile, acquire or retain biological weapons.
Article II	Undertaking to destroy biological weapons or divert them to peaceful purposes.
Article III	Undertaking not to transfer, or in any way assist, encourage or induce anyone to manufacture or otherwise acquire biological weapons.
Article IV	Requirement to take any national measures necessary to prohibit and prevent the development, production, stockpiling, acquisition or retention of biological weapons within a State's territory, under its jurisdiction, or under its control.
Article V	Undertaking to consult bilaterally and multilaterally and cooperate in solving any problems which may arise in relation to the objective, or in the application, of the BWC.
Article VI	Right to request the United Nations Security Council to investigate alleged breaches of the BWC, and undertaking to cooperate in carrying out any investigation initiated by the Security Council.
Article VII	Undertaking to assist any State Party exposed to danger as a result of a violation of the BWC.
Article X	Undertaking to facilitate, and have the right to participate in, the fullest possible exchange of equipment, materials and information for peaceful purposes.

For the purposes of prophylactic, protective, or other peaceful purposes, a country did not need to destroy their stockpiles after signing this convention. They could transfer biological and toxin weapons to storage or other research facilities, as long as they were for prophylactic, protective, or other peaceful purposes. Another loophole to the convention or at the very least a technical flaw.

So, here we are. Now the carefully parsed wording of ASD(NCB) Rosenbaum makes a lot more sense. Decoded and paraphrased, what she is saying is that, whatever they were doing, the US DoD together with the government of Ukraine was most definitely not in violation of the biowarfare convention.

One of the fascinating aspects about the Department of Defense admitting that they were "assisting" Ukraine with their biological weapons program using US taxpayer dollars is the reporting on the story itself. These days, when censorship and propaganda by US and other Western allies is rampant and actively defended as a necessity to "defend democracy" (the same justification deployed to rationalize censorship), the tell can be in how the propagandists at Wikipedia and the legacy media respond to a story. A basic Internet search on April 23, 2022, using Google reveals that only the *Washington Examiner* article mentioned above (Brest, 2022) comes up regarding the US biolabs. Duck Go searching on the Brave browser also brings up *Epoch Times* coverage [168]. Yet I have a clear recollection of reading and hearing about this in multiple mainstream news sources that day of the testimony. Those articles have apparently been "disappeared" from the Internet of things. *As if we needed yet another example to prove the point that we need a new Internet based on decentralized blockchain peer-to-peer technology* [173].

Furthermore, the "factchecker" group *Politifact* was quick to run a "factcheck" that refutes (without any actual evidence) that "The U.S. is not developing biological weapons in coordination with Ukraine, as Russian officials and far-right media outlets in the U.S. have claimed" (McCarthy, 2022). I do not believe it is a coincidence that this article was published on the *same exact day* as the congressional testimony by ASD(NCB) Rosenbaum, who said that there are no *offensive* biologic weapons in the Ukraine laboratories that the United States has been involved with. Subtle but key difference, as discussed above.

Also of note is the position of Wikipedia on this topic, which covers the subject only as a "conspiracy theory" and completely neglects to mention the actual documented fact that the US DoD/DTRA was funding a broad network of "biolabs" in Ukraine prior to the recent Russian invasion [174].

Note that the Wikipedia page was last edited by Philip Cross [175]. Philip Cross is most likely the pseudonym for a team of British civil servants (or at least all evidence leads to this hypothesis) who have consistently defamed me on the Wikipedia page for my CV, and for which there is absolutely no recourse. "Philip Cross" has literally made thousands and thousands of Wikipedia entries, all in favor of corporatism and neoconservative objectives. More evidence that government(s) and corporations are interfering with what is knowable on the web. Money and power are more important than free speech and transparency.

It appears based on the primary information documented above that the US Department of Defense, having admitted to their involvement in these biolabs to Congress, is evidently either busy erasing that admission by skewing Internet search results or is working with US intelligence community-associated "fact-checkers" (like *Politifact)* to rewrite a public record that is the opposite of what has already been admitted by US Government employees in congressional testimony. For the sake of trying to make sense out of this tangled web, it is useful for me to assume (or "hypothesize") that the long arm of the CIA or some other three-letter organization is involved in the systemic removal of all evidence of US involvement in bioweapons development. The competing theory is that big tech is censoring themselves for reasons unknown; you decide for yourself what you wish to believe.

The Biological Weapons convention (BWC) was written in 1972; that is fifty years ago. Gain-of-function research, molecular biology techniques, machine learning, and artificial intelligence are light-years more advanced relative to when the treaty was written. The famous "Asilomar Conference" that first defined ethical limits for recombinant DNA research occurred in 1975, three years later [176]. The ability to create truly horrific new viruses is no longer "rocket science." It is something scientists in most laboratories (and even those living in the dark world of "garage biology"), using reagents easily available worldwide, can readily achieve. The dystopian cyberpunk movie made in 1995 called *Twelve Monkeys,* directed by Terry Gilliam (whose vision was of a deadly virus that had been released upon the world with catastrophic consequences), could easily be our future. If anyone needs validation of the possibility of that vision coming to pass, the events involving SARS-CoV-2 post-September 2019 has clearly provided the necessary evidence.

Of note is that Israel has one of the most advanced biological warfare capabilities in the world [177, 178]. It is assumed by many that the Israel Institute for Biological Research in Ness Ziona is at the center of this program and is also developing vaccines and antidotes for chemical and biological warfare [179]. Israel is not a signatory to the Biological Weapons Convention (BWC).

But the question remains, what can be done?

The first and foremost way forward is to strengthen the BWC. There have been eight review conferences over the past fifty years to fix some of the more obvious issues of the original BWC. These have been largely unsuccessful. However, in the early review conferences a number of changes were made that enhanced the convention, including:

- Exchange data on high-containment research centers and laboratories or on centers and laboratories that specialize in permitted biological activities related to the convention.
- Exchange information on abnormal outbreaks of infectious diseases.
- Encourage the publication of biological research results related to the BWC and promote the use of knowledge gained from this research.
- Promote scientific contact on biological research related to the convention.
- Declare legislation, regulations, and "other measures" pertaining to the BWC.
- Declare offensive or defensive biological research and development programs in existence since January 1, 1946.
- Declare vaccine production facilities [172].

Unfortunately, these changes to the BWC have been unsuccessful, as the vast majority of states-parties have consistently failed to submit declarations on their activities and facilities [180]. In fact, reports indicate that virtually none of the signers of the convention have reported on their protective or peaceful activities.

Dr. Fillippa Lentzos, senior lecturer at King's College London, in Science & International Security in the Department of War Studies and in the Department of Global Health & Social Medicine, writes:

The treaty itself doesn't have any real penalties and given the difficulty of proving in an unclassified way that a country is in violation—it's challenging. That's been a major weakness in the whole bioweapons non-proliferation regime from the beginning [181].

There are many ways the BWC could be strengthened. The BWC receives minimal funding from member states and has a minimal staff. There are no processes for inspection of facilities. There are no penalties or consequences for not submitting declarations of offensive or defensive biological research programs. There is evidently no method for making the public aware of where these biological weapons research programs reside.

Furthermore, human augmentation, gain-of-function research, and cyber warfare are new technologies that need to be considered as part of the BWC or in a separate treaty. They have the potential to both revolutionize warfare and destroy civilizations.

As the ninth review conference of the BWC approaches, attention to these issues must be brought to the fore. First and foremost, the propaganda regarding this treaty must be addressed. For instance, take the following two passages from Wikipedia page on the Biological Weapons Convention (accessed 24 April, 2022):

> The Biological Weapons Convention (BWC), is a disarmament treaty that effectively bans biological and toxin weapons by prohibiting their development, production, acquisition, transfer, stockpiling and use [182].

This is the opening statement on Wiki about the BWC. Note that there no mention in Wiki that the treaty does allow biological weapons for prophylactic, protective, or other peaceful purposes.

> The BWC is considered to have established a strong global norm against biological weapons. This norm is reflected in the treaty's preamble, which states that the use of biological weapons would be "repugnant to the conscience of mankind." It is also demonstrated by the fact that not a single state today declares to possess or seek biological weapons, or asserts that their use in war is legitimate [182].

This last sentence is a mistruth or certainly a misrepresentation. As discussed above, most countries are in noncompliance with the reporting requirements added later to the BWC. They have not declared such use, because they are in noncompliance. The Arms Control Association writes: "These endeavors have been largely unsuccessful; the vast majority of states-parties have consistently failed to submit declarations on their activities and facilities"; see [172] for more details.

I believe that all of us have a role to ensure the safety of the world regarding biological weapons research. Succinct and bulleted ideas for creating a more durable and updated BWC are listed below:

- The public must be made aware that the BWC has significant loopholes regarding the development of biological weapons for prophylactic, protective, or other peaceful purposes, which might be used as a ruse to hide offensive weapon development.
- Changes must be made to Wikipedia by editors to correct the mistruths and misleading statements.
- General public interest in this issue must be driven by writing letters to the editors of major newspapers. People writing blogs, website articles, memes, and posting on social media posts will create pressure for the legacy media and the BWC review committee to respond.
- Israel is not a signatory to the BWC, and they have no plans to sign. There must be consequences for this. The fact that Israel is not a signatory and the significance of this must be a priority for global information distribution. It is reasonable to infer that, due to the close long-standing relationship between the US and Israel, Israel may be acting as a surrogate for US biological weapons research.
- Gain-of-function, human augmentation (a subset of transhumanism) and cyber warfare need to be addressed in either this convention, another existing convention, or a new convention or treaty.
- The propaganda and censorship surrounding the BWC must be stopped. Transparency is key to good governance at the national and global level. The public has a right to know that the treaty does not cover all biological weapons and that Israel has not signed the treaty.
- Pressure and legislation to stop Google and other search engines

from removing content that the "deep state" or three-letter agencies don't like must be applied. Public pressure on Congress to enact legislation to keep the Internet search engines from being manipulated by big tech, government national, or international intelligence community actors is critical.

- The UN is complicit in not being truthful about the BWC. The UN page on this subject does not mention that the treaty does allow for biological weapons for prophylactic, protective, or other peaceful purposes [182].

- The BWC has neither penalties for noncompliance nor mechanisms for inspection and verification of compliance. This should be immediately addressed at the next review committee. That noncompliance includes signers to declare legislation, regulations, and "other measures" pertaining to the BWC. These signatories must declare offensive or defensive biological research and development programs in existence since January 1, 1946, and declare related "vaccine" production facilities.

- The BWC does not have adequate reporting and investigative processes for infringements of the convention or the budget to do so. This should also be addressed at the next review committee. There should be an adequately budgeted standing committee that systematically inspects signers' facilities for biological weapons development and stockpiling.

The citizens of United States and the World Community deserve more transparency about the Biological Warfare Convention, and we must insist that it be revisited and updated to cover the current threat horizon and to close the loopholes.

CHAPTER 15
Most Journalists Are Scientifically Unqualified

Why does anyone rely on reporters to interpret scientific articles? They lack the necessary training, experience, and competence to interpret scientific publications and data, a skill that typically requires decades to master.

With few exceptions, corporatized media are not able to comprehend the complexities and ambiguities inherent in scientific discussions and so repeatedly fall back on the interpretations provided by those who are marketed as fair and accurate arbiters of truth—the US Government, the World Health Organization, the World Economic Forum, and various nongovernmental organizations that have an interest in promoting vaccines (Gates' Foundation, GAVI, CEPI, etc.) or other scientific agendas. But these organizations have political and financial objectives of their own and, in the case of the CDC, have clearly become politicized as previously discussed. When combined with the increasing prevalence of "advocacy journalism" (which has been actively promoted and funded by the Bill and Melinda Gates Foundation), the result has been that the corporate media have become willing vehicles for distribution of biased interpretations promoted by authority figures presented to the public as credible sources, but who actually practice the pseudopriesthood of scientism masquerading as science. As a consequence, corporate legacy media have largely become distributors and enforcers of government-approved (and composed) narratives and articles rather than objective and impartial investigators and arbiters of truth. This is particularly true of the perverse branch of scientific journalism that has ascended to prominence during the COVIDcrisis, the fact-checker organizations (some of which are sponsored by Thompson-Reuters, which has ties to Pfizer). But how does this propaganda ecosystem work, and what can be done about it?

To a large extent science and scientists are granted an exalted position in Western society due to an implied social contract. Western governments provide them support, and society grants elevated social status in exchange for valuable services. These services include performing their trade (doing "science") and teaching others both their craft and findings. Government-subsidized (noncorporate) scientists and science are trained and funded by citizens (through their taxes) to practice their craft objectively in a variety of technical domains including medicine and public health on behalf of the citizenry. This arrangement stands in contrast to corporate-funded scientists, who work to advance the interests of their employers, but who have often also been trained at taxpayers' expense.

The social contract between scientists and general citizenry assumes that those scientists employed *via* government funding act in a manner that is free of both political partisanship and external influence from corporations and nongovernmental advocacy organizations. This social contract is woven throughout federal government hiring and employment policies concerning the civilian science corps. These policies explicitly forbid these employees from engaging in partisan political activities while serving in an official capacity and forbid conflicts of interest stemming from influence of nongovernmental entities, whether for- or not-for-profit. When these terms and conditions are not upheld, the public justifiably objects to the breach of contract. This is why employees of the civilian scientific corps are protected from employment termination for political purposes by the executive branch, even though the Office of the President is tasked with managing the scientific enterprise. Failure of the civilian scientific corps to maintain personal and scientific integrity and/or political objectivity appears to have become a chronic condition, as evidenced by the politicization of the CDC. When politicization of scientific data and interpretation results in multiple policy decisions that fail to protect the interests of the general public, the public loses faith in both the scientists and the discipline that they purport to practice. This is particularly true when the breach of social contract is seen as advancing corporate or partisan interests.

There is an organizational paradox that enables immense power to be amassed by those who have risen to the top of the civilian scientific corps.

These bureaucrats have almost unprecedented access to the public purse, are technically employed by the executive, but are also almost completely protected from accountability by the executive branch of government that is tasked with managing them—and therefore these bureaucrats are unaccountable to those who actually pay the bills for their activities (taxpayers). To the extent these administrators are able to be held to task, this accountability flows indirectly from congress. Their organizational budgets can be either enhanced or cut during following fiscal years, but otherwise they are largely protected from corrective action including termination of employment absent some major moral transgression. In a Machiavellian sense, these senior administrators function as The Prince, each federal health institute functions as a semiautonomous city-state, and the administrators and their respective courtiers act accordingly. To complete this analogy, congress functions like the Vatican during the 16th century, with each Prince vying for funding and power by currying favor with influential archbishops. As validation for this analogy, we have the theater observed on C-SPAN each time a minority congressperson or senator queries an indignant scientific administrator, such as has been repeatedly observed with Anthony Fauci's haughty exchanges during congressional testimony.

Into this dysfunctional and unaccountable organizational structure comes the corporate media, which has become distorted and weaponized into a propaganda machine under the influence of multiple factors. The most overt driver of this co-optation has been that the Biden administration, through the CDC, made direct payments to nearly all major corporate media outlets while deploying a $1 billion taxpayer-funded outreach campaign designed to push only positive coverage about COVID-19 vaccines and to censor any negative coverage [183]. With this action, the corporate media behemoth has functionally become a fusion of corporate and state-sponsored media—a public-private partnership meeting the definition of corporatist fascism. According to the Associated Press [184], despite the 2013 legislation that changed the US Information and Educational Exchange Act of 1948 (also known as the Smith-Mundt Act) to allow some materials created by the US Agency for Global Media to be disseminated in the US, under the new law it is still unlawful for government-funded media to create programming and market their content to US audiences.

Nevertheless, this is precisely what was done in the case of the COVID-19 vaccine campaign.

Second, there has long been involvement of the intelligence community in domestic US media. Operation Mockingbird [185] is among the most well known of the incursions of the CIA into US media, but the extensive and long-standing influence of the spy agency in crafting domestic propaganda has been well documented by journalist Carl Bernstein in his article "The CIA and the Media" [186]. Among the corporate media outlets identified by Bernstein as having fallen under CIA influence is the *New York Times*, which is intriguing in light of the precise knowledge of (former) CIA officer Michael Callahan's CIA employment history inadvertently revealed by *NYT* reporter Davey Alba while interviewing me [187]. For further context, while speaking to me by cell phone early in 2020, Callahan specifically denied that there was any indication that the original SARS-CoV-2 virus sequence showed any evidence of intentional genetic modification, stating, "my guys have gone over that sequence in detail and there is no indication that it was genetically modified." In retrospect, it is now clear that was propaganda—or, speaking more plainly, an intentional lie. Disinformation. Many insiders now believe that the five-eyes spy alliance has been exploited during the COVIDcrisis to enable reciprocal domestic propaganda activities by participant states against the citizens of other member states that otherwise forbid their own intelligence agencies from domestic propaganda activities. Consistent with this is the malicious and aggressive editing of my own Wikipedia page (discussed by sardonic humorist "whatsherface" [188]) by an unusually prolific editor/pseudonym named Philip Cross, who apparently works for British intelligence services. Based on the totality of evidence, it is reasonable to infer that the US intelligence community has remained actively engaged in crafting and defending the COVIDcrisis narrative, either through direct influence with corporate media and specific reporters and/or indirectly via reciprocal "five-eyes" relationships.

In addition to the above, there are many specific examples of Dr. Anthony Fauci and colleagues acting to exploit corporate media to advance their bureaucratic and public policy agendas. Weaponization of his relationship with the media by Dr. Fauci during the time when AIDS was a major narrative is well documented in the book "The Real Anthony Fauci" [189]. During the COVIDcrisis, email exchanges using government servers and addresses (obtained by independent investigator Phillip Magness under Freedom of Information Request) concerning the Great Barrington Declaration demonstrate that Dr. Fauci continues to exert considerable influence over both lay and scientific press [190].

How does this work? How is Dr. Fauci able to influence corporate media and its reporters to compose and print articles about scientific and political issues that comport with his interests and perspectives as well as those of the Institute (NIAID) that he directs? The most straightforward of the ways that he influences corporate media and its reporters is through his proven ability to actually have reporters fired who write or broadcast stories that he does not like. In "The Real Anthony Fauci," Robert F. Kennedy Jr. documents how Dr. Fauci had journalists that he disapproved of fired.

More recently, Forbes fired journalist Adam Andrzejewski for revealing previously undisclosed details regarding Anthony Fauci's personal finances [191]. Fauci also repeatedly attacked Fox journalist Laura Logan for likening him to Joseph Mengele, which she had correctly identified as a characterization widely shared throughout the world [192]. Then there are the subtler reciprocal relationships that Dr. Fauci and his NIAID Office of Communications and Government Relations (OCGR) cultivate. The NIAID OGCR is organized into five different offices: the Director's Office, the Legislative Affairs and Correspondence Management Branch, the New Media and Web Policy Branch, the News and Science Writing Branch, and the Communications Services Branch. A search of the online HHS employee directory reveals that OGCR employs fifty-nine full-time employees, eight of whom staff the News and Science Writing Branch, and thirty-two of whom work for the New Media and Web Policy Branch. In contrast, only eight employees staff the Legislative Affairs and Correspondence Management Branch. It is important to recognize that NIAID is only one branch of the NIH, and these employees are dedicated to supporting the mission of that one single branch and its director, Dr. Fauci.

There is also a quid-pro-quo relationship between reporters and influential organizations or individuals. This relationship was nicely illustrated in the movie *The Big Short*, which documented the corruption that led to the "Great Recession" of 2007–2009 [193]. The movie included scenes involving investors and hedge-fund managers confronting financial industry journalists and bond ratings agency employees. In both cases, individuals whose structural role is typically seen as serving as a barrier to corruption and malfeasance were co-opted by the need to maintain good relationships with the industry and players whom they were tasked with overseeing. The same holds true in the case of the federal bureaucracy. Basically, if a journalist wishes to be granted timely access to press releases, OGCR-drafted content favorable to Dr. Fauci and the NIAID, or other insider information, he or she must not write critical or unflattering stories. The NIAID OGCR operation is much larger than most corporate media newsrooms, who have struggled to maintain staffing in the face of declining reader- and viewership, so maintaining good relations while avoiding retaliation is critical for any reporter who works a health and science beat.

A recent example involving the immunology, structural biology, and virology associated with evolution of SARS-CoV-2 Omicron escape mutants is useful for illustrating the problem of reporters interpreting complex scientific information. A group of Chinese scientists have recently had a tour-de-force study accepted for publication by the high-status scientific journal *Nature*. On June 17, 2022, an unedited preprint of a peer-reviewed article with the rather dry title "BA.2.12.1, BA.4 and BA.5 escape antibodies elicited by Omicron infection" was posted by *Nature* [194]. As an experienced reviewer with a reasonable level of understanding of the subject matter, I found this article to be one of the more challenging papers to read that I have encountered during the COVIDcrisis. Rich granular detail concerning the recent evolution of Omicron spike protein sequence and specifically the receptor binding domain (focused on BA.2.12.1 and BA.4/BA.5) is provided, and the Chinese team uses an array of the latest technologies to generate a mountain of data that are presented to the reader as a stream of condensed information with minimal supporting text (in part due to the word-length restrictions inherent in publication in *Nature*). This is a tough read, even for me, but clearly represents an amazing advance in understanding of the molecular evolution that is happening as Omicron continues to circulate in human populations who have received vaccines that fail to prevent infection, replication, and spread of the virus. There are even data that may support some of the hypotheses of Dr. Geert Vanden Bossche concerning the probability of shifts in glycosylation patterns as part of the antibody evasion evolution of the virus continues, shifts that he predicts may lead to markedly enhanced disease as well as immunological evasion.

This highly technical article was reviewed and presented to the world by Thomson-Reuters journalist Nancy Lapid, who writes a column titled "Future of Health." Her body of work, largely focused on the COVIDcrisis, now includes 153 such articles. She is a journalist, not a scientist. By way of full transparency, Thomson-Reuters has a variety of organizational leadership ties with Pfizer, a fact never disclosed in any of these articles. In Fact, Jim Smith—who is the president and CEO of Thompson-Reuters—is also a director of Pfizer, Inc. He also serves on the board of the World Economic Forum's Partnering Against Corruption Initiative [195]. Just a "small" conflict of interest!

Nancy Lapid's article covering this technically challenging *Nature* article is titled "Early Omicron infection unlikely to protect against current variants," which is a gross misrepresentation of the findings of the paper, providing no analysis of either clinical protection or of clinical samples obtained from a control set of patients who have been infected but not vaccinated [196]. The Reuters coverage goes on to say:

> People infected with the earliest version of the Omicron variant of the coronavirus, first identified in South Africa in November, may be vulnerable to reinfection with later versions of Omicron even if they have been vaccinated and boosted, new findings suggest.

This is a misrepresentation of the actual findings of this team. To take a page from the current vernacular, it is either "misinformation" (meaning an unintentional false representation of scientific data and interpretation), or "disinformation" (meaning an intentional false representation designed to influence thought or policy in some way). To complete the triad, "malinformation" is defined by the US Department of Homeland Security (DHS) as information that may be either true or false, but that undermines public faith in the US government. Propagation of any of these three types of information has been deemed grounds for accusations of domestic terrorism by DHS. As I try to avoid drawing conclusions about people's intentions (due to my inability to read their thoughts), I cannot distinguish between these different labels in the case of the (clearly false) interpretation that Thompson-Reuters has published with Nancy Lapid's story.

What the actual manuscript describes is detailed characterization of the evolution (including precise structural mapping of specific domain clusters of antibody-Spike protein interactions) of the new Omicron variants in relationship to both marketed and newly developed monoclonal antibodies as well as "neutralizing" naturally occurring antibodies obtained from patients who have either been vaccinated with the Chinese inactivated viral vaccine called "Coronavac" or "ZF2001" (an adjuvanted protein subunit vaccine), or were previously infected with an earlier variant of SARS-CoV-2 (or the original SARS!) and then vaccinated with "Coronavac" or "ZF2001" or both (Coronavac x2 first, then ZF2001 boost). The authors describe this clearly and precisely. This research does not involve any of the vaccines available in the United States, a key fact that Nancy Lapid fails to disclose. Whole inactivated or adjuvanted subunit vaccines are very different from mRNA or rAdV vectored genetic vaccines.

Important to understand in reading the paper is that the preponderance of information demonstrates that optimal acquired protection from infection by SARS-CoV-2 (via natural infection and/or vaccination) is not only provided by antibodies, but also requires a cellular (T-cell) adaptive immune response. This paper is only looking at one limited aspect of the rich and complex interactions between the innate and adaptive immune systems in human beings and infection by the virus SARS-CoV-2 (and also addresses previously SARS-infected individuals who have been boosted with "Coronavax"). Even in the abstract, the authors are quite precise in their summary of this fact that they are not assessing "protection," clearly demonstrating the inherent bias of the Nancy Lapid/Thompson-Reuters story. They are assessing and drawing conclusions regarding neutralization evasion of the currently circulating escape mutants regarding antibodies from patients as well as various monoclonal antibody preparations:

"Here, coupled with Spike structural comparisons, we show that BA.2.12.1 and BA.4/BA.5 exhibit comparable ACE2-binding affinities to BA.2. Importantly, BA.2.12.1 and BA.4/BA.5 display stronger neutralization evasion than BA.2 against the plasma from 3-dose vaccination and, most strikingly, from post-vaccination BA.1 infections."

This brief example illustrates the problem of untutored and unqualified reporters who reflect the biases of corporate media (and government) to serve as interpreters and arbiters of scientific truth. With few exceptions, they are just not qualified to perform the task of accurate reporting of complicated scientific findings. But both the general reader as well as government policy makers rely on corporate media to perform this task accurately and fairly.

Accurate presentation of scientific findings is necessary if the public and their elected representatives are to make both sound policy and medically informed personal choice decisions that are grounded in accurate and balanced quantifiable information obtained by best scientific practices. This is what they are paying for, and they deserve to have it delivered to them. If the public and policy makers wish to continue to rely on corporate legacy press to help them to understand complicated scientific and technical issues, "advocacy journalism" reporters need to get back in their lane and leave scientific and medical interpretation to experienced professionals. There are plenty of qualified scientists capable of reading and accurately communicating key findings from even such highly technical manuscripts as this recent *Nature* article [194]. The corporate press has the resources necessary to engage such specialists, and to be able to integrate and present multiple points of view that may include the perspective of the NIAID OGCR. But as is required for all peer-reviewed academic manuscripts in the modern era, the sources (and underlying data) should be disclosed in a transparent way, and potential conflicts of interest of those sources should also be disclosed.

In the interim, corporate media and their reporters should stop trying to spin that which they do not even comprehend.

CHAPTER 16
COVID-19 Vaccines and Informed Consent
By John Allison, JD

The author of this chapter, John Allison, JD, is a retired lawyer, licensed to practice in Washington State and the District of Columbia, with extensive private law firm and in-house experience. Most of John's law practice was devoted to the litigation of cases involving medical, toxicological, industrial hygiene, and product safety issues. Before retirement, he served as assistant general counsel in the legal department of a Fortune 100 company with overall responsibility for product liability and environmental and commercial litigation. He was also the lawyer for the company's Medical Department, including Corporate Toxicology, Epidemiology, and Product Responsibility. This chapter summarizes the results of his analysis, as a volunteer, of published information about the EUA-authorized genetic vaccines as well as his opinions related to the question of informed consent. The chapter is not intended to give legal advice. People who want legal advice on the issues raised in this chapter should consult with a lawyer licensed to practice in their jurisdiction.

COVID-19 Vaccines and Informed Consent: The fundamental right to make decisions about bodily health and medical treatments

Most Americans have long assumed that they have a fundamental right to make decisions about their own bodily health and the medical treatments they receive or decline. Informed consent is the ethical and legal principle by which that fundamental right is enforceable. To be able to give informed consent, a person needs to be informed about the risks and benefits of, and alternatives to, the proposed treatment.

The fundamental right to informed consent is particularly important with respect to the COVID-19 vaccines that are available in the United States pursuant to Emergency Use Authorizations (EUAs). Under the federal EUA statute, people are entitled to be informed about their right to accept or refuse administration of these vaccines, the consequences (if any) of refusing vaccination, and the benefits and risks of alternatives to the vaccines [197]. A different federal statute gives the manufacturers of EUA vaccines, and the people and organizations administering them, immunity from liability suits for damages [198]. Unless courts decide that the liability protection conferred to vaccine manufacturers cannot be enforced against people who did not give their informed consent to vaccination, people who suffer severe adverse effects after receiving a COVID-19 vaccine will not be able to recover compensation for their monetary and emotional distress damages, and the family members of people who die after receiving a COVID-19 vaccine will not be able to recover compensation for their loss, from the vaccine manufacturers or from the people who administered the vaccine.

Based on my analysis of published scientific studies, government reports, and other credible information, I have arrived at the following opinions with respect to the COVID-19 vaccines that are being widely used in the United States:

- Government misinformation about the safety and effectiveness of the COVID-19 vaccines, censorship of credible scientific and medical information about the risks of death and serious adverse effects of the COVID-19 vaccines, and vaccination coercion are depriving people of their ability to give informed consent to vaccination. Unless the limited effectiveness of the vaccines and the risks of death and serious adverse effects described in this chapter are disclosed to people before they are vaccinated, informed consent has not been obtained.

- Safe and effective drugs on the market for many years, such as Ivermectin and Hydroxychloroquine, have been proven by reputable doctors to be successful in the early treatment of COVID-19 [75, 199]. If those affordable drugs had been allowed to be more widely used in the United States before people needed to be hospitalized, many tens of thousands of people who died from COVID-19 would probably be alive today.

- The COVID-19 vaccines that are being widely used in the United States do not meet established criteria for establishing their short-term and long-term safety and efficacy. Serious safety signals—red flags—about these vaccines have been ignored, and continue to be ignored, by the FDA and the CDC [200]. The EUAs for the Pfizer-BioNTech, the Moderna, and the Johnson & Johnson/Janssen COVID-19 vaccines, and the FDA's approval of Pfizer's Comirnaty vaccine and Moderna's Spikevax vaccine, should be withdrawn. All of these vaccines should be taken off the market immediately.

Precautionary Principle Ignored in the COVID Era

SARS-CoV-2 is the coronavirus that causes COVID-19. Distinctive spike proteins on the surface of the virus enable it to penetrate cells and cause infection. The spike proteins mutate, producing the Delta variant, which became the dominant form of the virus by the middle of 2021 [201]. Continuing mutations of the spike protein produced the Omicron variant, which became the dominant form of the virus by the end of 2021 [202] [203]. We are now dealing with subvariants of Omicron.

The first confirmed case of COVID-19 in the United States was reported in mid-January 2020 [204]. The pandemic spread. COVID-19 vaccines were not available until the middle of December 2020, when the FDA granted emergency use authorization for the Pfizer-BioNTech and the Moderna vaccines [205, 206]. In February 2021 the FDA granted emergency use authorization for the Johnson & Johnson/Janssen vaccine [207]. In the spring of 2021, these vaccines became widely available in the United States, and mass vaccination programs began. By the end of 2021, millions of Americans, including workers in many different occupations, were fully vaccinated [208].

The COVID-19 vaccines do not produce immunity to COVID-19 because they are not designed to trigger an immune response to the SARS-CoV-2 virus. Instead, the vaccines are designed to trigger an immune response to the spike proteins on the surface of the original virus.

A number of studies demonstrate that the vaccines do not prevent infection or transmission of COVID-19. Fully vaccinated people can become infected and can also spread the SARS-CoV-2 virus to other vaccinated people and to unvaccinated people.

According to data on the CDC website, in the United States there were 384,536 deaths attributed to COVID-19 in 2020, before the vaccines were widely available [209]. In 2021, when vaccines were widely available and mass vaccination campaigns took place, there were 460,513 deaths attributed to COVID-19—an increase of 19.8% [210].

When the Delta and later the Omicron variants became the dominant form of the virus, government studies in different countries show that most COVID-19 hospitalizations and deaths occur among fully vaccinated people.

Now that the Omicron variant is the dominant form of SARS-CoV-2, the effectiveness of the mRNA vaccines (Pfizer and Moderna) diminishes significantly over just a few months. According to a Danish study, which has not yet been peer-reviewed, vaccinated people, more than ninety days after vaccination, are more likely than unvaccinated people to be infected by Omicron [211].

The COVID-19 vaccines contain genetic instructions that cause the body to produce enormous numbers of SARS-CoV-2 spike proteins in order to provoke an immune response. Unfortunately, it turns out that the spike proteins themselves are toxic to cells. For example, endothelial cells line the inside of arteries to make blood flow smoothly. Damage to the endothelial cells caused by spike proteins increases the potential for microscopic blood clots to form [212]. Those microscopic blood clots can travel to the lungs, increasing the risk of developing pulmonary arterial hypertension, which is a serious progressive condition that overtaxes and weakens the heart. There is no known cure for that condition.

In the mRNA COVID-19 vaccines manufactured by Pfizer and Moderna, the genetic instructions that cause the body to produce spike proteins are encapsulated in lipid nanoparticles. A preclinical study on laboratory animals conducted by Pfizer shows that the lipid nanoparticles and mRNA genetic instructions enter the bloodstream and accumulate in several organs, including the spleen, bone marrow, liver, and adrenal glands, and concentrate in the ovaries [200]. The body then starts producing spike proteins wherever the mRNA genetic instructions happen to land.

A number of serious medical conditions have been associated with the COVID-19 vaccines, including blood-clotting disorders, cardiac emergencies, myocarditis, Guillain-Barré Syndrome, autoimmune disease, spontaneous miscarriages, nervous system disorders, and female infertility [213].

The COVID-19 vaccines also interfere with the natural immune system, making a person more susceptible to viral infections and cancer [212]. This may be one of the reasons most COVID-19 symptomatic infections, hospitalizations, and deaths are now occurring among fully vaccinated people.

A recent laboratory study in Sweden indicates that the Pfizer-BioNTech COVID-19 vaccine is able to enter a human liver cell line, where it is reverse transcribed into DNA within a matter of hours [212]. As a result, the possibility that the COVID-19 vaccines affect DNA cannot be ruled out.

The mRNA COVID-19 vaccines also contain problematic ingredients. Both the Pfizer and the Moderna vaccines contain polyethylene glycol (PEG) as an active ingredient. An Expert Panel assessing the safety of PEG recommended against its use in ointments applied to damaged skin because some burn patients treated with a PEG-based antimicrobial cream experienced renal tubular necrosis and died of kidney failure [214]. The PEG used in the Moderna vaccine matches the description of a PEG product manufactured by Sinopeg, a company in China. According to the Sinopeg website, that product is for "research use only" [215]. The Moderna vaccine also contains a lipid known by the trade name SM-102. The Pfizer vaccine also contains a lipid known by the trade name ALC-0315. According to the safety information on the website of Cayman Chemical Company, which manufactures SM-102 and ALC-0315, both of those products are "for research use—Not for human or veterinary diagnostic or therapeutic use" [216, 217]. Yet, in the mRNA COVID-19 vaccines, PEG, SM-102 and ALC-0315 are being directly injected into people's bodies.

Because no long-term clinical studies were performed, there is no way of knowing whether or not vaccinated people will suffer severe adverse side effects in the future. This is a significant concern, since the vaccines increase the potential for developing cardiovascular disease and autoimmune disease, which can both take months or years to develop.

The Society of Actuaries collected and analyzed claims data from twenty life insurance companies that provide group term coverage in the United States representing roughly 90% of the employer-based group term life insurance industry [218]. All-cause mortality data for the pandemic period (April 1, 2020 through September 30, 2021) was compared to all-cause mortality data for the prepandemic baseline period (2017 through 2019). The analysis reveals a dramatic spike in deaths from all causes during the third quarter of 2021 (July 1 through September 30). During that quarter, excess mortality for all policyholders was more than 30% above baseline. The spike in deaths was even more dramatic for working-age people. Excess mortality for people ages 25 to 34 was 81% above baseline; excess mortality for people ages 35 to 44 was 117% above baseline; excess mortality for people ages 45 to 54 was 108% above baseline; and excess mortality for people ages 55 to 64 was 70% above baseline. The dramatic increase in deaths from all causes, particularly among working-age people, during the third quarter of 2021 when mass vaccination campaigns were well underway undermines the claim that the COVID-19 vaccines are safe and effective.

COVID Vaccine Efficacy: The Devil's in the Details

Pfizer's Comirnaty COVID-19 vaccine has received full FDA approval. However, that vaccine is not generally available in the United States. As Dr. Robert Malone pointed out in a sworn declaration filed in a case pending in the US District Court for the Middle District of Florida, the Pfizer-BioNTech vaccine and the Comirnaty vaccine are not interchangeable [219]. They are legally distinct products. Dr. Malone is an original inventor of the core mRNA technology used in the Pfizer-BioNTech and Comirnaty vaccines [1–8, 220, 221] . The potential for confusion when the approved vaccine is not generally available, and the available vaccines do not have FDA approval but are merely authorized for emergency use, was pointed out in a letter US Senator Ron Johnson sent to the Acting Commissioner of the FDA on August 26, 2021 [222]. The FDA has done nothing publicly to clear up that confusion.

Instead, the FDA perpetuated the confusion it created. On January 31, 2022, Moderna's Spikevax COVID-19 vaccine received full FDA approval. The FDA's press release announcing its approval claims that "Spikevax has the same formulation as the EUA Moderna COVID-19 Vaccine and is administered as a primary series of two doses, one month apart. Spikevax can be used interchangeably with the EUA Moderna COVID-19 Vaccine to provide the COVID-19 vaccination series [223]." However, Moderna's Spikevax COVID-19 vaccine is not generally available in the United States.

It has become clear that the vaccines do not prevent infection or transmission of COVID-19. Vaccinated people can become infected and can also spread the SARS-CoV-2 virus to other people. In July 2021 an outbreak of SARS-CoV-2 infections in Barnstable County, Massachusetts, led the CDC to reverse its position on the wearing of masks and to recommend that all people wear masks indoors when viral transmission is likely, regardless of their vaccination status [224]. The outbreak involved 469 people with COVID-19 infections; 79 percent of those people were symptomatic. Seventy-four percent of the symptomatic people were fully vaccinated. Five people needed to be hospitalized; four of the five were fully vaccinated.

At a symposium on December 10, 2021, Sucharit Bhakdi, MD, and Arne Burkhardt, MD, presented the results of their pathology analysis of the organs of fifteen people who had died after receiving a COVID-19 vaccine [225]. Both Dr. Bhakdi and Dr. Burkhardt have extensive backgrounds as academic medical researchers and professors in Germany. Based on their analysis, they concluded that the COVID-19 vaccines cannot protect against infection because the antibodies produced in response to the vaccines do not effectively protect the mucous membranes that line the respiratory tract. In their opinion, "the currently observed 'breakthrough infections' among vaccinated individuals merely confirm the fundamental design flaws of the vaccines." Drs. Bhakdi and Burkhardt also described the evidence of vaccine-induced autoimmune pathology that they found.

The results of a study published October 28, 2021, in the Lancet Infectious Diseases online confirm that "fully vaccinated individuals with breakthrough infections have peak viral load similar to unvaccinated cases and can efficiently transmit infection in household settings, including to fully vaccinated contacts" [226].

A study of hospital workers in Vietnam compared the SARS-CoV-2 viral load in the nostrils of people who had COVID-19 in 2020 (before vaccines were available) with the viral load in the nostrils of fully vaccinated people who were infected by the Delta variant in 2021 [227]. The study showed that fully vaccinated people infected by the Delta variant in 2021 had 251 times the SARS-CoV-2 viral load of the unvaccinated people who had COVID-19 in 2020 before the virus mutated to form the Delta variant. The study also shows that fully vaccinated people can transmit the Delta variant to other vaccinated people as well as to unvaccinated people.

The results of a study reported in the European Journal of Epidemiology on September 30, 2021, indicate that COVID-19 vaccination rates do not correspond with lower infection rates. The study found that "countries with higher percentage of population fully vaccinated have higher COVID-19 cases per 1 million people" [228].

A short video produced by Joel Smalley, a quantitative data analyst affiliated with the Health Advisory & Recovery Team in the UK, graphically shows dramatic spikes in COVID-19 deaths after the introduction of mass vaccination campaigns in each of forty countries [229].

On page 17 of an internal CDC document published in a July 30, 2021, article in the *Washington Post*, the CDC notes that "Delta variant breakthrough cases may be as transmissible as unvaccinated cases" [230]. The term "breakthrough cases" refers to people who get COVID-19 despite being fully vaccinated. The CDC makes that point again on page 22 of the document, and also notes that vaccines "may be less effective at preventing infection or transmission. . . . Therefore, more breakthrough and more community spread despite vaccination."

An August 6, 2021, technical briefing report by Public Health England is also instructive [231]. As of August 2, there were 300,117 confirmed and provisional COVID-19 cases in England attributable to the Delta variant, representing 56.7 percent of the total number of confirmed and provisional COVID-19 cases. Of the Delta variant cases, 655 patients died, and 402 of the patients who died, or 61 percent, were fully vaccinated.

The Delta variant also became the most dominant form of SARS-CoV-2 in Israel. As of August 15, 2021, of the patients with severe or critical COVID-19 who were hospitalized in Israel, 59 percent were fully vaccinated [232].

Project Salus, a Department of Defense and Joint Artificial Intelligence Center study, analyzed the effectiveness of mRNA COVID-19 vaccines against the Delta variant among Medicare beneficiaries sixty-five years and older. The project's September 28, 2021, report indicates, on page 7, that "In this 80% vaccinated >=65 population, an estimated *71% of COVID-19 cases occurred in fully vaccinated individuals*" [emphasis in the report]. The report also points out, on page 12, that "In this 80% vaccinated 65+ population, an estimated *60% of COVID-19 hospitalizations occurred in fully vaccinated individuals in the week ending August 7th*" [emphasis in the report] [233].

The COVID-19 Statistical Report published on November 10, 2021, by Public Health Scotland contains detailed information about COVID-19 cases, hospitalizations, and deaths by vaccination status during overlapping four-week periods in October and early November 2021. Based on the information presented in Tables 18, 19, and 20 of that Report: Fully vaccinated people accounted for 57% of the COVID-19 cases, 69% of the acute hospitalizations for COVID-19, and 87% of the "confirmed COVID-19 related deaths" [234].

A research paper published December 15, 2021, in the online Journal of the American Medical Association reports on the effectiveness of the Pfizer-BioNTech and Moderna COVID-19 vaccines among fully vaccinated male veterans ages sixty-five and older during the period from July to September 2021, when the Delta variant accounted for more than 70% of the new COVID-19 infections in the United States [235]. During that period, the researchers concluded that the effectiveness of the vaccines dropped from 60% one month after full vaccination to less than 20% by five months after full vaccination.

The COVID-19 vaccines appear to be far less effective against the Omicron variant than they were against the Delta variant. A December 31, 2021, technical briefing report by Public Health England indicates, on pages 10 to 12, that the effectiveness of the Pfizer and Moderna vaccines against symptomatic Omicron infection was between 65% and 70% during the first four weeks after the second dose [236]. By twenty weeks after the second dose, vaccine effectiveness against symptomatic disease fell to around 10%. For fully vaccinated people who also received a booster shot, vaccine effectiveness dropped to between 40% and 50% by ten weeks after the booster. Table 2 on page 8 of the technical briefing report indicates that, of the people infected by Omicron who were hospitalized, 43.2% were fully vaccinated and an additional 23.2% had also received a booster dose.

A January 13, 2022, report by the New South Wales COVID-19 Critical Intelligence Unit indicates that 68.9% of the hospitalized COVID-19 patients were double-vaccinated. New South Wales is an Australian state with 92.5% of its population ages twelve and older double-vaccinated.

January 19, 2022: A report by Public Health Scotland, in Table 16, indicates there were 218 confirmed COVID-19 related deaths in Scotland during the four-week period from December 11, 2021 through January 7, 2022.

Of the 218 people who died 160, or 73.4%, had received two or three doses of a COVID-19 vaccine.

The UK Health Security Agency's March 3, 2022 COVID-19 Vaccine Surveillance Report contains information in Table 12, on page 43, about deaths between January 31 and February 27, 2022, among patients who died within twenty-eight days of testing positive for COVID-19 or who had COVID-19 mentioned on their death certificate [237]. Of the 3,957 patients who died, 3,429, or 86.6%, had received two or three doses of a COVID-19 vaccine (725 people had received two doses, and 2,704 people had received three doses).

According to a more recent government report from New South Wales, there were eighty COVID-19 deaths during the week ending June 11, 2022. Table 1 of the report indicates that 81% of those deaths were in people who had received two or more doses of a COVID-19 vaccine.

A December 21, 2021, report on the status of the Omicron variant in Denmark indicates that 91,104 people were infected by Omicron between November 22 and December 14; 78.7% of those people were fully vaccinated. An additional 10.3% also had a booster shot.

A paper first printed in a preprint server and then published in PLoS Med presents data from Denmark about the effectiveness of COVID-19 vaccines against the Omicron variant [210]. According to the charts and the table on the last page of the article, vaccine effectiveness for two doses of the Pfizer-BioNTech COVID-19 vaccine was 55.2% for thirty days. Effectiveness fell to 16.1% during the second month postvaccination, and to 9.8% during the third month. After ninety-one days postvaccination, effectiveness fell to a negative 76.5%, meaning that fully vaccinated people with two doses of the Pfizer-BioNTech vaccine became 76.5% more likely than unvaccinated people to be infected by Omicron. The Moderna COVID-19 vaccine produced a similar pattern. Vaccine effectiveness for two doses was 36.7% for the first thirty days, falling to 30.0% during the second month and to 4.2% during the third month. After ninety-one days postvaccination, effectiveness fell to a negative 39.3%, meaning that fully vaccinated people with two doses of the Moderna vaccine became 39.3% more likely than unvaccinated people to be infected by Omicron.

A preview of a peer-reviewed article accepted for publication on December 23, 2021, in the journal *Nature* describes how the Omicron variant evades the COVID-19 vaccines [131]. The article mentions that recent reports show the efficacy of two doses of the Pfizer-BioNTech vaccine has dropped "to approximately 40% and 33% against [Omicron] in the United Kingdom and South Africa, respectively." The authors conclude that the "Omicron variant presents a serious threat to many existing COVID-19 vaccines and therapies, compelling the development of new interventions that anticipate the evolutionary trajectory of SARS-CoV-2."

A Swedish total population cohort study to evaluate the duration of vaccine effectiveness found "a progressive waning of vaccine effectiveness against SARS-CoV-2 infection of any severity during up to nine months of follow-up" [238]. Based on data from January 12 to October 4, 2021, the authors reported that, in the main cohort, "the estimated vaccine effectiveness was more than 90% in the first month, with a progressive waning starting soon thereafter, ultimately resulting in a non-detectable vaccine effectiveness after 7 months. Vaccine effectiveness waned across all subgroups, although differently according to vaccine schedule and type."

A September 2022 paper published in *JAMA* describes an analysis of the effectiveness of COVID-19 vaccines against infection with the Omicron variant in the province of Ontario, Canada [239]. The authors state that "Preventing infection due to Omicron and potential future variants may require tools beyond the currently available vaccines." For patients who had also received a third, or booster, dose of an mRNA vaccine, effectiveness against Omicron was 37% seven or more days after receiving the booster. The authors conclude that

> Two doses of COVID-19 vaccines are unlikely to protect against Omicron infection. [A] third dose of mRNA vaccine affords some protection against Omicron infection in the immediate term. However, the duration of this protection and effectiveness against severe disease is uncertain.

A February 15, 2022, preprint of a scientific paper, not yet peer-reviewed, describes the results of a study of healthcare workers in Israel who received four doses of an mRNA COVID-19 vaccine [240]. The authors found that, even after four doses, the Pfizer-BioNTech vaccine was only 30% effective against infection by Omicron, while adverse reactions were reported for 80% of the study participants who received that vaccine. The Moderna vaccine was found to be only 11% effective against infection by Omicron, while adverse reactions were reported for 40% of the study participants who received that vaccine.

More Harm Than Good

In addition to limitations on their effectiveness, the COVID-19 vaccines cause demonstrable harm to the human body. In a December 8, 2020, online comment to the FDA, Dr. Patrick Whelan, a pediatric rheumatologist at UCLA, expressed his concern that these vaccines "have the potential to cause microvascular injury to the brain, heart, liver, and kidneys in a way that does not currently appear to be assessed in safety trials of these potential drugs" [241].

Later scientific research has confirmed Dr. Whelan's concern. An article on the Salk Institute website discusses the results of a study published in Circulation Research that "shows conclusively that COVID-19 is a vascular disease, demonstrating how the SARS-CoV-2 virus damages and attacks the vascular system on a cellular level" [242, 243]. The critical finding of this research is that the spike protein by itself, without active virus material present, can damage endothelial cells and mitochondria. Endothelial cells line the inside of arteries, and mitochondria generate energy for cellular function.

The endothelial cells lining the inside of arteries are designed to help blood flow smoothly. Damage to the endothelial cells caused by spike proteins increases the potential for microscopic blood clots to form. Those microscopic blood clots can travel to the lungs, increasing the risk of developing pulmonary arterial hypertension, which is a serious progressive condition that overtaxes and weakens the heart. There is no known cure for that condition. All of the COVID-19 vaccines have been shown to cause other serious blood-clotting disorders, as well.

A study reported in the Annals of Diagnostic Pathology found diffuse "microvessel endothelial damage" in the brains and other organs of people who had died from COVID-19. The study also found that "endothelial damage is a central part of SARS–CoV-2 pathology and may be induced by the spike protein alone" [244].

A medical group with a preventive cardiology practice periodically administers a PULS (Protein Unstable Lesion Signature) test to its cardiac patients. The PULS test is a clinically validated cardiac test that measures the most significant protein biomarkers of the body's immune system response to arterial injury. The medical group uses the PULS test to calculate a "PULS score" that predicts the percentage chance for a patient to experience a new Acute Coronary Syndrome (ACS) within the next five years. ACS is a medical emergency that can be fatal. A heart attack is an example of ACS. The medical group's results for 566 patients are reported in an abstract published in *Circulation*, a journal of the American Heart Association. For each of those 566 patients, the patient's PULS score before the patient received a COVID-19 vaccine was compared with the patient's PULS score after receiving the second dose of an mRNA COVID-19 vaccine. The abstract reports an increase in PULS score from 11% prevaccination to 25% post vaccination, more than doubling the predicted risk of ACS within five years [245].

The UK Health Security Agency published its COVID-19 Vaccine Surveillance Report for Week 42 on October 21, 2021. That report contains information suggesting the vaccines weaken a person's natural immune system, perhaps permanently [246]. The report points out that nucleoprotein assays (Roche N) are used to detect postinfection antibodies. Page 23 of the Report notes "recent observations from UK Health Security Agency (UKHSA) surveillance data that N antibody levels appear to be lower in individuals who acquire infection following 2 doses of vaccination." This suggests that full vaccination interferes with the body's innate ability to produce antibodies, not only against the spike protein, but also against the shell of the virus, which is a crucial part of the natural immune response in unvaccinated people.

Dr. Ryan Cole, a board-certified pathologist who operates a diagnostic lab in Idaho, describes seeing a drop in CD8 "killer T-cells" after COVID-19 vaccination, indicating a weakened immune system. Since January 1, 2021, after the COVID-19 vaccines became available, Dr. Cole has seen a twenty-fold increase in endometrial cancers over what he sees on an annual basis. He has also seen an increase in melanomas and a significant increase in human papillomavirus when looking at the cervical biopsies of women. He has also seen an increase in herpes and shingles infections.

A group of research scientists in the Netherlands and Germany studied the effects of the Pfizer-BioNTech COVID-19 vaccine on the immune system. In a May 6, 2021, preprint article, which has not yet been peer-reviewed, the research scientists reported their conclusion that the Pfizer-BioNTech vaccine "induces functional reprogramming of innate immune responses, which should be considered in the development and use of this new class of vaccines" [247]. They also point out that "inhibition of innate immune responses may diminish anti-viral responses." A more recent peer reviewed article published June 2022 in Food and Chemical Toxicology explains how the mRNA COVID-19 vaccines subvert innate immunity and dysregulate the body's system "for both preventing and detecting genetically driven malignant transformation within cells [213]."

An article published in the *Journal of the European Academy of Dermatology and Venereology* describes the results of a study about reactivation of the varicella-zoster virus (VZV), which causes herpes zoster (shingles) in humans. The study found that people who received an mRNA COVID-19 vaccine were 1.8 times more likely to experience VZV reactivation within sixty days after vaccination than people who had not been vaccinated [248]. The authors point out that a "unifying thread" among other conditions that cause VZV to reactivate "is that they correspond to a decreased immune competence."

Dr. Harvey Risch, professor emeritus of epidemiology at the Yale School of Public Health and the Yale School of Medicine, analyzed COVID-19 infection rates reported by Public Health UK until March 2022. In an interview he said that Public Health UK was reporting infection rates per 100,000 people "according to vaccination status and by age. And they compared people who had been triple vaccinated, who had a booster with people who were completely unvaccinated by age group. And what they showed is above age 18 in every age group the rates of symptomatic infection in each age group were approximately threefold higher in the vaccinated people than the unvaccinated people." Dr. Risch expressed his opinion "that the vaccines have done damage to the immune system, such that it makes people more likely to get COVID over the longer term, not the short term vaccine benefit period, but after that, more likely to get COVID infections, more likely to get other respiratory infections" [249].

The Fact Sheet for the Johnson & Johnson/Janssen COVID-19 vaccine was amended on January 11, 2022, to add a warning about an increased risk of immune thrombocytopenia during the forty-two days following vaccination [250]. Immune thrombocytopenia usually occurs when the immune system mistakenly attacks and destroys platelets, which are designed to help the blood clot. The condition can make a person subject to easy and excessive bruising and bleeding.

A June 11, 2021, research article titled "Pathogenic Antibodies Induced by Spike Proteins of COVID-19 and SARS-CoV-2 Viruses," which has not yet been peer-reviewed, reports on the results of a study in China of the pathogenic roles and novel mechanism of action of certain antibodies specific to the spike protein [251]. According to the researchers: "The data indicate that certain anti-COVID-19 S1 antibodies [antibodies to the spike protein], such as REGN10987, are highly pathogenic because it has the high potential to bind to healthy human tissues, activating self-attacking immune responses and inducing serious adverse reactions in vivo." This means that one of the antibodies to the spike protein, REGN10987, has the potential to trigger autoimmune disease. The researchers went on to say that "the pathogenic antibodies can bind to unmatured fetal cells or tissues and cause abortions [spontaneous abortions, or miscarriages], postpartum labors, still births, and neonatal deaths of pregnant females."

DNA may very well be affected by the synthetic mRNA genetic instructions in the Pfizer and Moderna COVID-19 vaccines. Before 1975, it was generally accepted that genetic information was only transferred in one direction, from DNA to RNA. It was understood that an individual's DNA sequences are copied onto mRNA molecules, which in turn cause cells in the body to make the proteins that are essential for the individual's life. However, in 1975, Dr. Howard Temin was awarded a Nobel Prize for discovering that viruses with genomes consisting of RNA can be inserted into host cells' DNA through an enzyme known as reverse transcriptase. This means that genetic information can also be transferred from RNA to DNA. According to Dr. Daniel Nagase, a Canadian physician, children are particularly susceptible to this transfer of genetic information from RNA to DNA because they have higher levels of reverse transcriptase activity than adults [252].

According to a study published February 25, 2022, researchers at Lund University in Sweden found that the Pfizer-BioNTech COVID-19 vaccine is able to enter a human liver cell line in vitro (in the laboratory) [212]. Once inside the cell, the mRNA in the vaccine is reverse transcribed into DNA within as little as six hours.

No long-term clinical studies have been performed to evaluate the long-term safety of these COVID-19 vaccines. As a result, we have no way of knowing whether or not vaccinated people will suffer severe adverse side effects in the future. This lack of information is of particular concern, since the vaccines increase the potential for developing cardiovascular disease and autoimmune disease, which can each take months or years to fully develop [253]. Because clinical trial participants in the placebo (control) group were subsequently given the option of getting vaccinated, and a number of them chose to be vaccinated, there is no longer a statistically viable control group for a study of the long-term adverse effects of the vaccines.

As Dr. Peter McCullough pointed out in his October 15, 2021, sworn expert witness Declaration filed in the US District Court for the Middle District of Florida: "The COVID-19 genetic vaccines (Pfizer, Moderna, J&J) skipped testing for genotoxicity, mutagenicity, teratogenicity, and oncogenicity. In other words, it is unknown whether or not these products will change human genetic material, cause birth defects, reduce fertility, or cause cancer" [254].

Red Flags Rising

Researchers initially assumed that the mRNA vaccines would stay near the point of injection in the shoulder muscle. However, pursuant to a records request to the regulatory agency in Japan, Dr. Byram Bridle, a Canadian vaccine researcher and immunologist, obtained a copy of Pfizer's biodistribution study on laboratory animals [255]. According to that study, the lipid nanoparticles and the mRNA genetic instructions that cause the body to make spike proteins enter the bloodstream and accumulate in several organs, including the spleen, bone marrow, liver, and adrenal glands, and concentrate in the ovaries. The body then starts producing spike proteins wherever the mRNA genetic instructions happen to land.

Significantly higher rates of spontaneous miscarriages and menstrual cycle changes, including heavier bleeding, among women soon after receiving a COVID-19 vaccine have been reported.

Dr. Michael Yeadon, former vice president and chief science officer at Pfizer, Inc., has pointed out that the spike protein is faintly similar to a protein in the placenta that is essential for fertilization and a successful pregnancy [256]. He cited a study that found significantly elevated levels of antibodies against the placenta following administration of the Pfizer-BioNTech COVID-19 vaccine.

According to Dr. Richard Blumrick, a maternal-fetal medicine specialist, the SARS-CoV-2 virus does not cross the placenta [257]. However, the mRNA vaccines use lipid nanoparticles to encapsulate the mRNA genetic instructions that cause the body to produce spike proteins. The lipid nanoparticles are vectors, or carriers, that are likely to cross the placenta.

In their Expert Statement submitted in conjunction with a lawsuit in Italy, Drs. Bhakdi and Hockertz (Germany) and Dr. Palmer (Canada) "note that both the VAERS database and the EU drug adverse events registry (EudraVigilance) report fatalities in breastfed newborns after vaccination of their mothers" [258].

On June 25, 2021, the FDA announced revisions to the Fact Sheets for the Pfizer-BioNTech and Moderna COVID-19 vaccines adding a warning about the "increased risks of myocarditis (inflammation of the heart muscle) and pericarditis (inflammation of the tissue surrounding the heart) following vaccination" [259]. According to the FDA, the new warning in the Fact Sheets for vaccination providers "notes that reports of adverse events suggest increased risks of myocarditis and pericarditis, particularly following the second dose and with onset of symptoms within a few days after vaccination." Myocarditis causes permanent damage to the heart muscle and can ultimately result in premature death.

On July 13, 2021, the FDA approved revisions to the Fact Sheets for the Johnson & Johnson/Janssen COVID-19 vaccine to include warnings about an increased risk of Guillain-Barré syndrome following vaccination [230]. Guillain-Barré syndrome is a neurological disorder in which the body's immune system damages nerve cells. The syndrome can cause paralysis; there is no known cure.

In 1990, the government established the Vaccine Adverse Events Reporting System (VAERS), which is comanaged by the CDC and the FDA. It is intended to be a national early-warning system to detect possible safety problems with vaccines in the United States. A study performed for the US Department of Health and Human Services based on data collected from June 2006 through October 2009 found that only a very small percentage of adverse events are actually reported in VAERS [260]. This low reporting level is potentially catastrophic considering the volume of reports following COVID vaccination.

The number of serious adverse events and deaths that have been reported in VAERS for the COVID-19 vaccines is many times greater than the serious adverse events and deaths reported in VAERS for all other vaccines combined [261]. As of October 14, 2022, more than 31,470 deaths and more than 261,738 serious injuries including deaths have been reported to VAERS following administration of one of the COVID-19 vaccines [200]. Yet the CDC and the FDA continue to ignore these serious safety signals.

In contrast, in 1976 the federal government conducted a mass vaccination campaign against the swine flu. After roughly 25% of the population in the United States had been vaccinated, the government terminated the vaccination program due to reports of 25 deaths and 550 cases of Guillain-Barré Syndrome following vaccination [262].

In a July 13, 2021, letter to the director of the NIH, the director of the CDC, and the acting commissioner of the FDA, US Senator Ron Johnson pointed out that he had raised "the alarming safety signals emanating from VAERS" at a meeting with the director of the NIH on April 27, 2021 [263]. In his letter, Senator Johnson wrote, "By that date, the number of deaths following COVID-19 vaccination reported to VAERS had already reached 3,411, with 1,349 or 39.5 percent of those deaths occurring on Day 0, 1, or 2 following vaccination. I expected the director of NIH to share my concerns, but he, together with our other federal health agencies, has continued to downplay the significance of what VAERS is signaling."

Due to underreporting, the actual number of deaths following vaccination is likely to be far higher than the number reported in VAERS. On July 13, 2021, an expert witness signed a Declaration under penalty of perjury expressing her opinion that, as of July 9, 2021, "the deaths occurring within 3 days of vaccination are higher than those reported in VAERS by a factor of at least 5. This would indicate the true number of vaccine-related deaths was at least 45,000." That Declaration was filed in the US District Court for the Northern District of Alabama. Since July 9, 2021, thousands of additional deaths following COVID-19 vaccination have been reported in VAERS.

The World Health Organization maintains a VigiAccess database to make information about adverse drug reactions and adverse events following vaccination available to the public. The database can be accessed at www.vigiaccess.org, and information about adverse events following COVID-19 vaccination can be found by searching for: COVID-19 vaccine. Because the database is worldwide, it includes reports of adverse events associated with COVID-19 vaccines that are not available in the United States. According to the database, as of October 27, 2022, the World Health Organization received reports of 4,544,144 adverse events following COVID-19 vaccination. The data indicate that 66% are women, 33% are men; 41% are between 18 and 44 years old; and 28% are between 45 and 64 years old.

Craig Paardekooper, a researcher and computer programmer in the UK, and Alexandra Latypova, a researcher and biotech CEO, recently reported on the results of their analyses of VAERS data for the COVID-19 vaccines [264]. They found that some vaccine batches, identified by lot number, are many times more toxic than others. Roughly 80% of the vaccine batches accounted for one or two adverse events per batch reported in VAERS. Other batches accounted for hundreds or thousands of adverse events per batch reported in VAERS, and some of those adverse events involved death, disability, or serious illness. That should not be the case, as pharmaceutical manufacturers are required to follow good manufacturing practices to ensure their products are uniform. Paardekooper also found that the more toxic batches are not randomly distributed among lot numbers. Instead, he showed how the more toxic batches of the Pfizer-BioNTech vaccine have batch codes that are part of the same mathematical or alpha-numeric sequence.

In an October 11, 2021, article, the Health Advisory & Recovery Team in the UK reported that non-COVID-19 deaths of fifteen-to-nineteen-year-old males in England and Wales during the period May 1 to September 17, 2021, were significantly higher than the 2020 baseline [265]. The authors point out that the period coincides with the rollout of vaccinations for that age group. Also, the age adjusted mortality rate for vaccinated twelve-to-seventeen-year-olds "reached levels 60% higher than the peak mortality rate for unvaccinated people during the winter." The authors urged further investigation of the cause of the excess deaths.

On February 3, 2022, US Senator Ron Johnson sent a letter to Secretary of Defense Lloyd J. Austin III requesting information about "increases in registered diagnoses of miscarriages, cancer, or other medical conditions" in the Defense Medical Epidemiology Database (DMED) in 2021 compared to the five-year average from 2016 through 2020 [266]. Senator Johnson referred to testimony he heard from an attorney representing three Department of Defense whistleblowers "who revealed disturbing information regarding dramatic increases in medical diagnoses among military personnel. The concern is that these increases may be related to the COVID-19 vaccines that our servicemen and women have been mandated to take." The COVID-19 vaccines became widely available early in 2021, and the vaccination mandates went into effect later that year. Examples of increases in medical conditions reported in DMED for 2021 compared to the previous five-year average include: a 2,181% increase in hypertension; a 1,048% increase in diseases of the nervous system; increases from 474% to 894% in various types of cancer; and a 472% increase in female infertility.

In addition to the revelations of OneAmerica life insurance as detailed by Edward Dowd in Chapter 8, Lincoln National, the fifth largest life insurance company in the United States, reportedly experienced an even more dramatic increase in death benefits paid out under its group life insurance policies in 2021 compared to the two previous years. Group life insurance policies typically cover working-age people who receive coverage as an employee benefit from their employer. In 2019—before the COVID-19 pandemic came to the United States—Lincoln National paid out $500,888,808 in group death benefits. In 2020—before COVID-19 vaccines were widely available in the United States—the company paid out $547,940,260 in group death benefits. In 2021—when COVID-19 vaccines were widely available and mass vaccination programs were implemented—the company paid out $1,445,350,949 in group death benefits [267]. That is a 163.8% increase in group death benefits paid out by Lincoln National in 2021 compared to 2020.

Putting Healthy People at Risk

The SARS-CoV-2 virus rarely causes severe COVID-19 or death in people younger than eighteen years of age. A March 19, 2021, COVID-19 Pandemic Planning Scenarios document published by the CDC presented the agency's best estimate of the infection fatality ratio for COVID-19 patients in different age groups [268]. The infection fatality ratio represents the proportion of infected patients who die. In Table 1 of that document, the CDC provided its best estimate of infection fatality ratio of 20 deaths per 1 million infections for COVID-19 patients between 0 and 17 years of age. That is a ratio of 0.00002. An infection fatality ratio of 0.00002 means that 99.998% of COVID-19 patients ages 17 and younger are expected to survive.

Researchers at Yale and at the Albert Einstein College of Medicine reported on September 21, 2020, that children diagnosed with COVID-19 express higher levels of two specific immune-system molecules, which may explain why children infected with the SARS-CoV-2 virus tend to do much better than adults [269]. A more recent scientific article, published December 2021 in *Nature Immunology*, reported the authors' conclusion that "children generate robust, cross-reactive and sustained immune responses to SARS-CoV-2 with focused specificity for the spike protein [270]."

Dr. Ben Carson, who was the director of pediatric neurosurgery at Johns Hopkins for a number of years, has said that young children should "absolutely not" receive a COVID-19 vaccine. He is quoted as saying in an interview: "Do we want to put our children at risk, when we know that the risk of the disease to them is relatively small, but we don't know what the future risks are? Why would we do a thing like that? It makes no sense whatsoever" [271].

Myocarditis is currently known to be a serious adverse event for children and young people following COVID-19 vaccination. Myocarditis causes permanent damage to the heart muscle and can ultimately result in premature death. The authors of a February 14, 2022, report titled "SARS-CoV-2 Vaccination-Associated Myocarditis in Children Ages 12-17: A Stratified National Database Analysis" found that the "CAE [cardiac adverse event] cases in our investigation occurred a median of 2 days following vaccination, and 91.9% occurred within 5 days" [272]. Tellingly, 86% of the post-vaccination CAE patients in the study had to be hospitalized.

In a more recent study, researchers in France found "strong evidence of an increased risk of myocarditis and of pericarditis in the week following vaccination against COVID-19 with mRNA vaccines in both males and females, in particular after the second dose of the mRNA-1273 vaccine [273]." The mRNA-1273 vaccine is the Moderna COVID-19 vaccine.

A study of emergency medical services (EMS) calls in Israel analyzed data for emergency responses to patients experiencing cardiac arrest or acute cardiac syndrome between January 1, 2019, and June 20, 2021 [274]. The study found an increase of more than 25% in the incidence of each of those conditions among patients ages 16 through 39 after the rollout of Israel's COVID-19 vaccination program, compared to the earlier period before COVID-19 vaccines were available. No correlation was found between the incidence of those conditions and COVID-19 infection rates. The authors note that "the weekly emergency call counts were significantly associated with the rates of 1st and 2nd vaccine doses administered to this age group but were not with COVID-19 infection rates." They point out that their "findings raise concerns regarding vaccine-induced undetected severe cardiovascular side-effects and underscore the already established causal relationship between vaccines and myocarditis, a frequent cause of unexpected cardiac arrest in young individuals."

Don't Underestimate the Power of Natural Immunity

Researchers at Emory University in Atlanta and at the University of Washington in Seattle performed a longitudinal study of patients who had recovered from COVID-19 and achieved natural immunity to the virus [275]. The study results are reported in a July 20, 2021, scientific article published in Cell Reports Medicine. The researchers concluded that "broad and effective immunity may persist long-term in recovered COVID-19 patients."

According to a recent Israeli study, "natural immunity affords longer lasting and stronger protection against infection, symptomatic disease and hospitalization due to the Delta variant of SARS-CoV-2, compared to the BNT162b2 [Pfizer] two-dose vaccine induced immunity" [276]. More than 80 percent of the adults in Israel are fully vaccinated with the Pfizer-BioNTech COVID-19 vaccine. Fully vaccinated people in Israel are six to thirteen times more likely to be infected by the Delta variant than unvaccinated people who have natural immunity as the result of a previous COVID-19 infection.

Dr. Peter McCullough has pointed out that recovered COVID-19 patients who have natural immunity and later get a COVID-19 vaccine are at risk of experiencing more serious side effects from vaccination. The authors of an article published in the March 17, 2021, edition of the journal *Life* wrote that "our study links prior COVID-19 illness with an increased risk of vaccination side effects [277]." People who were vaccinated after recovering from COVID-19 were 56% more likely to experience a severe vaccination side effect requiring hospital care than people who were vaccinated without a previous COVID-19 infection.

Canceling Informed Consent

Pfizer recognizes that the long-term adverse effects of its COVID-19 vaccines (Pfizer-BioNTech and Comirnaty) are not currently known. In its Agreement to supply Pfizer-BioNTech COVID-19 vaccines to the government of a European country, Pfizer's subsidiary required the purchasing government, in paragraph 5.5 of the Agreement, to acknowledge "that the long-term effects and the efficacy of the Vaccine are not currently known and that there may be adverse effects of the Vaccine that are not currently known" [278]. It is reasonable to assume that the same provision was included in Pfizer's agreements with the governments of other countries, as well.

A patient cannot possibly give informed consent to receiving one of these COVID-19 vaccines when the patient is not warned in advance that the vaccines cause the body to produce an enormous number of spike proteins that are toxic to cells, impair bodily functions, and adversely affect the immune system.

Pfizer reportedly cut corners in the preclinical animal testing of the Pfizer-BioNTech COVID-19 vaccine. According to a May 28, 2021, article on the TrialSite News website, which reports on clinical trials, Pfizer performed its preclinical animal tests using "surrogate" mRNA instead of the mRNA that is actually in the vaccine [279].

Deficiencies in the clinical trials for the Pfizer-BioNTech COVID-19 vaccine are explained in a public document prepared by the Canadian Covid Care Alliance, a group of more than 500 independent Canadian doctors, other healthcare practitioners, and scientists [280]. The document also contains information from six months of follow-up in the Pfizer-BioNTech clinical trial, indicating that related adverse events were 300% higher in the vaccinated group than in the placebo (control) group, and severe adverse events were 75% higher in the vaccinated group than in the control group.

On November 19, 2021, pursuant to a Freedom of Information Act request, the FDA released some documents relating to its December 11, 2020, emergency use authorization of the Pfizer-BioNTech COVID-19 vaccine. One of the documents is an April 30, 2021, cumulative analysis of post-authorization adverse event reports that Pfizer submitted to the FDA [281]. The analysis covered adverse event reports in the Pfizer safety database through February 28, 2021. During the first three months the Pfizer-BioNTech vaccine was on the market, Pfizer received, and reported to the FDA, 42,086 relevant case reports containing 158,893 events. Demographically, 71% of the case reports involved women; 63% of the case reports involved patients between the ages of 18 and 64; 8869 "adverse events of special interest" were medically confirmed; and 1,223 fatalities were reported. Pfizer submitted this information to the FDA confidentially. It was not disclosed to the public until a court forced the FDA to do so, nearly two months after the FDA gave its approval to Pfizer's Comirnaty COVID-19 vaccine.

Pfizer's history is worth noting. In 2009, Pfizer, Inc. and one of its subsidiaries agreed to pay $2.3 billion to resolve criminal and civil liability arising from the illegal promotion of certain pharmaceutical products. According to the Department of Justice press release announcing the settlement, this was at the time "the largest healthcare fraud settlement in the history of the Department of Justice" [282]. As the assistant attorney general for the Civil Division pointed out, "Illegal conduct and fraud by pharmaceutical companies puts the public health at risk, corrupts medical decisions by healthcare providers, and costs the government billions of dollars." One of the federal prosecutors was quoted as saying, "The size and seriousness of this resolution, including the huge criminal fine of $1.3 billion, reflect the seriousness and scope of Pfizer's crimes."

SECTION 2: PUBLIC HEALTH – UTILITARIANISM AND INFORMED CONSENT

CHAPTER 17
Tyranny of the Modelers

Intersection of Utilitarianism, Geopolitics, Public Health, and Hubris

There are so very many factors that have contributed to the clear and compelling reality that the public health response to the global SARS-CoV-2 outbreak has been one of the greatest failures in public policy in modern history. But chief among those has been the grossly overestimated modeling projections of likely disease and death due to the virus.

Those well versed in the world of computer software coding are intimately familiar with the problem of "garbage in, garbage out" (GIGO), which is short slang for the real-world issue that the utility of any coded data set analysis is a function of the quality of the underlying data being analyzed and the assumptions engineered into the computer code. In retrospect, it is abundantly clear that the underlying data and assumptions that were used to develop the modeling that formed the basis for global public health policy decisions concerning the management of the outbreak were seriously flawed. These flawed analyses, which were promoted via a wide range of government policy analysis and media channels, almost universally wildly overestimated the risks of the virus.

At the core of both the national and globally coordinated public health policy COVID-19 response decisions lies a philosophical belief system known as "Utilitarianism." This is also the core philosophy often employed by Globalist organizations such as the World Economic Forum and can be found intertwined with another logical framework known as "Malthusianism." We are most familiar with the philosophy of Utilitarianism in the phrase "the greatest good for the greatest number."

Quoting from the Stanford Encyclopedia of Philosophy [283]

> Utilitarianism is one of the most powerful and persuasive approaches to normative ethics in the history of philosophy. Though not fully articulated until the 19th century, proto-utilitarian positions can be discerned throughout the history of ethical theory.
>
> Though there are many varieties of the view discussed, utilitarianism is generally held to be the view that the morally right action is the action that produces the most good. There are many ways to spell out this general claim. One thing to note is that the theory is a form of consequentialism: the right action is understood entirely in terms of consequences produced. What distinguishes utilitarianism from egoism has to do with the scope of the relevant consequences. On the utilitarian view one ought to maximize the overall good—that is, consider the good of others as well as one's own good.
>
> The Classical Utilitarians, Jeremy Bentham and John Stuart Mill, identified the good with pleasure, so, like Epicurus, were hedonists about value. They also held that we ought to maximize the good, that is, bring about "the greatest amount of good for the greatest number."
>
> Utilitarianism is also distinguished by impartiality and agent-neutrality. Everyone's happiness counts the same. When one maximizes the good, it is the good *impartially* considered. My good counts for no more than anyone else's good. Further, the reason I have to promote the overall good is the same reason anyone else has to so promote the good. It is not peculiar to me.

All of the features of this approach to moral evaluation and/or moral decision making have proven to be somewhat controversial, and subsequent controversies have led to changes in the Classical version of the theory.

Malthusianism is the idea that population growth is potentially exponential while the growth of the food supply or other resources is linear, which eventually reduces living standards to the point of triggering a population die-off. The theory is most clearly described in a 1798 treatise titled *An*

Essay on the Principle of Population, by English political economist Thomas Robert Malthus. This is the philosophy underlying the often-noted positions of Bill Gates and the World Economic Forum, which call for a drastic reduction in global human population, often referred to as the depopulation agenda. This illogic is examined in a succinct analysis published in *Scientific American* by Michael Shermer titled "Why Malthus Is Still Wrong. Why Malthus makes for bad science policy" [284]. As Mr. Schermer nicely summarizes,

"The power of population is so superior to the power of the earth to produce subsistence for man, that premature death must in some shape or other visit the human race," Malthus gloomily predicted. His scenario influenced policy makers to embrace social Darwinism and eugenics, resulting in draconian measures to restrict particular populations' family size, including forced sterilizations.

In his book *The Evolution of Everything* (Harper, 2015), evolutionary biologist and journalist Matt Ridley sums up the policy succinctly: "Better to be cruel to be kind." The belief that 'those in power knew best what was good for the vulnerable and weak' led directly to legal actions based on questionable Malthusian science. For example, the English Poor Law implemented by Queen Elizabeth I in 1601 to provide food to the poor was severely curtailed by the Poor Law Amendment Act of 1834, based on Malthusian reasoning that helping the poor only encourages them to have more children and thereby exacerbate poverty. The British government had a similar Malthusian attitude during the Irish potato famine of the 1840s, Ridley notes, reasoning that famine, in the words of Assistant Secretary to the Treasury Charles Trevelyan, was an 'effective mechanism for reducing surplus population.' A few decades later Francis Galton advocated marriage between the fittest individuals ("What nature does blindly, slowly, and ruthlessly man may do providently, quickly and kindly"), followed by a number of prominent socialists such as Sidney and Beatrice Webb, George Bernard Shaw, Havelock Ellis, and H. G. Wells, who openly championed eugenics as a tool of social engineering.

This is the philosophical basis of the depopulation agenda and policies that Mr. Gates and his Oligarch colleagues at the World Economic Forum seek to impose on all of us, for our own good of course. It is Malthusianistic theories that underlie the idea that the only way to prevent catastrophic global warming is by restricting carbon dioxide release into the atmosphere. This is a philosophy that completely disregards the amazing innovative, adaptive problem-solving capabilities of the human mind.

As taught in most universities, "Public Health" (as in the Masters of Public Health degree programs) is also largely based on these two 18th- and 19th-century philosophical theories (Utilitarianism and Malthusianism). As opposed to the disciplines of medicine and clinical research, which are grounded in the principles of the Hippocratic oath and beneficence as applied to the individual patient. Examples of beneficence in clinical research and medical practice include "Do no harm," "Balance benefits against risks," and "Maximize possible benefits and minimize possible harms."

And here is where we get to the crux of the issue. Medical hubris and the public health.

First a brief definition, so we are all on the same page:

> Hubris (/ˈhjuːbrɪs/; from Ancient Greek ὕβρις (húbris) 'pride, insolence, outrage'), or less frequently hybris (/ˈhaɪbrɪs/), describes a personality quality of extreme or excessive pride or dangerous overconfidence, often in combination with (or synonymous with) arrogance.

Apparently unaware of the irony, the WEF recognizes (in a very limited way) the problem of hubris and medicine—when they published the following article: "How hubris put our health at risk" [285].

The core thesis of modern public health is that a utilitarian approach can be used to generate a sort of spreadsheet of maximal public health benefit. To take an extreme example to illustrate the point, here is a sort of parable:

> A man walks into his doctor's office for a health checkup. After completion of the exam, he asks, "Doc, how am I doing?" His utilitarian MD-MPH turns and says, "You are in perfect health. Your heart is perfect, your liver is perfect, and your kidneys are

perfect. And I have four other patients who will die in the next week if they do not get transplants requiring a donated heart, liver, or kidney. So, I will be preparing you for surgery in one hour."

Four lives saved for one sacrificed. I think that we can all agree that, while this scenario may meet a Utilitarian standard, it fails to meet the fundamentals of Judeo-Christian belief systems regarding the Hippocratic oath and principle of beneficence. But if reports are correct, in the very utilitarian, Marxist reality that is modern China under the CCP, organ harvesting is a fact of life [286]. And I believe that the utilitarian bias of the WHO and US HHS, combined with the hubris of a belief system that assumes that the likes of Anthony Fauci and other bureaucrats have sufficient comprehension of the enormous complexity of the interactions of an emergent viral variant with a global human population, has led us to a very similar endpoint.

To a considerable extent, this has been driven and justified by the hubris of public health modelers who believe that they have sufficient knowledge to be able to identify all of the important interacting variables in this interaction of virus with human host population, to be able to reduce this complexity to a set of equations or a spreadsheet, and with this tool in hand, to be able to calculate the utilitarian "greatest good for the greatest number." And of those arrogant academic modelers whose hubris has led to massive suffering and avoidable loss of life, chief among them is Neil Ferguson, the physicist (!!) at Imperial College London who created the main epidemiology model behind the lockdowns.

Quoting from Phillip Magness's article "The Failure of Imperial College Modeling Is Far Worse than We Knew" [287]:

Ferguson predicted catastrophic death tolls back on March 16, 2020 unless governments around the world adopted his preferred suite of nonpharmaceutical interventions (NPIs) to ward off the pandemic. Most countries followed his advice, particularly after the United Kingdom and United States governments explicitly invoked his report as a justification for lockdowns.

Ferguson's team at Imperial would soon claim credit for saving millions of lives through these policies—a figure they arrived at through a ludicrously unscientific exercise where they purported to validate their model by using its own hypothetical projections as a counterfactual of what would happen without lockdowns. But the June hearing in Parliament drew attention to another real-world test of the Imperial team's modeling, this one based on actual evidence.

As Europe descended into the first round of its now year-long experiment with shelter-in-place restrictions, Sweden famously shirked the strategy recommended by Ferguson. In doing so, they also created the conditions of a natural experiment to see how their coronavirus numbers performed against the epidemiology models. Although Ferguson originally limited his scope to the US and UK, a team of researchers at Uppsala University in Sweden borrowed his model and adapted it to their country with similarly catastrophic projections. If Sweden did not lock down by mid-April, the Uppsala team projected, the country would soon experience 96,000 coronavirus deaths.

I was one of the first people to call attention to the Uppsala adaptation of Ferguson's model back on April 30, 2020. Even at that early date, the model showed clear signs of faltering. Although Sweden was hit hard by the virus, its death toll stood at only a few thousand at a point where the adaptation from Ferguson's model already expected tens of thousands. At the one-year mark, Sweden had a little over 13,000 fatalities from Covid-19—a serious toll, but smaller on a per-capita basis than many European lockdown states and a far cry from the 96,000 deaths projected by the Uppsala adaptation. The implication for Ferguson's work remains clear: the primary model used to justify lockdowns failed its first real-world test.

As we look back at the long list of public health lies and tragedies that have occurred since January 2020, I have been trying to think through what systemic changes should be implemented to help prevent such catastrophically poor decision making in the future. I suggest that at the top of the list we include jettisoning both the philosophical dependence of public health decision making (as taught in MPH programs) on utilitarian philosophy, and instead substitute a Judeo-Christian values-based public health decision-making process. We have let the MPH "Utilitarians" interject themselves in place of the traditional role of the Physician and have had to live through the consequences.

As Phil Magness eloquently summarizes in his definitive analysis of the impact of the Imperial College modeling on global public health COVID policy titled "The Failures of Pandemic Central Planning" [288]:

> Public health is identified both historically and in the present day as being acutely susceptible to knowledge problems, which in turn foster the conditions for a public choice trap that causes proposed policy measures to become ineffectual or even counter-productive in disease mitigation.

We need to stop letting arrogant physicist modelers generate garbage out from their inadequate models. Modeling results that are then hyped by the press and employed by public health bureaucrats to justify the globally deployed "solutions" that caused enormous suffering, avoidable death, and economic devastation.

CHAPTER 18

The Moral Right to Conscientious, Philosophical, and Personal Belief: Exemption to Vaccination

By Barbara Loe Fisher, Cofounder & President, National Vaccine Information Center (NVIC.org)

This article, originally written many years ago and recently updated by NVIC, came to my attention as I was doing research on informed consent. I was so taken with the way the issues were presented that I received permission from NVIC to republish it.

Barbara Loe Fisher is a longtime advocate of informed consent and free choice for any medical procedures, and in particular vaccines, that carry risk of injury or death. She has spoken on mainstream TV news, such as NBC and FOX; testified to state legislatures and the House of Representatives; and authored books on the topic of vaccine injury, vaccine death, and medical mandates.

* * * * *

The Moral Right to Conscientious, Philosophical, and Personal Belief

Many parents are not philosophically opposed to the concept of vaccination and do not object to every vaccine. However, they are philosophically opposed to government health officials having the power to intimidate, threaten, and coerce them into violating their deeply held conscientious beliefs in the event they conclude that either vaccination in general or, more commonly, a particular vaccine is not appropriate for their children.

The principle of informed consent to medical treatment has become a central ethical principle in the practice of modern medicine and is always applied to medical interventions that involve the risk of injury or death. Implicit in the concept of informed consent is the right to refuse consent or, in the case of vaccination laws, the right to exercise conscientious, personal belief or philosophical exemption to mandatory use of one or more vaccines.

Informed Consent: An Ethical Principle

The right to informed consent is an overarching ethical principle in the practice of medicine, for which vaccination should be no exception. We maintain this is a responsible and ethically justifiable position to take in light of the fact that vaccination is a medical intervention performed on a healthy person that has the inherent ability to result in the injury or death of that healthy person.

The Paternalistic Medical Model under Challenge

The reason that informed consent has been increasingly adopted since World War II as the guiding ethical principle governing the patient-physician relationship is as deeply rooted in the comparatively new discipline of political science as it is in more ancient philosophies. At the heart of medicine's struggle to come to grips with a human being's right to informed consent to medical intervention is a challenge to one tenet of the Hippocratic philosophy in the practice of medicine. That being that the physician and the physician alone should determine which medical intervention will benefit the patient.

This traditional paternalistic medical model is increasingly being rejected by today's more educated healthcare consumers. The logic of medical informed consent represents a historic challenge to the supremacy of the allopathic medical model as the only means of maintaining health and preventing disease. The movement toward a more diversified, multidimensional model healthcare system is a phenomenon occurring not only in the United States, but in many other technologically advanced countries.

These are contentious and sometimes frightening days, both for consumers and healthcare providers fighting for the right to have better information and more healthcare choices, as well as for medical doctors and the institutions they deserve, who understandably do not like the intrusion or disruption of the status quo. While social change is never easy for the challenger or the challenged, in an enlightened society change can often present a remarkable opportunity for growth and renewal for everyone if perspective is maintained and neither side engages in a take-no-prisoners mentality.

Together with a general rejection of the historically paternalistic character of the patient-physician relationship in favor of one based on truth-telling and a more equal decision-making partnership, the post-World War II concept of the right to informed consent has centered on an acknowledgment of the inviolability of the individual's human right to autonomy and self-determination. This ethical concept, born out of unparalleled tragedy during WW II German forced human experimentation, has emerged as the single most important force in shaping modern bioethics.

From Aristotle to Kant: Defining Moral Virtue

In the centuries prior to World War II, religious scriptures as well as some of the greatest philosophers in history have acknowledged that the very meaning of life itself hinges on the ability of the individual to choose his own fate. Aristotle, that masterful defender of empirical knowledge and creator of virtue ethics, insisted that wisdom and moral virtue comes from within each individual, from cultivating the feelings that cause us to act in compassionate, truthful, and noble ways. Aristotle's respect for man's unique ability to reason and choose to be virtuous convinced Thomas Aquinas, who in turn convinced a threatened Catholic Church, that religion did not have to be afraid of acknowledging man's ability to discover truth through reason and sense experience as well as through spiritual revelation.

After the Protestant Reformation led by Martin Luther, when individual responsibility began to be considered more important than obedience to religious doctrine, the 16th and 17th centuries saw dramatic scientific discoveries such as those by Galileo and Isaac Newton that spawned a new breed of philosophers like Thomas Hobbes, who developed a scientific system of ethics emphasizing organized society, the state, and political structures.

Toward the end of the 18th century, the great German philosopher Immanuel Kant maintained that the ultimate moral principle, known as the categorical imperative, is the golden rule in its logical form, that being:

Act as if the principle on which your action is based were to become by your will a universal law of nature.

Kant insisted that no human being should ever treat another human being as a means to an end no matter how good or desirable that end may appear to be.

Utilitarianism: A Political Doctrine Turned Into a Pseudo-Ethic

Kant was challenged by British philosopher Jeremy Bentham, a contemporary of Dr. Edward Jenner. Bentham developed an ethical and political doctrine known as utilitarianism. Utilitarianism, which is a consequentialist theory, judges the rightness or wrongness of an action by its consequences and holds that an action that is moral or ethical results in the greatest happiness for the greatest number of people. With its emphasis on numbers of people, Bentham created utilitarianism primarily as a guide to state legislative policy, and, according to Arras and Steinbock, modern cost-benefit analyses "are the direct descendants of classical utilitarianism."

Utilitarianism, which was a major philosophical influence on Marxism, was implemented in its most extreme and tragic form by those in control of the German state during World War II. In a remarkable series of articles by physician bioethicists and lawyers published in a November 1996 issue of *JAMA*, there is a compelling description of how physicians in service to the state employed the utilitarian rationale that a fewer number of individuals can be sacrificed for the happiness of a greater number of individuals. Physicians and public health officials played a leading role in scientific experiments designed to find ways to cleanse the German state of all infection of it by individuals the state had decided harmed the public good, including physically and mentally handicapped children and adults, as well as those suffering from serious diseases.

The Nuremberg Code: The Rights of Individuals Must Come First

Out of the Doctors' Trial in Nuremberg came the Nuremberg Code, of which Yale law professor, physician, and ethicist Jay Katz has said, "if not explicitly then at least implicitly commanded that the principle of the advancement of science bow to a higher principle: protection of individual inviolability. The rights of individuals to thoroughgoing self-determination and autonomy must come first. Scientific advances may be impeded, perhaps even become impossible at times, but this is a price worth paying."

In another article, Dr. Katz said that the judges of the Nuremberg tribunal, overwhelmed by what they had learned, "envisioned a world in which free women and men, after careful explanation, could make their own good or bad decisions, but not decisions unknowingly imposed on them by the authority of the state, science, or medicine."

Bioethicist Arthur Caplan concurred when he said, "The Nuremberg Code explicitly rejects the moral argument that the creation of benefits for many justifies the sacrifice of the few. Every experiment, no matter how important or valuable, requires the express voluntary consent of the individual. The right of individuals to control their bodies trumps the interest of others in obtaining knowledge or benefits from them."

The First Principle of the Nuremberg Code is:

> The voluntary consent of the human subject is absolutely essential. This means that the person involved should have legal capacity to give consent; should be so situated as to be able to exercise free power of choice, without the intervention of any element of force, fraud, deceit, duress, overreaching, or other ulterior form of constraint or coercion; and should have sufficient knowledge and comprehension of the elements of the subject matter involved as to enable him to make an understanding and enlightened decision.

The Nuremberg Code, which speaks most specifically to the use of human beings in medical research, but also has been viewed by bioethicists and US courts as the basis for the right to informed consent to medical procedures carrying a risk of injury or death, was followed by the passage in 1964 of the Helsinki Declarations by the World Medical Association. Like the Nuremberg Code, the Helsinki Declarations emphasized the human right to voluntary, informed consent to participation in medical research that may or may not benefit the individual patient, science, or humanity.

Judeo-Christian Ethical Tradition Protects Freedom of Conscience

But even if the Nuremberg Code and Helsinki Declarations had never been promulgated and pointed us toward the morality of accepting the human right to informed consent to medical interventions that can kill or injure us, there is the strong Judeo-Christian ethical tradition that protects the sacred right of the individual to exercise freedom of conscience even if it conflicts with a secular law of the state.

This freedom is considered so inviolable in Catholic canon that the definition of moral conscience is discussed in detail in the catechism of the Catholic Church, which holds that "Conscience is a judgment of reason whereby the human person recognizes the moral quality of a concrete act that he is going to perform, is in the process of performing or has already completed. In all he says and does, man is obliged to follow faithfully what he knows to be just and right. It is by the judgment of his conscience that man perceives and recognizes the prescription of the divine law." In even stronger terms, the Catholic Church warns that "a human being must always obey the certain judgment of his conscience. If he were deliberately to act against it, he would condemn himself."

In the Old Testament of the Bible, which is the basis for Jewish law and the guide for each believer in Jewish law to discover the will of God, Abraham is asked by God to sacrifice his son to demonstrate his faith. Although Abraham is willing, God does not force Abraham to sacrifice his son. In fact, God makes it clear that human sacrifice to demonstrate allegiance is not appropriate. Why should physicians in a modern state have the power to ask more of a parent than God asked of Abraham?

Bioethics: Humans Are Not Objects or Means to an End

Bioethicists George Annas and Michael Grodin have stated that "Whenever war, politics or ideology treat humans as objects, we all lose our humanity." Or, as Elie Weisel said, "When you take an idea or a concept and turn it into an abstraction, that opens the way to take human beings and turn them, also, into abstractions."

In any war, whether it be a war using humans armed with guns in an attempt to defeat other humans, or a war using humans injected with vaccines in an attempt to eliminate microorganisms, it is easy for those in charge to view the instruments of that war—human beings—as objects and a means to an end. But the great moral tradition of Judeo-Christian Western thought does not support this dangerous concept.

David Walsh, an ethicist and political scientist who spoke at the May 1996 Institute of Medicine Risk Communication Workshop, made it clear that the only time the state has the moral authority to override a human being's inviolable right to autonomy and force him to risk his life for the state is when the very survival of the community is at stake. When, during a workshop break, several participants asked him to define what that means in terms of communicable disease, Dr. Walsh replied, "when the number of deaths caused by a disease in a community outweigh the number of births." It is interesting to note that no plague in history, not even the Black Plague and certainly not any vaccine preventable disease we have today, nor the AIDS epidemic, meets that standard.

Philosopher Hans Jonas, in one of the most brilliant and moving essays ever written on the subject of bioethics, reminds us that a state may have the right to ask an individual to volunteer to die for what the state has defined as the common good, but rarely, if ever, does a state have the moral authority to command it. Like Dr. Walsh, Jonas warned of the extraordinary emergency circumstances that should be in effect before the state can ethically override individual autonomy. He concluded:

> Let us not forget that progress is an optional goal, not an unconditional commitment, and that its tempo in particular, compulsive as it may be, has nothing sacred about it. Let us also remember that a slower progress in the conquest of disease would not threaten society, grievous as it is to those who have to deplore that their particular disease be not yet conquered, but that society would indeed by threatened by the erosion of those moral values whose loss, possibly caused by too ruthless a pursuit of scientific progress, would make its most dazzling triumphs not worth having.

Even Bertrand Russell, a confirmed agnostic and sometime devotee of the utilitarian ethic, warned that "our conduct, whatever our ethic may be, will only serve social purposes in so far as self-interest and the interests of society are in harmony."

He added, "It is the business of wise institutions to create such harmony as far as possible."

Mandatory Vaccination Laws Force Violation of Moral Conscience

It is not in the best interest of the citizens of a free society or of public health officials in positions of authority in the federal or state government to use the heel of the boot of the state to crush all dissent to mandatory vaccination laws and force individuals to violate their deeply held conscientious beliefs. It is not in the best interest of those of you, who deeply believe in the rightness of using vaccines to eliminate microorganisms, to be mistrusted and feared by the people being forced to use the vaccines you create and promote for universal use.

It is very hard for people to trust government officials who track and hunt children down to ensure compliance with mandatory vaccination laws that are now equating chicken pox with smallpox and hepatitis B with polio. It is terrible when Americans live in fear of state officials who show up on parents' doorsteps with subpoenas charging them with child abuse for failing to vaccinate; who threaten parents for refusing to vaccinate their surviving children with the same vaccine that injured or killed another one of their children; who strip, handcuff, and imprison a teenager for failing to show proof he got a second MMR shot; who deny children the right to go to school; who deny poor pregnant mothers the right to get food or welfare unless all their children are vaccinated with all government recommended vaccines.

How can the people believe or want to do what public health officials say when they live in fear of them?

We as parents, who know and love our children better than anyone else, we, by US law and a larger moral imperative, are the guardians of our children until they are old enough to make life-and-death decisions for themselves. We are responsible for their welfare, and we are the ones who bear the grief and the burden when they are injured or die from any cause. We are their voice, and by all that is right in this great country and in the moral universe, we should be allowed to make a rational, informed, voluntary decision about which diseases and which vaccines we are willing to risk their lives for—without fearing retribution from physicians employed by the state.

Argue with us. Educate us. Persuade us. But don't track us down and force us to violate our moral conscience.

CHAPTER 19
Bioethics and the COVIDcrisis

To provide context for the following examination of the bioethical foundations of current policy and practice that underpin experimental COVID vaccine deployment in many Western nations, allow me to begin by a personal anecdote from June 2021.

I was on a call with a Canadian primary care physician, and we were discussing the vaccine adverse side effects and what was happening in Canada. The physician has since been harassed, threatened with loss of medical license, and had his office broken into and his computer destroyed by the Canadian government, and therefore wishes to remain nameless. He related the story of the six (in his mind) highly unusual clinical cases of postvaccination adverse events that he had personally observed in his practice involving vaccination of his patients with the Pfizer mRNA vaccine product, all within the first couple of months from the initial vaccine rollout. What was most alarming to me was that my clinical primary practice physician colleague told me that each of these cases was reported as per the proper channels in Canada, and each was summarily determined to not be vaccine-related by the authorities without significant investigation. Furthermore, he reported to me that any practicing physician in Canada who goes public with concerns about vaccine safety is subjected to a storm of derision from academic physicians and also threatened with potential termination of employment (state-controlled socialized medicine) and loss of license to practice. This is one face of censorship in the time of COVID. This is in addition to the censorship that has been happening not only on social media platforms, but from within search engines, such as Google.

Although the censorship on social media may seem an efficient and immediate solution to the problem of medical and scientific misinformation, it paradoxically introduces a risk of propagation of errors and manipulation. This is related to the fact that the exclusive authority to define what is 'scientifically proven' or 'medically substantiated' is attributed to either the social media providers or certain institutions, despite the possibility of mistakes on their side or potential abuse of their position to foster political, commercial or other interests [289].

But what are official public health leaders afraid of? Why is it necessary to suppress discussion and full disclosure of information concerning mRNA reactogenicity and safety risks? One would expect that the governmental public health and regulatory affairs infrastructure would be committed to analyzing the vaccine-related adverse event data rigorously. Is there information or patterns that can be found, such as the cardiomyopathy signals, or the latent virus reactivation signals? An objective and unbiased HHS enterprise should be enlisting the best biostatistics and machine learning experts to examine these data, and the results should—no must—be made available to the public promptly.

Please follow along and take a moment to examine the underlying bioethics of this situation. I believe that adult citizens must be allowed free will, the freedom to choose. This is particularly true in the case of clinical research. Although there are no licensed vaccine products available in the USA at this time, the mRNA and recombinant adenovirus vaccine products that are being distributed in the USA remain experimental. Furthermore, we are supposed to be doing rigorous, fact-based science and medicine. If rigorous and transparent evaluation of vaccine reactogenicity and treatment-emergent postvaccination adverse events is not done, we (the public health, clinical research, and vaccine developer communities) play right into the hands of "anti-vaxxer" memes and validate many of their arguments. The suppression of information, discussion, and outright censorship concerning these current COVID vaccines, which are based on gene therapy technologies, casts a bad light on the entire vaccine enterprise. It is my opinion that the adult public can handle information and open discussion. Furthermore, we must fully disclose any and all risks associated with these experimental research products.

In this context, the adult public are basically research subjects that are not being required to sign informed consent due to a governmental EUA (and liability) waiver granted to the vaccine manufacturer/distributors, doctors, and hospitals. Furthermore, as these are products developed during a pandemic, they are not under the same liability rules that other vaccine products are. But that does not mean that they do not deserve the full disclosure of risks that one would normally require in an informed consent document for a clinical trial. And now some national authorities have called for widespread dosing of EUA vaccines to adolescents and the young, which by definition are not able to directly provide informed consent to participate in clinical research—written or otherwise.

The key point here is that what is being done by suppressing open disclosure and debate concerning the profile of adverse events associated with these vaccines violates fundamental bioethical principals for clinical research and the doctrine of informed consent. This goes back to the Geneva convention and Helsinki declaration [290]. There must be informed consent for experimentation on human subjects and for medical procedures. The human subjects—you, me, and the citizens of these countries—must be informed of risks. Censorship and social media banning contribute to preventing full disclosure and discussion of risks, and therefore violate the fundamental right of all of us to consent to medical procedures based on fully informed consent. As a community, we have already had a discussion and made our decision; we cannot compel prisoners, military recruits, or any other population of humans to participate in a clinical research study. For example, see the Belmont report, which provided the rationale for US federal law Code of Federal Regulations 45 CFR 46 (subpartA), referred to as "The Federal Policy for the Protection of Human Subjects" (also known as "The Common Rule").

According to the Office for Human Research Protections,

> The *Belmont Report* was written by the National Commission for the Protection of Human Subjects of Biomedical and Behavioral Research. The Commission, created as a result of the National Research Act of 1974, was charged with identifying the basic ethical principles that should underlie the conduct of biomedical and behavioral research involving human subjects and

developing guidelines to assure that such research is conducted in accordance with those principles. Informed by monthly discussions that spanned nearly four years and an intensive four days of deliberation in 1976, the Commission published the *Belmont Report*, which identifies basic ethical principles and guidelines that address ethical issues arising from the conduct of research with human subjects.

Quoting from the *Belmont Report*:

> Informed Consent.—Respect for persons requires that subjects, to the degree that they are capable, be given the opportunity to choose what shall or shall not happen to them. This opportunity is provided when adequate standards for informed consent are satisfied.
>
> While the importance of informed consent is unquestioned, controversy prevails over the nature and possibility of an informed consent. Nonetheless, there is widespread agreement that the consent process can be analyzed as containing three elements: information, comprehension and voluntariness [291].

The doctrine of informed consent is based upon the right of every individual to determine what shall be done to his or her body in connection with medical treatment, and every patient is entitled to receive the information from their physician needed to allow him or her to make an informed decision on whether or not to consent or refuse treatment. As patients are entitled to this information, physicians must make reasonable disclosures to their patients about the risks associated with a proposed medical procedure or treatment. In the USA, the doctrine of informed consent is applied at the state level. It is at the state level that mass vaccination policies are made.

During this pandemic, the federal government has abused this relationship with states by removing true informed consent at the state level, from the vaccine rollout and by hiding data [150]. As the federal government is and was responsible for the vaccine distribution and processes, they have managed to circumvent the doctrine of informed consent for this procedure.

The adverse event profile blackout of the mRNA and adenovirus vaccine by the CDC in mainstream media, social media platforms, and big tech (such as Google, Twitter, and Facebook) makes the current situation whereby the state governments are responsible for informed consent both immoral and untenable.

In response to a FOIA request filed by TheBlaze, HHS revealed that through the CDC, it purchased advertising from major news networks including ABC, CBS, and NBC, as well as cable TV news stations Fox News, CNN, and MSNBC; legacy media publications including the *New York Post*, the *Los Angeles Times*, and the *Washington Post*; digital media companies like BuzzFeed News and Newsmax; and hundreds of local newspapers and TV stations.

> These organizations published articles and video segments regarding the vaccine that were nearly uniformly positive about the vaccine in terms of both its efficacy and safety. Hundreds of news organizations were paid by the federal government to advertise for the vaccines as part of a "comprehensive media campaign," according to documents The Blaze obtained from the Department of Health and Human Services. The Biden administration purchased ads on TV, radio, in print, and on social media to build vaccine confidence, timing this effort with the increasing availability of the vaccines. The government also relied on earned media featuring "influencers" from "communities hit hard by COVID-19" and "experts" like White House chief medical adviser Dr. Anthony Fauci and other academics to be interviewed and promote vaccination in the news [183].

The advertising and positive articles regarding the vaccine do not mention or discuss in a negative light the adverse events profile of the vaccine, the irregularities of clinical trials research, clinical trial data, and criticisms regarding the vaccine [183]. Combined with the fact that the CDC has been hiding data regarding the vaccines means that even physicians have not been privy to the data that would allow true informed consent [150].

Though virtually all of these newsrooms produced stories covering the COVID-19 vaccines, the taxpayer dollars flowing to their companies were not disclosed to audiences in news reports, since common practice dictates that editorial teams operate independently of media advertising departments and news teams felt no need to make the disclosure (as some of the media outlets reached for comment explained).

To my eyes, it appears that public health leadership has stepped over the line and has violated every one of the bedrock principles: information, comprehension, and voluntariness, which form the foundation upon which the ethics of clinical research are built. I believe that this must stop. We must have transparent public disclosure of risks, in a broad sense, associated with these experimental vaccines and with these gene therapy-based medical procedures. It is either that, or the entire modern bioethical structure that supports human subjects research will have to be rethought.

Furthermore, as these vaccines available to the US public are not yet market-authorized (licensed), coercion of human subjects (including military personnel) to participate in medical experimentation is specifically forbidden. Therefore, public health policies that meet generally accepted criteria for coercion to participate in clinical research are forbidden.

For example, if I were to propose a clinical trial involving children and entice participation by giving out ice cream to those willing to participate, any Institutional Human Subject's Safety Board (IRB) in the United States would reject that protocol. If I were to propose a clinical research protocol wherein the population of a geographic region would lose personal liberties unless 70% of the population participated in my study, once again, that protocol would be rejected by any US IRB based on coercion of subject participation. No coercion to participate in the study is allowed. In the case of human subject clinical research, in most countries of the world this is considered a bright line that cannot be crossed. So, now we are told to waive that requirement without even so much as open public discussion being allowed?

In conclusion, I hope that you will take a moment and consider for yourself what is going on. The logic seems clear to me.

1. An unlicensed medical product deployed under emergency use authorization (EUA) remains an experimental product under clinical research development, even after licensure by the FDA of these vaccines.

2.   EUA authorized by national authorities basically grants a short-term right to administer the research product to human subjects without written informed consent.

However, the doctrine of consent is still in effect. This doctrine of informed consent requires that every patient is entitled to receive the information from their physician needed to allow him or her to make an informed decision on whether or not to consent or refuse treatment. The doctrine of informed consent is mostly regulated at the state level in the United States, as the regulation of medical practice is a responsibility of each state (and is not a federal responsibility). With the rollout of the vaccine, the federal government overlooked the doctrine of informed consent and ignored state laws. Physicians and their patients could not participate in full informed consent because information was censored or hidden [150]. The Geneva Convention, the Helsinki declaration, and the entire structure that supports ethical human subjects research require that research subjects be fully informed of risks and must consent to participation without coercion.

In my opinion, a line has been crossed.

In response to this, the public must continue to lobby legislators and politicians. That includes strengthening our bioethics laws, including the common rule, which must not be ignored and disregarded in the future. The federal government actions since the rollout of the EUA vaccines cannot be allowed to become an accepted precedent or norm. And interested organizations and people like you and me need to remind members of Congress and the Senate of the need for such legislation. The doctrine of informed consent must be codified into federal law in such a way that it acts as an umbrella over all other federal legislation regarding informed consent.

The good news is that at the state level, where most of the vaccine regulations actually matter, positive legislation to control vaccination mandates and to stop the erosion of informed consent has been making good progress. In the United States, our state laws supersede federal law unless it is an item that has been specifically written into the Bill of Rights or the US Constitution. States can regulate drugs, vaccines, medical licenses, and insurance. The federal government is authorized to regulate interstate commerce, but not the practice of medicine.

CHAPTER 20
The Illusion of Evidence-Based Medicine

How the government stopped worrying and learned to love propaganda

> "Evidence-based medicine is 'the conscientious, explicit and judicious use of current best evidence in making decisions about the care of individual patients.' The aim of EBM is to integrate the experience of the clinician, the values of the patient, and the best available scientific information to guide decision-making about clinical management."

In 1990, a paradigm shift occurred in the development of new medicines and treatments. An idea so big that it was supposed to encompass the whole of medicine. It was to start initially at the level of preclinical and clinical trials and work all the way through the system to the care and management of individual patients. This new concept for how medicine would be developed and conducted is called *evidence-based medicine (EBM)*. Evidence-based medicine was to provide a more rigorous foundation for medicine, one based on science and the scientific method. Truly, this was to be a revolution in medicine—a nonbiased way of conducting medical research and treating patients. Sounds great, right?

So, what happened? Like many things that start out with the best of intentions, the initiative appears to have been subverted, co-opted, and weaponized to advance the financial and political interests of those who can pay for large, expensive randomized controlled trials. Big Pharmaceutical companies and their government/administrative state partners.

There is a fundamental flaw in the logic of evidence-based medicine as the basis for the practice of medicine as we know it, a practice based on science; one that determines care down to the level of the individual patient. This flaw is nestled in the heart and soul of evidence-based medicine, which (as we have seen during the COVIDcrisis) is not free of politics. It is naive to think that data and the process of licensure of new drugs are free from bias and conflicts of interest. In fact, this couldn't be any further from the truth. The COVID-19 crisis of 2020 to 2022 has exposed for all to see how evidence-based medicine has been corrupted by the governments, hospitalists, academia, Big Pharma, academic journals, tech, and social media. They have leveraged the processes and rationale of evidence-based medicine to corrupt the entire medical enterprise.

Evidence-based medicine depends on data. For the most part, the data gathering and analysis process is conducted by and for the pharmaceutical industry, then reported by senior academics. The problem, as laid out in an editorial in the *British Medical Journal*, is as follows [292]:

> The release into the public domain of previously confidential pharmaceutical industry documents has given the medical community valuable insight into the degree to which industry-sponsored clinical trials are misrepresented. Until this problem is corrected, evidence-based medicine will remain an illusion.
>
> This ideal of the integrity of data and the scientific process is corrupted as long as financial (and government's) interests are prioritized over the common good.
>
> Medicine is largely dominated by a small number of very large pharmaceutical companies that compete for market share but are effectively united in their efforts to expanding that market. The short-term stimulus to biomedical research funding has been celebrated by free market champions, but the unintended, long-term consequences for medicine have been severe. Scientific progress is thwarted by the ownership of data and knowledge because industry suppresses negative trial results, fails to report adverse events, and does not share raw data with the academic research community. Patients die because of the adverse impact of commercial interests on the research agenda, universities, and regulators.

The pharmaceutical industry's responsibility to its shareholders means that priority must be given to their hierarchical power structures, product loyalty, and public relations propaganda over scientific integrity. Although universities have always been elite institutions prone to influence through endowments, they have long laid claim to being guardians of truth and the moral conscience of society. But in the face of inadequate government funding, they have adopted a neoliberal market approach, actively seeking pharmaceutical funding on commercial terms. As a result, university departments become instruments of industry: through company control of the research agenda and ghostwriting of medical journal articles and continuing medical education, academics become agents for the promotion of commercial products. When scandals involving industry-academe partnership are exposed in the mainstream media, trust in academic institutions is weakened and the vision of an open society is betrayed.

The modern embodiment of the "corporate university" also compromises the concept of academic leadership. No longer are positions of leadership due to distinguished careers. Instead, the ability to raise funds in the form of donations, grants, royalty revenue, and contracts dominates the requirements for university leaders. These "leaders" now must demonstrate their profitability or show how they can attract corporate sponsors.

The US government, particularly NIAID, controls a significant amount of the grants and contracts of most academic institutions in the USA. Together with the Bill and Melinda Gates Foundation and Wellcome Trust, they also can determine what research is conducted and who is funded to conduct that research.

The US government also controls the narrative. Take for example the use of the media, CDC, and the FDA to control the narrative about early treatment for COVID-19, and the growing evidence demonstrating routine and direct collusion between a wide range of US Government agencies and technology/social media companies to control and censor information (in clear violation of the First Amendment of the US Constitution). By now we should all know about the corruption of the early clinical trials of hydroxychloroquine [293]. On the basis of these faked studies, one of the safest drugs in the world was recommended to not be used in an outpatient setting—most likely, in order to increase vaccine acceptance and avoid emergency use authorization clauses that would otherwise create a barrier to widespread deployment of experimental products including vaccines. Or, in another example, how our government used propaganda to control the use of Ivermectin by such tactics as calling it unfit for human use and labeling it as a "horse wormer." All indications are that these efforts by the US government were to dissuade early treatment and thereby to stop vaccine hesitancy or otherwise complicate the Emergency Use Authorization (EUA) pathway.

Beyond our government and the pharmaceutical companies skewing evidence-based medicine for their own purposes, then there is the university system, which is more interested in generating income than creating a research program that is free from bias.

> Those who succeed in academia are likely to be key opinion leaders (KOLs in marketing parlance), whose careers can be advanced through the opportunities provided by industry. Potential KOLs are selected based on a complex array of profiling activities carried out by companies, for example, physicians are selected based on their influence on prescribing habits of other physicians. KOLs are sought out by industry for this influence and for the prestige that their university affiliation brings to the branding of the company's products. As well-paid members of pharmaceutical advisory boards and speakers' bureaus, KOLs present results of industry trials at medical conferences and in continuing medical education. Instead of acting as independent, disinterested scientists and critically evaluating a drug's performance, they become what marketing executives refer to as "product champions."
>
> Ironically, industry sponsored KOLs appear to enjoy many of the advantages of academic freedom, supported as they are by their universities, the industry, and journal editors for expressing their views, even when those views are incongruent with the real evidence. While universities fail to correct misrepresentations of the science from such collaborations, critics of industry face

rejections from journals, legal threats, and the potential destruction of their careers. This uneven playing field is exactly what concerned Popper when he wrote about suppression and control of the means of science communication. The preservation of institutions designed to further scientific objectivity and impartiality (i.e., public laboratories, independent scientific periodicals and congresses) is entirely at the mercy of political and commercial power; vested interest will always override the rationality of evidence. [292].

Regulators (ergo, the FDA and CDC in the USA) receive funding from industry and use industry funded and performed trials to approve drugs, without (in most cases) actually reviewing the raw data. What confidence do we have in a system in which drug companies are permitted to "mark their own homework" rather than having their products tested by independent experts as part of a public regulatory system? Unconcerned governments and captured regulators are unlikely to initiate necessary change to remove research from industry altogether and clean up publishing models that depend on reprint revenue, advertising, and sponsorship revenue.

Some proposals for reforms include:

- Regulators must be freed from drug company funding. This includes the FDA funding—which must come directly from the government, as opposed to pharma fees, as now is the case. Tying employee salaries to pharma fees creates a huge conflict of interest within the FDA.
- The revolving door between regulators like the FDA, the CDC, and Big Pharma (as well as tech/media) must stop. Employment contracts for regulatory government positions must have "non-compete" clauses whereby employment opportunities are limited upon leaving these regulatory agencies. Likewise, Big Pharma executives should not fill leadership positions at regulatory agencies.
- Taxation imposed on pharmaceutical companies to allow public funding of independent trials; and, perhaps most importantly,

anonymized individual patient-level trial data posted, along with study protocols. These data to be provided on suitably accessible websites so that third parties, self-nominated or commissioned by health technology agencies, could rigorously evaluate the methodology and trial results.

- Clinical trial data must be made public. Trial consent forms are easily changed to make these anonymized data freely available.

- Publication of data must be open and transparent. The government has a moral obligation to trial participants, real people who have been involved in risky treatment and have a right to expect that the results of their participation will be used in keeping with principles of scientific rigor.

- The government has a moral obligation to the public to conduct clinical trials in ways that are nonbiased by industry.

- The Foundation for the CDC and the Foundation for the NIH, which run clinical trials and studies for these organizations (while their boards are made up of pharma industry executives and employees), must be decommissioned. We have laws in this country whereby the government does not accept volunteer labor, or direct donations to influence government decisions. These NGOs are doing just that. These practices must be stopped. They are intentionally using these organizations to bypass federal laws concerning exertion of undue influence on federal decision making.

- Off-label drugs must continue to be used by the medical community. The early treatment protocols, which have saved countless lives, have documented the important role that physicians have played in finding cheap and effective treatments for COVID as well as many other diseases. Let doctors be doctors.

- If the scientific and medical journals are to function as the arbiters of medical truth, they must be stopped from taking monies from Big Pharma. This includes the sales of reprints, banner ads, print ads, etc.

- Government must stop interfering with the publishing of peer-reviewed papers and discussion on social media. A free press must remain free from coercion from government. We all know countless examples, such as the Trusted News Initiative (TNI) and

White House meetings with big tech to influence what is allowed to be printed. And the billion dollars spent by the US Government to promote these EUA/unlicensed "vaccine" products that do not prevent infection or transmission of the SARS-CoV-2 virus. This type of practice is a direct assault on our First Amendment rights. It also skews the utility and validity of evidence-based medical decision making.

- Informed consent, one of the foundations of modern medicine, has been stymied by the FDA, NIH, the CDC hospitalists, big tech, and social media. They have been hiding data and skewing results. When people cannot get the information they need to make an informed decision, evidence-based medicine cannot function correctly.
- The government must stop picking winners and losers. Evidence-based medicine requires a nonbiased playing field.
- Industry concerns about privacy and intellectual property rights should not hold sway.

If we are ever to trust and support the concept of evidence-based medicine again, significant changes to the system must be enacted. The only question is . . . is our government up to the job?

CHAPTER 21
ARPA-H, Intelligence Community within NIH

The intelligence community infiltration into the health research bureaucracy continues.

I really did not begin to understand the Washington DC/Bethesda-based National Institutes of Health (NIH) research bureaucracy until my research laboratory was recruited and relocated from the University of California, Davis, to the University of Maryland, Baltimore School of Medicine in 1997. Before then, I had a vague notion that the NIH intramural (Bethesda/ Rockville-based research campus) and extramural (mostly Rockville admin- istrative campus) infrastructure was a sort of research paradise, where all the really important government-funded biomedical research work was done. For the lucky few who were good enough, the elite of the elite, they were able to work unencumbered by the daily grind of the endless begging-for-dollars grant and contract writing (and associated funding politics) that has come to dominate the lives of most academic biomedical researchers.

After Jill and I relocated the lab to Baltimore's inner harbor, I realized that the University of Maryland Medical School, Baltimore campus, had become sort of a satellite to the Bethesda NIH complex, and I started being asked to make presentations and participate in "study sections"—one step along the "peer review" process of selecting which research gets funding and which will wither on the vine. The process of how that sausage-making machine actually works is another story, but suffice to say that it is nothing like what people are led to believe. "Peer review" is largely a sham, and what and who actually gets funded is pretty much completely at the whim of the top bureaucrats who run the place (Dr. Fauci is just one example)—who are both untouchable and unaccountable. Absent a major moral transgression scandal, they literally cannot be fired by the executive branch (which is supposed to provide oversight), and their internal government customer is essentially congress. Which gives rise to the strategy of rigorously tracking and reporting NIH institute funding allocation by state and congressional district.

Here's how this works. Congress allocates money to the NIH, which money then gets divided into the intramural activities (internal government researchers) and extramural activities (the money that goes back to congressional districts in the form of grants and contracts). Over time, this has resulted in a "you scratch my back, I'll scratch yours" feedback loop of taxpayer-funded (or borrowed) money. Not at all what the administrative and technical genius Vannevar Bush had envisioned when he led the creation of the Office of Scientific Research and Development (OSRD) in 1941. OSRD, born of the WWII war effort, essentially became the granddaddy of the entire federal scientific research enterprise—including NIH, and neither congress nor the executive branch have ever really looked back to assess if the American people are getting good value.

The objectivity of the sycophant journalists and corporate press (the *Washington Post* and *New York Times* provide notable examples) that should be examining and exposing this circle jerk is compromised by the same problem that led to their not reporting on the mortgage loan malfeasance that resulted in the financial meltdown of the "Great Recession" of 2007–2009. The conflict of interest that drives the curious myopia of the fourth estate in such matters has been elegantly described in the blockbuster movie *The Big Short*. Here's how that one works. If you are a journalist covering the healthcare beat and wish to have access to the inside "scoops" concerning what is going on or coming up within the US Government "health" bureaucracy, you need to have insider access. You want to be on the list of people who are contacted concerning an upcoming press release or emerging issue. But the quid pro quo is that you have to play nice with the big boys, and not be too critical (for a deeper dive, look up "controlled opposition" or "Hegelian dialectic") or you will lose access to the centers of power and associated information stream. For an excellent example illustrating how Drs. Fauci, Collins, and peers have learned to manipulate this reciprocal relationship with the press to advance their own power and political agendas, see their emails concerning the authors of the Great Barrington Declaration [294, 295].

Over the ensuing eighty years since Vannevar Bush, the whole thing has become a self-perpetuating and ever-expanding bureaucratic monster that is so deeply woven into the fabric of government that it may never be possible to reimagine or reform the beast. After all, who does not support better healthcare? Who knows better than the "experts" who lead the bureaucracy? Any congressperson or executive branch political appointee (or heretic physician/scientist) who dares to question is attacked, gaslighted, and vilified by both the bureaucrats themselves and the corporate-controlled press. This bureaucratic "healthcare" enterprise has become untouchable and has been further entrenched by building "public-private partnerships" (essentially corporatism or really fascism by another name) with the medical-industrial complex. Never mind the fact that this feedback loop of self-interest has spawned one of the most expensive healthcare systems in the world, and that the overall health and longevity of the American taxpayers who fund it continues to slip, year by year, down the ladder of world health outcome ranking.

I am searching my brain for the right metaphor to express the reality of the NIH that I actually encountered in moving from the academic epicenter of California agriculture to the East Coast belly of the medical-industrial complex beast—an astronomic Black Hole comes closest. Like the effects of a Black Hole on spacetime, the massive amount of money allocated to the NIH bureaucracy by the US federal government (year, after year, after year) distorts every aspect of modern medical research, across the United States medical research enterprise and beyond throughout the world.

With that prelude and context, let's examine the new NIH program called ARPA-H (Advanced Research Projects Agency for Health). The 6.5 billion US dollar (initial budget) program was conceived of during the Trump administration but was created by Dr. Francis Collins and White House Office of Science and Technology Policy Director Eric S. Lander during the current Biden administration. ARPA-H is so new that it has yet to be assigned to one of the twenty-seven NIH institutes and centers and in the interim during this formation phase resides within the Office of the Director of NIH (Director, Dr. Elias A. Zerhouni) [296]. A cool billion dollars for 2022 has already been transferred into the program, even though it is not really operating yet. As they say in DC, a billion here, a billion there, and pretty soon it adds up to real money ($45 billion this year for NIH, give or take). Of course, that does not include the monies for related government organizations like the Biomedical Advanced Research and Development Authority (BARDA) (a mere $1.6 billion) or the entire FY2022 budget of the office of the assistant secretary for Preparedness and Response (ASPR- $3.3 Billion including BARDA).

Office of the Secretary - Advanced Research Projects Agency for Health **Operating Plan for FY 2022** *Dollars in Millions*		
Activities	**FY 2021** Final	**FY 2022** Final
Advanced Research Projects Agency for Health......................	-	-
Discretionary Budget Authority/1...	-	1,000.000
Total, Program Level..	-	1,000.000

This amount has been transferred to the National Institutes of Health.

("Nice" detail on their budget breakdown, by the way!)
Now, just to provide context, ASPR has operational responsibilities for the advanced research, development, and stockpiling of medical countermeasures as well as the coordination of the federal public health and medical response to emergencies and disasters. ASPR/BARDA funds development and purchase/stockpile of all the vaccines, drugs, respirators, etc., to meet the nation's biodefense needs, for a cost of $3 billion plus change. So, what is the mission that ARPA-H is going to fulfill for a bit less than 1/3 of that budget? According to its own website [297]:

> The proposed mission of ARPA-H *could be* to make pivotal investments in break-through technologies and broadly applicable platforms, capabilities, resources, and solutions that have the potential to transform important areas of medicine and health for the benefit of all patients and that cannot readily be accomplished through traditional research or commercial activity.
> ARPA-H will:

- Speed application and implementation of health breakthroughs to serve all patients
- Foster breakthroughs across various levels—from the molecular to the societal
- Build capabilities and platforms to revolutionize prevention, treatment, and cures in a range of diseases
- Support "use-driven" ideas focused on solving practical problems that advance equity and rapidly transform breakthroughs into tangible solutions for all patients.
- Focus on multiple time-limited projects with different approaches to achieve a quantifiable goal.
- Use a stage-gate process, with defined metrics, and inject accountability through meeting these metrics.
- Overcome market failures through critical solutions or incentives
- Use the Defense Advanced Research Projects Agency (DARPA) as a model to establish a culture of championing innovative ideas in health and medicine.

Could be. In the investment community, that would be called a "nonbinding forward-looking statement." As I read it, the mission of ARPA-H is (maybe? could be?) to do what NIH, with its' $45 billion per year in funding, is failing to accomplish. How, you ask? Apparently by mimicking the way that the infamous Defense Advanced Research Projects Agency (DARPA) is run and trying to transplant a bit of entrepreneurial spirit into NIH. And to be able to employ alternative contracting mechanisms to engage with the Medical-Industrial complex such as the "Other Transactional Authority" (OTA) mechanism used to bypass the Federal Acquisition Regulations. Per NIH: "An Other Transaction (OT) is a unique type of legal instrument other than a contract, grant, or cooperative agreement. Generally, this awarding instrument is not subject to the FAR, nor grant regulations unless otherwise noted for certain provisions in the terms and conditions of award." By the way, Pfizer has nicely exploited the OTA to build yet another way to shield themselves from any liability incurred with their COVID mRNA vaccine program [298]. Sounds great—what could possibly go wrong <sarcasm>?

A Bloomberg interview of Francis Collins may be among the most helpful in trying to figure out what is really being planned for the initial $6.5B ask [299]. Seeking clarification, when I called in to the Office of the Director of NIH to try to get information on which part of the bureaucracy was going to get dominion over the budget, I was told that the decision had not yet been made. But in this interview, Francis Collins seems pretty sure he knows the answer:

What kind of person are you looking for to lead ARPA-H?
A very entrepreneurial person who has experience in moving forward projects that are high risk, but high reward, and quickly. Maybe somebody who also is experienced with failure, because we want to be sure we know how to see that when it's coming and make decisions quickly. Most likely, this will be somebody from the private sector, or at least somebody who's had significant private sector experience.

The ARPA-H head would have a reporting line to the NIH director, who would need to be pretty hands-off as far as interfering with the decision process and what projects to pursue, but very hands-on in terms of providing the kind of administrative support that's going to be necessary to get this agency started as quickly as possible.

There is a debate going on about whether this should be a presidential appointment. The weight of evidence would say no because then it starts to seem political. And there might be risks involved there, so they will probably be appointed by the health and human services secretary.

Cutting through the chaff here, what I read is basically that 1) NIH is way too slow and kludgy (it often takes five years from initial scientific funding concept to getting money out the door), 2) DoD/DARPA, which Fauci and others have criticized as not getting the biodefense job done (but remember that DARPA would not fund the EcoHealth Alliance coronavirus gain-of-function research at the Wuhan Virology lab which Fauci funded!), is running circles around NIH, 3) BARDA is not able to get the job done either (there is a big overlap between ARPA-H and the BARDA mission), 4) Collins wants this money to come to the Office of the Director, NIH, 5) Collins does not want the director of ARPA-H to be accountable to the President/

Executive Branch, 6) Collins wants the NIH to be more able to compete with private sector efforts ("this will be somebody from the private sector").

Furthermore, based on the ARPA-H website, it looks to me like biometric identification is going to be a big focus here [300]. It is a reasonable possibility that this will be the center that will have ownership of driving forward various aspects of the Transhumanism agenda for the civilian sector, potentially to include DNA-based identification technologies.

This basically overlaps with the same mission and logic used to justify the NIAID Vaccine Research Center (VRC), which is the group that partnered with Moderna to create that mRNA COVID vaccine (resulting in a nice patent royalty stream from the federal investment for all concerned) [301]. The general wisdom applies yet again. When the current funding and bureaucracy is not getting the job done, add more money and bureaucracy. How is ARPA-H different?

To get a peek under the bureaucratic skirts so that we can better understand what is really going on here ("you will know them by their actions, not by their words"), it could be useful to examine who is the brilliant entrepreneurial spirit who has been grabbed out of a leading biomedical innovator corporation to inject a bit of chutzpah into the ossified NIH bureaucracy.

But it turns out, what the White house was "really" looking for was another spook. This time, they zeroed in on and specifically recruited a "DARPA veteran" [302]. In fact, a former director of DARPA who believes that ARPA-H should be should be fully separated from the National Institutes of Health and its vision should be something "bigger" [303]. The new director appointed is Dr. Renee Wegrzyn, who has held important positions at the Defense Advanced Research Projects Agency (DARPA) and the Intelligence Advanced Research Projects Agency (I-ARPA). As mentioned, behavioral and brain sciences are to be incorporated into the research mission of ARPA-H [304].

In addition to the appointment of Dr. Wegrzyn, HHS Secretary Becerra issued a directive establishing an organizational structure for ARPA-H in October 2022. In the establishment notice found in the Federal Register, the structure for ARPA-H has grown broader than initially conceived by members of congress and will now stand as a separate entity inside the NIH structure. Although it is hard not to believe that this was the plan at the White House all along.

Frankly, what I see is yet another spook being strategically placed into the federal side of the "public-private partnership" that exists between the global medical-biodefense-academic-industrial complex and the US federal government and given a nice juicy $6.5B birthday gift with no strings attached and no ability of the executive branch to provide oversight. ARPA-H appears to me to be an intelligence community operational research arm that has been embedded into the Office of the Director of NIH. What could possibly go wrong?

SECTION 3: CORPORATE MEDIA, CENSORSHIP, PROPAGANDA, AND POLITICS

CHAPTER 22
Inverted Totalitarianism

The ideal subject of totalitarian rule is not the convinced Nazi or the dedicated Communist, but people for whom the distinction between fact and fiction . . . and the distinction between true and false . . . no longer exist.
—Hannah Arendt, *The Origins of Totalitarianism*

Because science, medicine, and politics are three threads woven into the same cloth of public policy, we have to work to fix all three simultaneously. The corruption of political systems by corporatists and others committed to globalism has filtered down to our science, medicine, and healthcare systems and must be exposed. Furthermore, the perversion of science and medicine by corporate interests is expanding its reach; it is pernicious and intractable. Regulatory capture by corporate interests runs rampant throughout our politics, governmental agencies, and institutes. The corporatists have infiltrated all three branches of government, and it is up to us, the people, to take control back. Corporate-public partnerships that have become so trendy have another name; that name is fascism—the technically correct political science term for the fusion of the interests of corporations and the state to yield a hybrid governance structure. Basically, the tension between the interest of the republic and its citizens (which Jefferson felt should be primary) and the financial interests of business and corporations (Hamilton's ideal) has swung far too far toward the interests of corporations and their billionaire owners at the expense of the general population. The antitrust laws including the Sherman act have become toothless tigers and pose little or no barrier to racketeering, collusion, and the rampant war profiteering that has characterized the COVIDcrisis.

The nation and its governing arms (including the intrenched bureaucracy often referred to as the "deep state" or "administrative state") now primarily serve the interests of multinational corporations as well as their managers and owners, instead of the other way around—serving the general citizenry. The term that best describes this system of government is called *inverted totalitarianism*. This political science terminology was first coined in 2003 by the political theorist and writer Dr. Sheldon Wolin. *Inverted totalitarianism* is what the government of the United States has devolved into, as Dr. Wolin had warned might happen in the book *Democracy Incorporated* [305]. The United States has been co-opted into a "managed democracy." The American republic was placed into the hands of oligarchs by bureaucratic imperatives and managerial principles and practices, which have created a creeping form of totalitarianism. Now we can clearly see the liberties and freedom guaranteed by the US Constitution being eroded rapidly just as Wolin predicted. The consequence is the establishment of this new form of totalitarianism, which (unlike classical totalitarianism) does not have an authoritarian leader. Instead, inverted totalitarian governments are run by a nontransparent group of managers and elites who manage the country from within. What President Trump might call the "deep state." Or what Steve Bannon originally called the "Uniparty." In effect, our democracy has been turned upside down while being captured by corporate interests that endorse authoritarian policies—hence "inverted totalitarianism" [305].

The infiltration of this version of fascism has gone so far that even routine aspects of the political sphere are determined by corporate interests. This was cemented in the Supreme Court's ruling in Citizens United v. the Federal Election Commission, a decision that reversed century-old campaign finance restriction laws. This ruling has enabled corporations and other outside groups to spend unlimited amounts of money on elections.

As a nation, we are once again confronted by a historic conflict that goes back to Jefferson and Hamilton and the founding of this country. Simply stated, this is the old issue of whether capitalism is a tool of (representative, constitutional republic) democracy or is democracy a tool of capitalism. Will democracy control capitalism or will capitalism control democracy? In this context, it is useful to think of two of our founding fathers, Jefferson and Hamilton, as representing the two sides of the coin in this tension, and awareness of this dynamic conflict was built into our Constitution. In theory, the people, through democratic representation, are empowered to elect leaders and pass laws, but when corporate financial interests become too powerful, they are able to overthrow this relationship and capture political control by exerting financial power and influence over the political process.

In the 21st century, a new threat to democracy and the people's rule has emerged. That is the party of Davos, the alliance of transnational corporations (TNC) and their representatives as the leaders and managers of global governance as embodied in the World Economic Forum and other transnational nongovernmental structures including the United Nations. This has yielded an emerging system of inverted totalitarianism on a global scale. Transnational corporate rule is not limited by national rules and regulations, and its tentacles are everywhere. At the national level, we see its effects on the judicial, legislative, and executive branches. On the international level, the money and power from the transnational organizations have bought off entities like the World Health Organization. The nominal head and coordinating body of this globalized effort to control the world through capitalism has become the World Economic Forum, whose primary belief is that national boundaries are less important than global connectivity and management through corporatism.

Here we are today. In many ways, the hidden head of this unelected corporatist government structure is now the leadership of the World Economic Forum, which meets in the winter in Davos-Klosters, Switzerland, and in summertime in either Tianjin or Dalian China, where the heads of corporations and the apex predator financial firms come together to decide the governing decisions of the world (including the practice of medicine), and politicians and other influencers travel as supplicants akin to courtiers approaching an Emperor or a Pope.

So, if inverted totalitarianism is smashed together with the development of such a global corporatist ruling elite, the question becomes who is running the United States? With the largest brushstrokes possible, it seems to come down to the World Economic Forum, the party of Davos, the large global banks and investment funds, and the central banks (Federal Reserve in the US) including the Bank of International Settlements (BIS). And what we have seen develop, worldwide during the time of COVID, is gross overreach, managerial incompetence, and a reflexive tendency toward authoritarianism that have resulted in an appalling loss of life, liberty, and ability to pursue happiness here in America.

What somehow must be accomplished, if we wish to retain the US

Constitutional form of government and the freedoms guaranteed by the founding documents, is to return to the primacy of the individual: "*That government of the people, by the people, for the people, shall not perish from the earth.*" It is people who should control the levers of government—we must return balance between the Jeffersonian and Hamiltonian ideals. Capitalism in the service of a representative democracy, not the other way around. That is what this great nation was founded on.

Resistance is almost the only way that this can be fought. It has to be organized; funding and the effort must have professional organizers. Resistance has begun, largely in church communities and self-assembling social media aggregates, with people organizing to form autonomous groups outside of the formal power structures that include the two main political parties—both of whom have been significantly captured by corporate interests (giving rise to the Uniparty). In order for this resistance effort to be successful, these groups will have to remove themselves from corporate influence and funding.

This is why the government, corporate interests, and "mainstream" (corporate-captured legacy) media find alternate social media platforms that they can't control to be so threatening. They know these forums are a principal threat, and that control of these alternative information streams are key if the power structure is to be kept from flipping back to control by the people as originally envisioned in the US Constitution.

A powerful popularist who is not tied to the global elite or narrative can also work to break the power of inverted totalitarianism. That is why President Trump, with all his flaws, was so popular. People responded to his messaging that the primacy of representative democracy, the idea that is so deeply connected to the American Enlightenment, is broken. That the "deep state" that Trump speaks of is real. And he is right, it is real. It has become self-evident that the bureaucratic imperatives and managerial principles and practices that now make up the United States governing body are outside of the boundaries of what the founding fathers intended.

Beyond the role of the current populist movement, the corporate elite and managers have silenced many of the great American thinkers and doers of the last fifty years. To frame this regarding the specific case of science and medicine, Sheldon Wolin presciently wrote:

"One of the things we have seen over the last 30 or 40 years is a gradual silencing of people who are doctors or scientists," Saul said. "They are silenced by the managerial methodology of contracts. You sign an employment contract that says everything you know belongs to the people who hired you. You are not allowed to speak out. Take that [right] away and you have a gigantic educated group who has a great deal to say and do, but they are tied up. They don't know how to untie themselves. They come out with their Ph.D. They are deeply in debt. The only way they can get a job is to give up their intellectual freedom. They are prisoners."

Regulatory and operational capture of the hospitals, health insurance, and physicians is almost complete now. The government, Big Pharma, big tech, and hospital administrators have worked in consort to silence scientists and physicians. This is not something new. Sheldon Wolin defined the problem a decade ago. What is new is the rise of the Internet and social media to become a "town hall" within which populists can interact and organize. It is this tool that we can use to resume control of the country as originally intended. Each of us must take back our power regarding our own and our family's healthcare. We must be responsible for our own bodies and not give that right away to others. We must choose our physicians carefully and conscientiously. Corporatist medicine and science must be avoided as primary sources for information. New, noncorporation-influenced medicine and science sources need to be developed. That includes medical information aggregator sites and journals that are free from corporate influences.

Independent social media platforms and alternative news sources are key to building communities and organizations that can resist inverted totalitarianism. And because they are key to the resistance, governments and the corporate elite will work to censor and remove all such platforms. That will include outright purchase, as well as a variety of governmental controls placed on alternate media sources. The current move to embrace censorship by the legacy corporate media, big tech, and their hypnotized minions are examples of what our future holds. Be ready to jump platforms. Archive email lists and be ready to activate new accounts quickly. It is the resistance and our communities that are important, not the platform.

This is a war to save our great nation. Be prepared, because this new form of global fascism, a new form based on inverted totalitarianism, is an enemy like none of us have seen before. As we move forward on this fight to save the world, remember we have one big tool. That is our ability to resist. There are many people who are willing to resist, to change, and to not accept a world where corporations rule by fiat. People who will work to create a world whereby medicine is more than just numbers and a way to sell new drugs and vaccines.

Do not forget the words of Abraham Lincoln:

We the people are the rightful masters of both Congress and the courts, not to overthrow the Constitution but to overthrow the men who pervert the Constitution.

CHAPTER 23
Behavioral Control and the End of the American Dream

Remember those Johns Hopkins Pandemic war games that occurred over the span of decades (including Event 201)? Where the outcome usually ended in a need to control the populace, and in which behavioral modification techniques are used to enforce cooperation from citizens?

Right now, one can (virtually) wander over to Johns Hopkins Center for Health Security and see that their current projects include an analysis of "antimisinformation actions," which they call the Environment of Misinformation.

World leaders, governments, big media, Big Pharma, and tech giants are busy planning out the next pandemic response. For those of us who believe there is a better way to live than to be controlled by these organizations, it is well past time to plan out responses to all these draconian measures. And to begin to develop biothreat countermeasure strategies that do not involve the use of heavy-handed censorship, propaganda, mandates, and behavioral modification techniques. You know, the old-fashioned way where the government relies on people to use their own critical thinking skills to assess what is best for themselves and their families, after helping citizens to obtain all the relevant information that is necessary and available? As opposed to the view that it is the government's job to completely control both information and citizenry as if citizens are livestock to be managed in a battlefield while waging war against an armed hostile opponent.

So, let's review their planning and consider how "we" have been controlled, nudged, censored, and lied to during this pandemic, and review and learn from what governments are already planning for how to control us better during the next "public health emergency."

Definition: psyops (From the US Department of Defense)

> Psyops are the use of propaganda and psychological tactics to influence emotions and behaviors.

The US Department of Defense (DOD) 2004 and 2010 Counterinsurgency Operations Reports define "psyops" as the following:

> The mission of psychological Operations (psyops) is to influence the behavior of foreign target audiences to support US national objectives. A psyop accomplishes this by conveying selected information and advising on actions that influence the emotions, motives, objective reasoning, and ultimately the behavior of foreign audiences. Behavioral change is at the root of the psyop mission.

Read that last sentence again. "Behavioral change is at the root of the psyop mission." Sound familiar?

At the heart of a psyops operations are behavioral tools or mind control techniques such as hypnosis, mass formation, censorship, security theater, use of fear to drive anxiety and propaganda.

How does our military use psyops? Here are some key terms.

DELIBERATELY DECEIVE

Military deception missions use psychological warfare to deliberately mislead enemy forces during a combat situation.

INFLUENCE WITH INFORMATION

Military Information Support Operations (MISO) missions involve sharing specific information to foreign audiences to influence the emotions, motives, reasoning, and behavior of foreign governments and citizens. This can include cyber warfare and advanced communication techniques across all forms of media.

ADVISE GOVERNMENTS

Interagency and government support missions shape and influence foreign decision making and behaviors in support of United States' objectives.

PROVIDE COMMUNICATIONS FOR RESCUE EFFORTS
Civil Authorities Information Support (CAIS) missions aid civilian populations during disaster relief situations by sharing critical information to support the rescue effort.

What is the history of Army Psychological Operations?
Founded during World War I to devastate opposing troops' morale, the PSYOPS unit has played a critical role in World War II, the Vietnam War, and recent operations in Afghanistan and Iraq, where unconventional warfare provided by PSYOPS has been crucial to national security.

What is nudging?
"A nudge is any attempt at influencing people's judgment, choice or behavior in a predictable way that is motivated because of cognitive boundaries, biases, routines, and habits in individual and social decision-making posing barriers for people to perform rationally in their own self-declared interests, and which works by making use of those boundaries, biases, routines, and habits as integral parts of such attempts."

And now—for the fly in the toilet. . .
To help illustrate the concept, one of the simplest examples of "nudging" is the use of the image of a "fly" placed in urinal toilet bowls. First introduced at Schiphol airport in Amsterdam back in 1999, the idea was simple: etch the image of a fly in the urinal, and men will aim for it almost every time, well, at least over 80% of the time. Urinal cleaning costs went way down without forcing anyone to do anything.

Now that the terms nudging and pysops are defined and understood, think about these terms in context of mass formation and the approved COVIDcrisis narrative, and the remarkable overreaction of corporate media and big tech in response to the Joe Rogan/Malone podcast (Episode #1757 Joe Rogan Experience). Here is the quote that prompted such an overwrought response:

> Basically, [there was a] European intellectual inquiry into what the heck happened in Germany in the '20s and '30s—very intelligent, highly educated population and they went barking mad.

The answer is mass formation psychosis. When you have a society that has been decoupled from each other and has free-floating anxiety, and a sense that things don't make sense, we can't understand it, and then their attention gets focused by a leader or a series of events on one small point (just like hypnosis), they literally become hypnotized and can be led anywhere. And one of the aspects of that phenomenon, is that people that they identify as their leaders, the ones that typically that come in and say "You have this pain and I can solve it for you. I, and I alone, can fix this problem for you"—they will follow that person through hell. It doesn't matter whether they lie to them or whatever. The data are irrelevant, and furthermore anyone who questions that narrative is to be immediately attacked. They are the other. This is central to mass formation psychosis, and this is what has happened.

The following article illustrates one example of this amazingly coordinated response to the terms "mass formation psychosis" [306]:

Van Bavel (the AP source) seems to believe that he is not guilty of either spreading or believing propaganda, manipulating people, or being manipulated.

In a *Nature* article in 2020, Van Bavel posited that "insights from the social and behavioral sciences can be used to help align human behavior with the recommendations of epidemiologists and public health experts." What is this if not an attempt to push people to do what they're told?

The article addresses using fear as a means to control people, in the right doses: "A meta-analysis found that targeting fears can be useful in some situations, but not others: appealing to fear leads people to change their behavior if they feel capable of dealing with the threat but leads to defensive reactions when they feel helpless to act. The results suggest that strong fear appeals produce the greatest behavior change only when people feel a sense of efficacy, whereas strong fear appeals with low-efficacy messages produce the greatest levels of defensive responses."

So, the expert opinion that was cited to fact-check the term "Mass Formation" is actually from an expert on the use of behavioral techniques and fear to ensure compliance within a population in support of public health objectives. So, denying that mass formation exists, while writing about it in the scientific literature. It doesn't get more Orwellian, does it?

> Political warfare is the art of heartening one's friends and disheartening one's enemies. It makes use of ideas, words, images, and deeds to compel or convince friends, foes, or neutrals into cooperation or acquiescence. Effective political warriors know that the best way to prevail in modern ideological conflict is not through killing, but through persuasion, cooption, and influence.
>
> *Frank Gaffney, Jr.*

Clinical Research to develop Vaccination PsyOps Messaging

Another example of the development and deployment of psyops and nudge technology during the COVIDcrisis involves messaging designed to address "vaccine hesitancy." Even though the testing of the vaccine products was highly abbreviated and did not meet regulatory norms for either vaccines or genetic therapies, the development and clinical testing of psyops and nudge messaging was carefully performed using a prospective randomized, controlled clinical trial structure with short-term three- and six-month follow-up.

During July of 2020, Yale University initiated a clinical trial (#NCT04460703) to develop and optimize means to psychologically manipulate people to overcome "vaccine hesitancy" to the COVID genetic vaccines via message control and content [307]. This interventional clinical trial, titled "Persuasive Messages for COVID-19 Vaccine Uptake: a Randomized Controlled Trial, Part 1," enrolled 4,000 human subjects aged eighteen and older and was designed to assess one primary and four secondary endpoints. Although the sources of funding for most clinical trials of this size and complexity are usually clearly stated, in the case of this study the funding sources have been carefully hidden behind a veil of academic research institutes that do not disclose their sources of funding, which is very unusual. The sources of support listed in the final publication summarizing study results are the

Tobin Center for Economic Policy at Yale University, the Yale Institute for Global Health, the Institution for Social and Policy Studies, and the Center for the Study of American Politics at Yale University. None of which state their funding sources on their publicly available web pages.

This study tested different messages about vaccinating against COVID-19. Participants were randomized to one of twelve arms, with one control arm and one baseline arm. The study was designed to compare the reported willingness to get a COVID-19 vaccine at three and six months of it becoming available between the ten intervention arms relative to the two control arms.

The Primary outcome endpoint was self-reported intention to get COVID-19 vaccine immediately after the intervention message, and the likelihood of getting a COVID-19 vaccination within three months and then six months of it becoming available.

Secondary Outcome Measures included:

1. Vaccine confidence scale using a validated scale to assess the impact of the messages on vaccine confidence.
2. Persuade others. This is a measure of a willingness to persuade others to take the COVID-19 vaccine.
3. Fear of those who have not been vaccinated. This is a measure of a comfort with an unvaccinated individual visiting an elderly friend after a vaccine becomes available.
4. Social judgment of those who do not vaccinate.

A group of different messages were tested to determine which would be the most effective for achieving primary and secondary outcomes. In other words, specific propaganda messaging was experimentally tested, conclusions drawn, and then these results were used to guide the subsequent federal US (and global) propaganda campaign to promote uptake of a poorly tested, unlicensed, experimental use authorized medical procedure and product.

The experimental messages that were tested included the following, each of which are familiar to all who were subjected to the subsequent propaganda campaign:

1. **Personal freedom message:** How COVID-19 is limiting people's personal freedom, and by working together to get enough people

vaccinated, society can preserve its personal freedom. The specific message tested was as follows: "COVID-19 is limiting many people's ability to live their lives as they see fit. People have had to cancel weddings, not attend funerals, and halt other activities that are important in their daily lives. On top of this, government policies to prevent the spread of COVID-19 limit our freedom of association and movement. Remember, each person who gets vaccinated reduces the chance that we lose our freedoms or government lockdowns return. While you can't do it alone, we can all keep our freedom by getting vaccinated."

2. **Economic freedom message**: How COVID-19 is limiting people's economic freedom, and by working together to get enough people vaccinated, society can preserve its economic freedom. "COVID-19 is limiting many people's ability to continue to work and provide for their families. People have lost their jobs, had their hours cut, and lost out on job opportunities because companies aren't hiring. On top of this, government policies to prevent the spread of COVID-19 have stopped businesses from opening up. Remember, each person who gets vaccinated reduces the chance that we lose our freedoms or government lockdowns return. While you can't do it alone, we can all keep our ability to work and earn a living by getting vaccinated."

3. **Self-interest message:** COVID-19 presents a real danger to one's health, even if one is young and healthy. Getting vaccinated against COVID-19 is the best way to prevent oneself from getting sick. "Stopping COVID-19 is important because it reduces the risk that you could get sick and die. COVID-19 kills people of all ages, and even for those who are young and healthy, there is a risk of death or long-term disability. Remember, getting vaccinated against COVID-19 is the single best way to protect yourself from getting sick."

4. **Community interest message:** A message about the dangers of COVID-19 to the health of loved ones. The more people who get vaccinated against COVID-19, the lower the risk that one's loved ones will get sick. Society must work together and all get vaccinated. "Stopping COVID-19 is important because it reduces the risk that members of your family and community could get sick and

die. COVID-19 kills people of all ages, and even for those who are young and healthy, there is a risk of death or long-term disability. Remember, every person who gets vaccinated reduces the risk that people you care about get sick. While you can't do it alone, we can all protect every-one by working together and getting vaccinated."

5. **Economic benefit message:** A message about how COVID-19 is wreaking havoc on the economy and the only way to strengthen the economy is to work together to get enough people vaccinated. "COVID-19 is limiting many people's ability to continue to work and provide for their families. People have lost their jobs, had their hours cut, and lost out on job opportunities because companies aren't hiring. On top of this, government policies to prevent the spread of COVID-19 have stopped businesses from opening up. Remember, each person who gets vaccinated reduces the chance that we lose our freedoms or government lockdowns return. While you can't do it alone, we can all keep our ability to work and earn a living by getting vaccinated."

6. **Guilt message:** About the danger that COVID-19 presents to the health of one's family and community. The best way to protect them is by getting vaccinated, and society must work together to get enough people vaccinated. Then it asks the participant to imagine the guilt they will feel if they don't get vaccinated and spread the disease. "Imagine how guilty you will feel if you choose not to get vaccinated and spread COVID-19 to someone you care about."

7. **Embarrassment message:** The danger that COVID-19 presents to the health of one's family and community. The best way to protect them is by getting vaccinated and by working together to make sure that enough people get vaccinated. Then it asks the participant to imagine the embarrassment they will feel if they don't get vaccinated and spread the disease. "Imagine how embarrassed and ashamed you will be if you choose not to get vaccinated and spread COVID-19 to someone you care about."

8. **Anger message:** The message is about the danger that COVID-19 presents to the health of one's family and community. The best way to protect them is by getting vaccinated and by working together to make sure that enough people get vaccinated. Then it asks the

participant to imagine the anger they will feel if they don't get vaccinated and spread the disease. Message 3) + "Imagine how angry you will be if you choose not to get vaccinated and spread COVID-19 to someone you care about."

9. **Trust in science message:** A message about how getting vaccinated against COVID-19 is the most effective way of protecting one's community. Vaccination is backed by science. If one doesn't get vaccinated that means that one doesn't understand how infections are spread or who ignores science. "Getting vaccinated against COVID-19 is the most effective means of protecting your community. The only way we can beat COVID-19 is by following scientific approaches, such as vaccination. Prominent scientists believe that once available, vaccines will be the most effective tool to stop the spread of COVID-19. The people who reject getting vaccinated are typically ignorant or confused about the science. Not getting vaccinated will show people that you are probably the sort of person who doesn't understand how infection spreads and who ignores or is confused about science."

10. **Not bravery message:** A message that describes how firefighters, doctors, and frontline medical workers are brave. Those who choose not to get vaccinated against COVID-19 are not brave. "Soldiers, fire-fighters, EMTs, and doctors are putting their lives on the line to serve others during the COVID-19 outbreak. That's bravery. But people who refuse to get vaccinated against COVID-19 when there is a vaccine available because they don't think they will get sick or aren't worried about it aren't brave, they are reckless. By not getting vaccinated, you risk the health of your family, friends, and community. There is nothing attractive and independent-minded about ignoring public health guidance to get the COVID-19 vaccine. Not getting the vaccine when it becomes available means you risk the health of others. To show strength get the vaccine so you don't get sick and take resources from other people who need them more, or risk spreading the disease to those who are at risk, some of whom can't get a vaccine. Getting a vaccine may be inconvenient, but it works."

The final peer-reviewed scientific article summarizing the findings of this prospective randomized clinical trial has been published in the journal *Vaccine* under the title "Persuasive messaging to increase COVID-19 vaccine uptake intentions" [308]. The peer-reviewed and published findings include the following:

Abstract

Widespread vaccination remains the best option for controlling the spread of COVID-19 and ending the pandemic. Despite the considerable disruption the virus has caused to people's lives, many people are still hesitant to receive a vaccine. Without high rates of uptake, however, the pandemic is likely to be prolonged. Here we use two survey experiments to study how persuasive messaging affects COVID-19 vaccine uptake intentions. In the first experiment, we test a large number of treatment messages. One subgroup of messages draws on the idea that mass vaccination is a collective action problem and highlighting the prosocial benefit of vaccination or the reputational costs that one might incur if one chooses not to vaccinate. Another subgroup of messages built on contemporary concerns about the pandemic, like issues of restricting personal freedom or economic security. We find that persuasive messaging that invokes prosocial vaccination and social image concerns is effective at increasing intended uptake and also the willingness to persuade others and judgments of non-vaccinators. We replicate this result on a nationally representative sample of Americans and observe that prosocial messaging is robust across subgroups, including those who are most hesitant about vaccines generally. The experiments demonstrate how persuasive messaging can induce individuals to be more likely to vaccinate and also create spillover effects to persuade others to do so as well.

Discussion

Overall, the results point both to a set of effective messages and the potential efficacy of specific messages for some particular subgroups. On average, a simple informational intervention is effective, but it is even more effective to add language framing vaccine uptake as protecting others and as a cooperative action. Not only does emphasizing that vaccination is a prosocial action increase uptake, but it also increases people's willingness to pressure others to do so, both by direct persuasion and negative judgment of non-vaccinators. The latter social pressure effects may be enhanced by highlighting how embarrassing it would be to infect someone else after failing to vaccinate. The Not Bravery and Trust in Science messages had substantial effects on other regarding outcomes and for some subgroups, but do not appear to be as effective as the Community Interest messages in promoting own vaccination behavior. Importantly, in distinct samples fielded several months apart, the Community Interest, Community Interest + Embarrassment, and the Not Bravery messages produced substantively meaningful increases for all outcomes measures relative to the untreated control, and in some instances did so in comparison to the Baseline information condition.

Our findings are consistent with the idea that vaccination is often treated as a social contract in which people are expected to vaccinate and those who do not are sanctioned. In addition to messages emphasizing the prosocial element of vaccination, we observed that messages that invoked reputational concerns were successful at altering judgment of those who would free ride on the contributions of others. This work could also help explain why social norm effects appear to overwhelm the incentive to free ride when vaccination rates are higher. That is, messages that increased intentions to vaccinate also increased the moralization of non-vaccinators suggesting that they are fundamentally linked to one another. These messages will need to be adapted in specific cultural contexts with relevant partners, such as community leaders.

The robust effect of the Community Interest message advances our current understanding of whether public health messaging that deploys prosocial concerns could be effective at increasing COVID-19 vaccine uptake. The results of both experiments presented here support prior work that demonstrated the effectiveness of communication that explains herd immunity on promoting vaccination. It also suggests that a detailed explanation of herd immunity may not be necessary to induce prosocial behavior.

Beyond the theoretical contribution, the results have practical implications for vaccine communication strategies for increasing COVID-19 vaccine acceptance. We identified multiple effective messages that provide several evidence-based options to immunization programs as they develop their vaccine communication strategies. Importantly, the insights into differential effectiveness of various messages by subgroup (e.g., men vs women) could inform messaging targeted to specific groups. Understanding heterogeneous treatment effects and the mechanisms that cause differential responses to persuasive messaging strategies requires additional testing and theoretical development. We view this as a promising avenue for future work.

Some may still harbor illusions about the general citizen population (including the medical community) having been subjected to a focused, planned, and field-tested psyops campaign designed to promote compliance with acceptance of unlicensed vaccines. Hopefully this paper will help dispel some of the confusion and controversies that are currently circulating. We have all been psychologically manipulated by our governments. Intentionally. Manipulated to accept an unlicensed, poorly tested medical product that is neither safe nor effective under any previously accepted definition of effectiveness as a vaccine.

Behavior Control of Civilians by Their Own Government Is the New Normal

What we have experienced during the COVIDcrisis is full-on modern political and information warfare, which consists of effectively using social and MSM media to control the narrative. But this is asymmetric warfare. The practitioners have the resources, the power, the money, mainstream media, Big Pharma, tech giants, and social media supporting their efforts. When Biden decided by executive order to enforce mandates, which are political in nature (the science does not support mandates as the vaccines do not stop spread of the disease and may create vaccine escape mutants), then mandates became a censored topic. When the government decided that they would not support the use of Ivermectin and HCQ, despite our laws that allow such usage, these also became censored topics and taboo to discuss. Messages about Ivermectin being dangerous horse-paste were planted thoughout the Internet and TV. These are examples of censorship and propaganda.

Now, there is a small, but growing guerrilla army of supporters of freedom. Strangely enough, the resistance has come from "conservatives," and now many in the younger generation are taking up the mantle. Indications are that it is becoming hip to be conservative. Will Democrats lose the next generation because of their draconian public health responses to COVID-19?

If this informal freedom resistance campaign speaks too loudly or speaks truth to power too often, they are taken out by warnings and de-platforming. They are losing their right to free speech on a daily basis. Even some conservative politicians are no longer allowed to use such platforms as YouTube or Twitter. Senator Ron Johnson, the ranking member for the Permanent Subcommittee on Investigations, has lost the right to publish on YouTube. That should scare and shock anyone with a functioning brain. Our politicians are being banned from free speech because big tech, working with the current Democratic administration, doesn't like their messaging, their interference with the approved narrative. This goes beyond censorship to use of the power of the state to eliminate opposition, and it is interfering with both our right to a free press and our elections. Articles on such subjects as the lab leak are removed from search engines and are not allowed to be republished by the Trusted News Initiative. Professionals are losing their jobs, being investigated, and losing their licenses for speaking out and/or treating patients with COVID-19 early in the course of the disease. Some might call these tactics defenestration. As this happens, more and more people are realizing that the very freedoms that made our country what it is are at stake.

The use of fear by governments and cooperating media to control behavior in the COVIDcrisis is gradually being acknowledged, and considered "totalitarian by Members of Scientific Pandemic Influenza Group on Behavior, and that some express regret about "unethical" methods shows that governments have gone beyond nudging and are working in the realm of pysops and totalitarian measures" [309]. More recently, one of the largest newspapers in Denmark apologized for its journalistic failure during COVID-19, for only publishing the official government narrative without question. This newspaper continued to go along with this plan long after it was clear that the government narrative was crumbling. The link between governments and the media to control the population has become normalized.

Where as a nation, as a society, and as individuals do we go from here?

For me, this is a battle that has completely changed my life, my way of thinking, and my perspective on my government and world leaders. There is no going back for me. I will not let this great nation, this world, go down the road of totalitarianism if I can help it. That is a recipe for the end of the American dream.

CHAPTER 24

Propaganda, Corporatism, Journalism, Advertising, and the Noble Lie

Knowledge of the theory and practical implementation of mass formation psychology (also known as "mass psychosis") can and is being used by propagandists, governments, nongovernmental organizations, and the World Economic Forum to sway large groups of people to act for the benefit of the propagandists' objectives [310–314]. Although a major crisis of some sort can be extremely useful when seeking to advance propaganda and manipulate the beliefs of populations of people (war, hyperinflation, or public health for example—or all three at once), these psychological theories can be applied even without strong evidence of a compelling crisis. A propagandist-leader just has to be sufficiently compelling to manufacture a crisis. Such leaders are often identifiable by a surrounding cult of personality, frequently supported and amplified by sycophants who benefit from close association with the leader.

One current example illustrating this involves the almost global acceptance of mask use by the general population over the past two years, despite ample evidence that surgical and cloth masks are basically ineffective [315–317]. Because Fauci and his acolytes at the CDC insisted that masks work, public acceptance of a very intrusive element into people's lives (and children's education) was almost universal. Data demonstrating lack of effectiveness of masks for preventing spread of the SARS-CoV-2 virus were largely irrelevant, and either rejected or unable to be either published in the scientific literature or acknowledged by those who have become hypnotized by the mass formation process. Even the logic of masking children was accepted without question despite the clear and compelling evidence of harm. The globally propagated six-foot social distance rule, which was completely arbitrary, provides another example.

Paul Joseph Goebbels was the chief German propagandist for the Nazi Party and was then promoted to the Reich Minister of Propaganda from 1933 to 1945. He was truly a master and arguably the creator of the concept that the State can control people by introducing propaganda into print and broadcast news (and moving pictures) to enable State-based control of entire populations. Goebbels's wicked brilliance was to exploit racism as a tool to promote German nationalism to the point of mobilizing and motivating Germany to engage in a globalized war for political, military, and economic dominance. His writings and speeches on propaganda have been studied by leaders and governments ever since, much as the Italian Renaissance writings of Niccolò Machiavelli continue to be a cornerstone of modern interstate realpolitik. Examples of Goebbels's insights include the following:

> There was no point in seeking to convert the intellectuals. For intellectuals would never be converted and would anyway always yield to the stronger, and this will always be 'the man in the street.' Arguments must therefore be crude, clear and forcible, and appeal to emotions and instincts, not the intellect. Truth was unimportant and entirely subordinate to tactics and psychology.
>
> If you tell a lie big enough and keep repeating it, people will eventually come to believe it. The lie can be maintained only for such time as the state can shield the people from the political, economic, and/or military consequences of the lie. It thus becomes vitally important for The State to use all of its powers to repress dissent, for the truth is the mortal enemy of the lie, and thus by extension, the truth is the greatest enemy of The State.

Goebbels applied the theories behind what is now described by Dr. Mattias Desmet as "mass formation" or "mass psychosis" to practical politics within a nation-state. Academic writings concerning the formation of a "mass" or a crowd, otherwise known as mass formation, was an accepted discipline during the time when Goebbels was developing his insights, with many scholars including Gustave Le Bon, Freud, McDougal, and Canetti making intellectual contributions to his thinking.

To expand further upon the point, in a related article titled "Silencing the Lambs—How Propaganda Works," renowned journalist and author John Pilger writes: "Leni Riefenstahl said her epic films glorifying the Nazis depended on a 'submissive void' in the German public. This is how propaganda is done" [318]. He describes how, in the 1970s, he met Leni Riefenstahl (one of Hitler's leading propagandists), whose epic films glorified the Nazis. She told him that the "patriotic messages" of her films were dependent not on "orders from above," but on what she called the "submissive void" of the German public. "Did that include the liberal, educated bourgeoisie?" Pilger asked. "Yes, especially them," she said.

Pilger then proceeded to discuss observations of the playwright Harold Pinter, who was a personal friend.

> "U.S. foreign policy," Pinter said, is "best defined as follows: kiss my arse or I'll kick your head in. It is as simple and as crude as that. What is interesting about it is that it's so incredibly successful. It possesses the structures of disinformation, use of rhetoric, distortion of language, which are very persuasive, but are actually a pack of lies. It is very successful propaganda. They have the money, they have the technology, they have all the means to get away with it, and they do."

In accepting the Nobel Prize for Literature, Pinter stated:

> The crimes of the United States have been systematic, constant, vicious, remorseless, but very few people have actually talked about them. You have to hand it to America. It has exercised a quite clinical manipulation of power worldwide while masquerading as a force for universal good. It's a brilliant, even witty, highly successful act of hypnosis.

Pilger asked Pinter if the "hypnosis" he referred to was the "submissive void" described by Leni Riefenstahl.

> "It's the same," he replied. "It means the brainwashing is so thorough we are programmed to swallow a pack of lies. If we don't

recognize propaganda, we may accept it as normal and believe it. That's the submissive void."

Le Bon, a French social psychologist, is often seen as the founder of the field of study focused on crowd (group) psychology. Le Bon defined a crowd as a group of individuals united by a common idea, belief, or ideology, and he believed when an individual becomes part of a crowd, he/she undergoes a profound psychological transformation. The individual ceases to think independently and instead relies on the group synthesis of a set of simplified ideas. According to this theory, crowd formation requires a set of simplified ideas that the group incorporates, at which point an individual who has become integrated into the group ceases to psychologically exist as an independent mind and functionally becomes hypnotized. Le Bon maintained that a group typically forms around an influential idea that unites a number of individuals, and this idea then propels the group (or mass) to act toward a common goal. However, he also concluded that these influential ideas are never created by members of the crowd. Instead, they are most often given to the crowd by a leader or set of leaders. According to Le Bon, in order for an idea to unite and influence a crowd, it must first be dumbed down to the level that the entire crowd can understand it. It must be easily understood by all within the crowd.

Just to provide a current example to illustrate the point, a scientific discipline could develop a new type of vaccine as a solution to a public health crisis. That complex research and resulting technology may have required decades of effort. On average, the crowd as a whole would be incapable of comprehending such complex theories or technologies, so socially engineering acceptance of the vaccine (by a crowd or mass) would require this new concept for vaccination to be thoroughly simplified before the idea could become the focus of a hypnotic, single-minded belief in the solution (the new type of vaccine). Le Bon proposed that this is where group leaders come in. Under the "Le Bon model," the leader of a crowd (for example, someone like Dr. Anthony Fauci) will enable this process by distilling these complicated concepts (or technologies) down to a small set of simplified ideas that the crowd can accept, incorporate, and act upon as their own. One of the most important elements of this is the requirement for a "trusted leader" to be accepted by the crowd, a process that can be actively advanced through propaganda and censorship/information control. Once a crowd truly accepts a leader, it is almost impossible for them to reject that leader, whether or not the lies that he or she may tell are actually done with "noble" intent or purpose.

Over the last two years, we have seen clear evidence that both our government as well as those of Great Britain and many other Western democracies have learned and actively apply the lessons of Gustave Le Bon and Joseph Goebbels quite well.

Going back in time, in a book titled *Propaganda and Persuasion*, historians Jowett & O'Donnell wrote about Hitler's basic principles of propaganda, which were based upon Goebbels's work and advice. They are:

Hitler's Basic Principles
(abstracted from Jowett & O'Donnell's *Propaganda and Persuasion*

- Avoid abstract ideas. Appeal to the emotions.
- Constantly repeat just a few ideas. Use stereotyped phrases.
- Give only one side of the argument.
- Continuously criticize your opponents.
- Pick out one special "enemy" for special vilification.

Unfortunately, national and world governmental organizations as well as corporate mass media have learned more than just the lessons of mass psychosis and propaganda. World governments and large financial interests have now united to produce harmonized propaganda through a wide variety of media outlets, such as big tech, social media, and mainstream media. We have entered a new era of total thought control exerted on a global scale, which is often referred to as psychological operations or psyops.

Before proceeding further, it is important to provide examples to illustrate what is going on in the modern psyops operations led by governments, nongovernmental organizations, global forums such as the United Nations and World Health Organization, and the World Economic Forum. Helpful examples include the following:

Operation Mockingbird: Operation Mockingbird was organized by Allen Dulles and Cord Meyer in 1950. The CIA spent about one billion dollars a year in today's dollars, hiring journalists from Corporate Media including CBS, the *New York Times*, ABC, NBC, *Newsweek*, Associated Press, and others to promote their point of view. The original operation reportedly involved some 3,000 CIA operatives and hired over 400 journalists. In 1976, the domestic operation supposedly closed, but less than half of the media operatives were let go. Furthermore, documentary evidence shows that much of the Operation Mockingbird was off-shored at that time. It is rumored that British Intelligence picked up many of the duties of Operation Mockingbird on behalf of the US intelligence community (see the Trusted News Initiative, for example).

The Trusted News Initiative (TNI): TNI is a British Broadcasting Corporation (BBC)-led organization that has been actively censoring eminent doctors, academics, and those with dissenting voices that contravene the official COVID-19 narrative. Anything contrary to this narrative is considered disinformation or misinformation and will be deleted, suppressed, or deplatformed. Misinformation and disinformation are considered anything not aligned with the World Health Organization and/or the regional Public Health Authority-approved "truth." In the case of the USA—that "truth" is established by Anthony Fauci, the CDC, and the FDA. The TNI uses advocacy journalism and journals to promote their causes. The Trusted News Initiative is more than this, though; if you go back to Hitler's basic principles, the members of the TNI are using these core principles to control the public. The known TNI partners include: Associated Press, AFP; BBC, CBC/Radio-Canada, European Broadcasting Union (EBU), Facebook (whose founders fund articles being written for *The Atlantic*), *Financial Times*, First Draft, Google, *The Hindu*, Microsoft, *New York Times*, Reuters, Reuters Institute for the Study of Journalism, Twitter, You Tube, *Wall Street Journal, Washington Post*. In many ways, the TNI currently functions as a monopolistic trade organization designed to suppress competition from nontraditional information sources such as "new media" and so may be subject to the United States' Sherman Antitrust Act, which reads, "Every contract, combination in the form of trust or otherwise, or conspiracy, in restraint of trade or commerce among the several states, or with foreign nations, is declared to be illegal."

World Economic Forum: The World Economic Forum (WEF) is one of the key think tanks and meeting places for the management of global capitalism and is arguably coherent enough to qualify as the leading global "deep state" organization. Under the operational leadership of professor Klaus Schwab, it has played an increasingly important role in coordinating the globalized hegemony of large pools of transnational capital and associated large corporations over Western democracies during the last three decades. Many of its members are active in using COVID-19 to carry out a "Great Reset" (as described in the writings of Klaus Schwab and initially announced by King George III) to dispossess and implement digital tracking and control of people as a step toward what many believe will institute a technofeudalism as well as the WEF objective of a fourth industrial revolution incorporating technologies collectively referred to as "transhumanism." Genetic mRNA vaccines have been identified by both Western governments and the WEF as a first step toward an inevitable "transhumanism" agenda. A case can also be made that the organization and operation of the WEF also violates the United States' Sherman Antitrust Act.

Social Credit systems: China's social credit system is a combination of government and business surveillance that gives citizens a "score" that can restrict the ability of individuals or corporations to function in the modern world by limiting purchases, acquiring property, or taking loans based on past behaviors. Of course, how one uses the Internet directly impacts the social credit score. This is the origin of the social credit system that appears to be evolving in the United States. Environmental, social, and governance (ESG) metrics are a kind of social credit system designed to coerce businesses—and, by extension, individuals and all of society—to transform their practices, behaviors, and thinking. Many government leaders in USA who have been trained by the WEF Young Leaders or Influencers programs are pushing this scoring system and actively promoting the idea in the USA. Already, financial institutions such as PayPal and GoFundMe, as well as some more mainstream banking systems, are actively deciding who can use their services based on a social credit scoring system [319].

WEF Young Leaders Program: This is a five-year World Economic Forum training program that handpicks individuals most likely to succeed in politics, corporate governance, and as key influential royalty. The WEF helps connect these graduates with leaders and capital to ensure that they rise in the ranks of national or world politics and/or corporate governance. The training program agenda is kept secret, and it is very rare that a graduate will discuss the program publicly. However, the benefits of global corporatism, combined with social engineering, are main components of the program. The Young Leaders program started in 1992 (under a different name) and has graduated close to 4,000 people (the full list can be found at the MaloneInstitute.org website). They include a "who's who" of leaders and influencers in politics, big tech, media, the pharmaceutical industry, and finance.

A small subset of the graduates from the Young Leaders Program in the United States include:

Politics and Policy: Jeffrey Zients (White House Coronavirus Response Coordinator since 2021), Jeremy Howard (cofounder of lobby group "masks for all"), California Governor Gavin Newsom, Peter Buttigieg (candidate for US President in 2020, US secretary of transportation since 2021), Chelsea Clinton, Huma Abedin (Hillary Clinton aide), Nikki Haley (US ambassador to the UN, 2017–2018), Samantha Power (US ambassador to the UN, 2013–2017, USAID Administrator, since 2021), Ian Bremmer (founder of Eurasia Group), Bill Browder (US-British financier), Jonathan Soros (son of George Soros), Kenneth Roth (director of Human Rights Watch), Paul Krugman (economist), Lawrence Summers (US Secretary of the Treasury, 1999–2001), Black Lives Matter cofounder Alicia Garza, Ivanka Trump, and Tulsi Gabbard.

Legacy Media: CNN medical analyst Leana Wen, CNN's Sanjay Gupta, Covid Twitter personality Eric Feigl-Ding, Andrew Ross Sorkin (*New York Times* financial columnist), Thomas Friedman (*New York Times* columnist), George Stephanopoulos (ABC News), Lachlan Murdoch (CEO of Fox Corporation, cochair of News Corp), Justin Fox (Bloomberg), Anderson Cooper (CNN).

Technology and Social-Media: Microsoft founder Bill Gates, former Microsoft CEO Steven Ballmer, Jeff Bezos. Google cofounders Sergey Brin and Larry Page, Elon Musk, former Google CEO Eric Schmidt, Wikipedia cofounder Jimmy Wales, PayPal cofounder Peter Thiel, eBay cofounder Pierre Omidyar, Facebook founder and CEO Mark Zuckerberg, Facebook COO Sheryl Sandberg, Moderna CEO Stéphane Bancel, Pfizer CEO Albert Bourla (a WEF Agenda Contributor), and Pfizer VP Vasudha Vats.

Note: The agenda of the WEF and their graduates and affiliates must be publicized. When the reader encounters people on this list, please remember that they have been trained by the WEF. It is my opinion that the alliance of these graduates is not to the USA, but rather to corporatist globalism.

The Great Reset: Economist Peter Koenig is a geopolitical analyst and a former senior economist at the World Bank and the World Health Organization (WHO), where he has worked for over thirty years on water and environment around the world. He lectures at universities in the US, Europe, and South America. He describes a plan, commonly known as "The Great Reset," which the Rockefellers, the Gateses, the WEF, the Windsors, and the Rothschilds are implementing as the domination of a small minority of corporate globalists over the majority of the population. They are using the cover of anti-COVID measures and an overstated public health crisis to push through these measures.

Koenig claims to have coined the term in October 2020 as shorthand to represent a plan designed as "the antidote to democracy." The Great Reset would be the total corporate takeover of all aspects of life. As defined by Koenig, the Great Reset involves using the global technocratic biosecurity state (otherwise known as the global public health system) to implement these changes. The end results will mean extensive restrictions on the physical environment around people, a forced digitization, and a loss of bodily autonomy (having a say in your own health decisions).

The Great Reset was officially launched in 2020, not by Klaus Schwab or Bill Gates, but by Charles, Prince of Wales, at the time, and now King of Britain and the United Kingdom. Charles's official website announced on June 3, 2020: "Today, through HRH's Sustainable Markets Initiative and the World Economic Forum, The Prince of Wales launched a new global initiative, The Great Reset" [320].

A royal tweet then declared [320]:

#TheGreatReset initiative is designed to ensure businesses and communities 'build back better' by putting sustainable business practices at the heart of their operations as they begin to recover from the coronavirus pandemic.

On face value, "The Great Reset" is also the title used for the 50th annual meeting of the World Economic Forum, which was held during June 2020. The event brought together high-profile business and political leaders. Convened by Charles, Prince of Wales, and the WEF, the focus was on rebuilding society and the economy following the COVID-19 pandemic. The above description is what one finds on your basic search engine, but the motives are less than pure. A less flattering definition of The Great Reset would be Capitalism with Chinese characteristics: A two-tiered economy, with profitable monopolies and the state on top and socialism for the majority below.

With these basic terms in our toolbox, let's return to the central topic. In the coordinated propaganda and censorship response to the COVID-19 public health crisis, globalists and corporatists are directly incorporating Hitler's own principles for crowd control. If we look closer, we can clearly see coordinated actions by the BBC-led Trusted News Initiative, various members of the Scientific-Technological elite, large financial groups (such as Bank of America, Vanguard, BlackRock, and State Street), and the World Economic Forum acting in real time to suppress a growing awareness by the general public of having been actively manipulated. It is increasingly becoming clear that these organizations and aligned nation-states have been using crowd psychology tools to generate significant fear and anxiety of COVID-19 to advance their agendas on a global scale. They have used COVID to drive a planned and coordinated agenda known as The Great Reset.

Multiple governments have now admitted to actively using fear, "Nudge" technology, and "Mass Formation"—related theories as a tool for totalitarian population control during this outbreak. These are basically psychological operations aimed at populations of nations. One glaring example has been operating in the UK:

> Scientists on a committee that encouraged the use of fear to control people's behavior during the Covid pandemic have admitted its work was "unethical" and "totalitarian."
>
> SPI-B warned in March last year that ministers needed to increase "the perceived level of personal threat" from Covid-19 because "a substantial number of people still do not feel sufficiently personally threatened."

Gavin Morgan, a psychologist on the team, said: "Clearly, using fear as a means of control is not ethical. Using fear smacks of totalitarianism. It's not an ethical stance for any modern government. By nature I am an optimistic person, but all this has given me a more pessimistic view of people" [321].

This has been occurring at the same time that the various versions of SARS-CoV-2/Omicron are destroying the legitimacy of government and WHO propaganda concerning the "Safe and Effective" mRNA vaccines and associated mandates.

A Canadian COVIDcrisis case study

A leading hypothesis to explain the obsession with vaccination is that vaccine passports are a strategic portal to achieving a key World Economic Forum objective. That objective being development of a digital identity-based social credit scoring system that will enable management of human behavior by weaponizing banking and access to funds based on behavior and speech. The core hypothesis here is that Western democratic governments (which in normal times have personal privacy constraints) can leverage the emergency to mandate a digital identity and proof of vaccination with associated QR codes and cell phone-based contact tracing in the interests of "public health." This despite the proven fact that the mRNA SARS-CoV-2 vaccines not only do not stop viral infection, replication, and transmission, but may even increase infection and disease risk. The logic apparently is that the Western democracies, which are increasingly acting like infiltrated client states of the World Economic Forum, can both economically and politically benefit by implementing a universal social credit system akin to that which has been pioneered and gradually implemented (think boiling frog) in the People's Republic of China by the Chinese Communist Party. For further on that topic, see [322] and [323].

And now, thanks to WEF young leader program-trained Justin Trudeau and his WEF-trained Finance Director Chrystia Freeland (a member of the WEF board of trustees), we have a peek under the covers, a foreshadowing of sorts, about the potential blowback issues with the whole "manipulate people to do what you want using social credit scores and weaponizing access to banking" strategy. It turns out that WEF darlings Justin and Chrystia were supposed to be the tip of the spear for piloting the WEF digital ID system

[324]. So, faced with the threat of peaceful truckers occupying Ottawa and making like Canadian Geese with their horns, they decided to just go all the way and weaponize the banking system to meet the "enormous threat" to Canadian national security posed by the truckers [325]. Well, that did not turn out quite as planned [326]. Apparently, when those holding both large and small bank accounts realized that TD Bank and other large Canadian banks were no longer a safe harbor, they decided to withdraw their funds.

Canadian Finance Director Chrystia Freeland, although proudly listed by the WEF as an exemplar young leaders program graduate [327], was not trained as an economist, but rather her background is in Russian history and literature and in fact was previously posted to Russia as a journalist for the *Financial Times* [328].

Back on point, it seems that both working class as well as high net worth people really do not like the threat of having their money stolen or frozen by government bureaucrats on a political crusade. So, what to do when what you previously thought was a safe harbor turns out to be plagued by arbitrary and capricious currents? Leave, and withdraw your money from the bank. When the geniuses at the WEF meet to game out the "lessons learned" from Justin and Chrystia's short sharp shock, what are they going to conclude? Will they conclude that *"The fault, dear Brutus, is not in our stars, But in ourselves"*? That is highly doubtful. More typical will be that they will decide that the fault lies in having moved too fast, and that the Overlords should have told their apparatchiks to stick to the incrementalistic "boiling frog" strategy.

The fallback of the WEF globalists running totalitarian Canada apparently will be to "build back better" in part by seeking to develop a cashless society. One that involves 100% centralized digitized currency tied to a digital ID. Henceforth, a social credit system that is bulletproof from the power of the people and their money. If the WEF has their way, it is only time before this concept envelops the world.

We now face a serious challenge to take back control of our sovereignty, both in the USA and abroad. This is not going to be an easy battle, but it is one that we all have a part to play in. The first item of business is to expose the WEF for what it is, along with their graduates, acolytes, and camp followers. Then make sure that these people do not hold office or leadership positions until they clarify where their loyalty is: their country or their global alliances. We must not concede our fundamental rights of privacy and speech, guaranteed by the First Amendment. That means no digital passports for vaccination or otherwise. No government tracking of our movements, tied to said passports. No social credit scores allowed.

Now, was there any foreshadowing that could have alerted them to this little problem? Well, to be blunt, yes. Have you noticed how bicoastal real estate (and farmland) prices seem really inflated in the last decade? What's going on there? A case can be made that what is happening is that foreign investors, particularly private money with origins in the People's Republic of China (PRC), is looking for shelter in real estate markets [329]. See, if you are a Chinese billionaire (and there are lots of those), you have a problem. At any moment the PRC can decide that your social credit score (or that of your company) is in default, and the Chinese Communist Party can seize your funds. What's a rich Chinese billionaire to do? The solution is simple: seek shelter offshore, and in fact, much of that money has been parked in Canada and the USA. Even if you take a 50% loss due to an "adjustment" in valuation of inflated offshore real estate, you are still ahead of having all of your money stolen by some arbitrary bureaucrat on a mission to impress supervisors. So, when your portfolio manager discovers that Canada is now going full CCP by weaponizing their banking system for political purposes? Well, what would you do? I think that *Get the Heck out of Dodge* is pretty much the universal answer. Resulting in a cascade of economic badness that threatened to completely crash the new Totalitarian state of Canada. That one did not even require a Harvard exchange program degree in Russian history and a background in journalism to figure out.

The silver lining in this whole mess is that the WEF young leaders program seems more interested in turning out compliant bureaucrat functionaries than razor-sharp, highly trained intellects. And Justin Trudeau and Chrystia Freeland certainly fit the mold. Let's give thanks that they have alerted us to where Klaus Schwab, the King of England, and their WEF minions want to quietly take us. Now, if you care about your freedom and the autonomy of your nation, it is time to act. Or forever hold your peace. There is no medical emergency [330]. It is time for congress to act to shut down the illegal suspension of Constitutional rights [331] using a fraudulent justification [332]. And let those of us who actually do an honest day's work for a living (like physicians, farmers, and truckers) get back to work.

So where do we go from here? The damage done has been profound. Lives unnecessarily lost, businesses decimated, careers destroyed, and I do not even know where to begin with our entire healthcare system and associated federal health and human services bureaucracy.

Long ago (before I was interviewed by Joe Rogan) when I still had Twitter and LinkedIn accounts, before being deplatformed for posting a very accurate video [333] documenting the Pfizer clinical research malfeasance, I warned of the potential crisis of credibility that public health would face if data revealed that Ivermectin was safe and had some effectiveness against COVID, that the "lab leak hypothesis" had merit, and that the genetic vaccines were not completely safe and effective. Instead of coming clean, the propagandists doubled down. And now we have not only profoundly damaged the vaccine mission and public health in general, but we have destroyed public trust in physicians, academic medicine, and the entire hospital system. That damage is going to take decades to rebuild, and I hardly know where to start or how to think through what steps will be required.

I have a colleague who happens to be a lawyer and a mother. I think her last text to me provides an excellent summary of my position. There must be accountability.

> Well, we are not going to let them "just move on." They will pay for what they did. We will legally challenge everything they did to prevent it from happening again, and then we will reform education so that our children know how to recognize the signs of tyranny in the future.

Roger that.

Censorship: the true name for cancel culture

> Whoever controls the narrative controls the world. Humans are storytelling creatures, so whoever can control the stories the humans are telling themselves about what's going on in the world has a great deal of control over the humans. Our mental chatter tends to dominate such a large percentage of our existence that if it can be controlled the controller can exert a tremendous amount

of influence over the way we think, act, and vote. The powerful understand this, while the general public mostly does not . . .

—Caitlin Johnstone

Cancel Culture - definition: "Cancel culture or call-out culture is a modern form of ostracism in which someone is thrust out of social or professional circles—whether it be online, on social media, or in person. Those subject to this ostracism are said to have been 'cancelled'"

Censorship - definition: "Censorship is the suppression of speech, public communication, or other information. This may be done on the basis that such material is considered objectionable, harmful, sensitive, or 'inconvenient.' Censorship can be conducted by governments, private institutions, and other controlling bodies. Governments and private organizations may engage in censorship."

Calling acts of censorship "cancel culture" or "cancelled" only makes the act of deplatforming someone and taking away free speech rights more palatable. It is laughable to equate taking away our fundamental rights under the Constitution with being "thrust out of a social or professional circle." When the term "cancel culture" is used—those who use this language are giving the censors a free pass. Basically, if censorship is done for the "right reason(s)," then it is "okay," right? No. Just no. Stop. Think. Let's make them use the correct words. We have to make people, corporations, and governments accountable for their actions.

Censorship and propaganda as tools for population control are being normalized. We should call it what it is, using real words and avoiding soft euphemisms.

The slippery slope of advertising pharmaceutical products

Promoting Unlicensed Vaccines is Lawbreaking. US Federal law as well as FDA policy is quite clear on this point. Unlicensed healthcare products: No advertising or promotion allowed.

At the time I am writing this, the Western nations are rapidly moving toward deployment and mandates involving yet more inadequately tested, Emergency Use Authorized genetic vaccine products (in other words, not licensed or market authorized). In the case of the US HHS EUA authorized bivalent (Wuhan-1/BA.4/5 product), which will be deployed without any clinical testing (according to Anthony Fauci, because there is insufficient time), it seems an appropriate time to review the law concerning marketing of unlicensed medical products.

This is all about reducing the risk of hospitalization and death from Omicron BA.5, which (based on data from all over the world) appears to be most significant in the highly genetic "vaccinated" population, but which is being rapidly replaced by new Omicron variants that are likely to become dominant before the new bivalent vaccines are deployed to any significant extent. The US HHS public health official concerns that appear to be motivating this extraordinary action include the combination of impending fall/winter (which has been associated with higher rates of infection, morbidity,

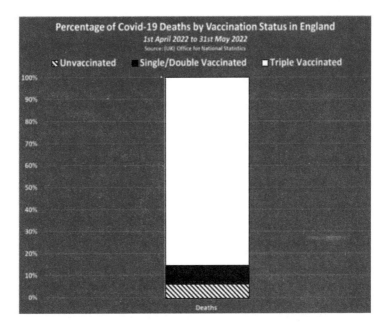

Official figures published by the UK government reveal the fully/triple vaccinated population have accounted for nine of every ten COVID-19 deaths over the last year, 91% of all COVID-19 deaths in England since the beginning of 2022, and 94% of all COVID-19 deaths since the beginning of April 2022. This data set developed from the last official disbursement of data made available to the public, and it appears that the official UK website now only has the full dataset available as a raw Excel file. However, The Free Republic website analyzed those data, and the full analysis is found on their website, https://freerepublic.com/focus/f-bloggers/4089325/posts.

and mortality from both Influenza and the Novel Coronaviruses), as well as the unexplained and unaddressed global inconvenient truth that those who have received three or more mRNA vaccine doses are at highest risk for hospitalization and death. A lower-cost, scientifically and clinically proven method for reducing fall and winter risks for RNA respiratory disease and death would be to facilitate awareness and appropriate uptake of Vitamin D (see Chapter 12), but this seems to be a forbidden topic in US HHS public health messaging.

The FDA and CDC (and Dr. Barney Graham, on interview with the *Washington Post*) appear to have conceded that these vaccines are unlikely to do much for even slowing down infection, replication, and spread of this virus. And in the opinion of many (including myself), these new products may well supercharge the development of even more sophisticated "vaccine escape mutant" viruses.

> We may have gotten about as much advantage out of the vaccine, at this point, as we can get," said Barney Graham, an architect of coronavirus vaccines. "We can tweak it and maybe evolve it to match circulating strains a little better," Graham said. "It will have a very small, incremental effect.

Even Dr. Paul Offit, who seems to have never met a vaccine that he did not think should be administered to children, raised concerns in his interview with *Time* magazine:

> Dr. Paul Offit, a member of the advisory committee, says this strategy makes him "uncomfortable" for several reasons. He notes that the data presented from Pfizer-BioNTech and Moderna in June involving their BA.1 booster shot, which focused on the levels of virus-fighting antibodies the vaccine generated, were underwhelming. "They showed that the neutralizing antibody titers were between 1.5- and two-fold greater against Omicron than levels induced by a booster of the ancestral vaccine," he says. "I'd like to see clear evidence of dramatic increase in neutralizing antibodies, more dramatic than what we saw against BA.1, before launching a new product. We're owed at least that.

And yet, here we are.

AXIOS: FDA authorizes Omicron boosters [334]

Why it matters: The updated shots, retooled to target the BA.5 strain that accounts for most cases in the US today, are expected to become available after Labor Day.

Between the lines: The reformulated mRNA shots got regulators' blessings without first being tested in humans.

- They are also the first to move ahead without an FDA advisory committee weighing in, marking a shift that more closely mirrors the annual flu shot approval process.
- The Biden administration is prioritizing speed over having all the data on how the vaccines work in real life. Some experts warn that this could make some people leery about getting the reformulated shots [335].

According to Dr. Peter Marks, as we now know was previously the case with Dr. Deborah Birx, apparently the modern standard for granting Emergency Use Authorization for medical countermeasure products is "hope," although I am unable to find that as a condition for granting EUA in the current regulatory guidance documents. Perhaps that criterion will be added in upcoming revisions to the congressional authorization language.

Echoing prior similar comments by Dr. Birx, Dr. Marks has clearly stated that he "hopes" the updated "booster" injection will hold and not require "lots of vaccines" each year [336]. The FDA justification for the new bivalent Pfizer/BioNTech and Moderna products does not include clinical trials of this particular vaccine, but instead relies on previous clinical trials conducted for other strains of the virus [337].

Note the FDA language buried at the very bottom of the statement:

The *amendments to the EUAs* were issued to Moderna TX Inc. and Pfizer Inc. These products have been added on by amendment to existing Emergency Use Authorizations by the FDA without any external review or comment. These products are NOT "licensed," they are experimental EUA authorized. If the US Federal COVID medical emergency declarations are lifted, these products can no longer be distributed. They are only authorized while there is a relevant declared medical emergency, and only while there remains no licensed alternative.

With the previously Emergency Use Authorized genetic COVID-19 / SARS-CoV-2 vaccines, a concerted and coordinated marketing campaign was deployed and was funded by the US Government and may have included funding from non-governmental sources.

In the case of the infamous "Big Bird/Sanjay Gupta" CNN marketing campaign, according to Ad Week dot com, organizers responsible for this marketing to children and elderly of an unlicensed medical product included Sesame Workshop, Warner Media, and the Ad Council [338].

What does US federal and Canadian law have to say about marketing of unlicensed healthcare products?

A nice summary below defines the legal landscape [339]:

Internationally, regulations exist to prohibit the advertising or promotion of unlicensed healthcare products [340]. In Canada, Section 9(1) of the *Food and Drugs Act* states that, "*no person shall label, package, treat, process, sell or advertise any drug in a manner that is false, misleading or deceptive or is likely to create an erroneous impression regarding its composition, merit or safety.*" Since the terms and any proposed indication of unlicensed healthcare products have not been established, advertising such products is not permitted [341].

Similar provisions are laid out in the US Code of Federal Regulations (CFR), Title 21, sections 312.7(a) and 812.7(a): the promotion of any investigational drug or medical device (including a new use under investigation for an existing device) is expressly prohibited [342, 343].

To continue with the example, *AdWeek* has this to say about the "Campaign," which was clearly and explicitly an advertising campaign prepared and distributed by the advertising industry group "The Ad Council" [344]:

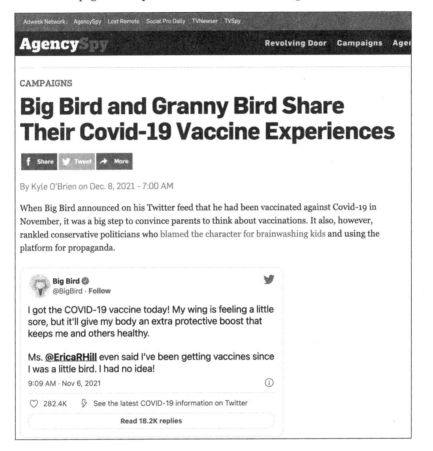

Well, an 8'2" bird doesn't back down after a few negative comments and the large yellow beaked one even brought Granny Bird along to talk about their vaccine experiences and assure parents and children that getting the vaccine is the right thing to do.

In a new campaign by Sesame Workshop, WarnerMedia and the Ad Council, Big Bird and Granny Bird talk about how getting the shot may have made his wing hurt a little, but it was the right thing to do to keep everyone safe and healthy.

It's part of the Ad Council's massive Covid-19 Vaccine Education Initiative.

All indications are that we going to see more lawless marketing of these "new" EUA-authorized and unlicensed medical products in the near future.

Why the promotion or advertising of unlicensed healthcare products is prohibited

The primary concern about the promotion or advertising of unlicensed healthcare products or off-label uses is that a healthcare provider may form an opinion about a product's use on the basis of the claims made by its company before it receives regulatory approval, and that opinion may be incorrect relative to the pending regulatory approval. Such an erroneous opinion on the part of the healthcare provider could lead to incorrect use of the licensed product, thus using the product off-label.

When disseminating information on unlicensed healthcare products may be deemed acceptable

Although companies that develop healthcare products are not allowed to promote unlicensed products or off-label uses, disseminating information about an investigational product may be acceptable under certain circumstances, as outlined below.

Scientific information: Medical conferences and continuing medical education

The US 21 CFR section 213.7(a) recognizes that the prohibition of promoting investigational healthcare products is not intended to restrict the full exchange of scientific information concerning such a product, including presenting scientific findings in scientific or lay media. Additionally, it is recognized by the EU Advertising Directives that without industry sponsorship of scientific meetings and attendance by doctors at such meetings, the medical community would be less well informed.

Companies that develop healthcare products commonly sponsor medical conferences, continuing medical education (CME), or events for the exchange of scientific information (for example, a poster presentation of a disease state at a medical conference). All sponsorships should be developed in line with the following:

> Distinguish the critical difference between the provision of information, and promotional material (advertising). Evaluate whether the material is informational or promotional. Do not distribute information if it is promotional in nature.

Avoid "unduly influencing" speakers to disseminate off-label information.

Clearly label the marketing status of the product so that you do not mislead your audience. (For example, you may encounter a situation where a product is approved in some jurisdictions but not in the country of a conference, or that in that particular country it has a different approved indication of use.)

To ensure compliance, companies may establish internal programs, procedures, and policies that follow industry guidelines regarding CME events and the exchange of scientific information. For instance, the Accreditation Council for Continuing Medical Education (ACCME) has strict accreditation requirements to which CME providers must adhere when holding an event and providing related educational materials [345, 346].

From a regulatory/legal standpoint, these EUA products remain under "clinical investigation." What are the rules for that?

Clinical investigation
When a new drug or medical device, or a new use of a licensed product, is under investigation, any claims of safety and effectiveness about such healthcare products are prohibited unless the company is seeking to recruit clinical investigators or enroll patients in a study [345]. Permissible activities may include:

"Institutional ads" in which a company states that it is conducting research in a certain therapeutic area to develop a new product, but does not mention the proprietary or established name of the product
"Coming soon" advertisements, which state the name of the product, but make no representation about the new product's safety, efficacy or intended use

Dr. Peter Marks, director of the formerly respected FDA Center for Biologics Evaluation and Research (CBER), "*hopes*" that these modified bivalent vaccines will work, but my understanding of the literature concerning immune imprinting/original antigenic sin in the context of the current

SARS-CoV-2 genetic vaccines strongly suggests that the approach being employed with these products will only exacerbate the problems of immune imprinting. Based on their comments (or lack thereof) concerning these new, inadequately tested genetic "vaccine" products, Dr. Marks and his colleagues appear to be willfully ignorant of this literature and the associated implications.

Just to be completely clear, this peer-reviewed literature, published in many of the highest profile scientific journals, is easily found with simple PubMed search terms. I have previously reviewed many of these studies in two articles and have testified about the risks and issues in my testimony to the Texas State Senate on June 26, 2022 [347–349]. I mention these things just to make that point that this information has been readily available for FDA, CDC, and anyone else to review for quite some time.

In my opinion, the FDA and CDC have continued to bypass well-established regulatory norms and have permitted outright lawless activities (including prohibited marketing of unlicensed medical products) in their rush to deploy genetic vaccine products for a disease that is readily managed using a wide variety of early clinical treatment protocols. "Basically, we're playing in a sandbox of unknown benefits and unknown risks, because they don't have clinical trials," Dr. Vinay Prasad, an epidemiologist at the University of California, San Francisco, has stated. "The United States should have required randomized, controlled trials before clearing the boosters."

I hope my predictions about these rushed and largely untested "new" bivalent vaccine products are wrong, and they do not make the immune imprinting/original antigenic sin problems worse. However, I fear that this is what the current data appear to indicate is likely to happen. But at a minimum, the recent political spin angle that this "vaccine" mess is all the fault of the Trump administration based on operation warp speed bypassing normal vaccine development procedures can no longer carry water. With deployment of these bivalent vaccines based on "hope" rather than clinical data demonstrating safety and effectiveness, the current executive branch administration is now guilty of precisely the sin of which they accuse the prior Republican administration.

I "hope" that we will not see yet more lawless marketing of unlicensed medical products. I suspect that my "hope" will be as worthless as that of

Dr. Birx and Dr. Marks has been. Unfortunately, the new FDA marketing campaign for the reformulated bivalent booster COVID-19 vaccines demonstrates that we are to be subjected to yet another marketing campaign for an untested experimental product. In this case, we are being enticed to treat our bodies and immune systems like a computer software product. [350].

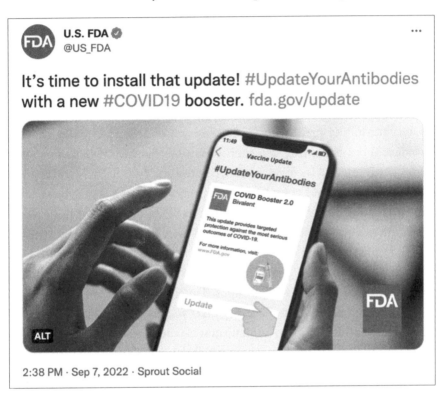

Advocacy (or Solutions) Journalism

Over the last two years, I have come to realize that "journalism" and "journalists" seem to have changed in some fundamental way. I used to believe that there were standards and bedrock ethics that all journalists working for major publications ascribed to. I guess I had thought that the stereotype of the intrepid journalist toiling away in a brave and unending quest for truth was the norm. That what was exemplified in the classic movie *All the President's Men* was how the journalistic process worked. But no longer. My eyes are now opened that this is not how modern journalism works. Frankly, I feel so naive for believing that. What I have personally experienced, again and again, is something very different. What my colleagues have experienced is very different.

It turns out that journalism has morphed into something new. That the teaching of journalism as a nonbiased exercise is no longer in vogue. Instead, what is often being taught is called "advocacy" journalism. To some extent, this type of journalism has always existed, but now it has been normalized and codified into the norm in most university-level journalism schools.

Allow me three general examples to illustrate the point:

First example. Many years ago, when I was working for the "Aeras Global Tuberculosis Vaccine Foundation," which was one of the early Bill and Melinda Gates Foundation nonprofit vaccine companies, the CEO hired a media consulting firm that mainly consisted of a Pulitzer Prize-winning journalist and a marketing manager. To ensure that favorable stories about the organization and its mission were printed, the "journalist" and the marketing specialist would consult with their clients (in this case "Aeras") and learn what story the organization wanted to be told in a major print publication. An article pushing the story would then be crafted, all of the necessary background assembled to meet whatever editorial review standards were likely to be encountered. Then, this prebaked work product would be fed to some "journalist" working for the targeted publication. Free work product, no labor required, what's not to like? My first "you are not in Kansas anymore" moment concerning modern journalism was when I saw this process used to "place" an article into the *Economist*, which I had naively believed operated as an independent arbiter of truth. Silly me. Even then, I thought—well, this can't be the norm, can it?

The **second example** comes from having repeatedly been on the receiving end of "gotcha" journalism as it is currently practiced. "Journalists," particularly many of the younger ones, seem to use a variety of ploys to draw out information that they can weaponize in some manner to support a predetermined storyline that they wish to promote. Often it is a sort of confidence game, like a con artist might employ, where they flatter or use phrases like "I just want to help you to get your story out" to get the subject to let down their guard and agree to an interview. After establishing a relationship with the subject, they then draw out details using increasingly aggressive questions focused on supporting the true agenda. These personal details are often woven into a story line designed to delegitimize someone or otherwise reveal some salacious character flaw. Then the article drops, and the naive subject suddenly finds that they have been duped into revealing personal details that have been weaponized to support a predetermined narrative. Having experienced this myself a few times, I now often see this strategy (and various versions of this con) repeatedly play out with colleagues. As for me, lesson learned is to vet the "journalist" by reading prior work, and just say no when it becomes clear that they are a specialist in this type of strategy. But in the end, a journalist is there to tell a story, and it is most likely not the one that the subject might have wished to have been told and there is a good chance it is a poor representation of the truth. That "story" is the one that the journalist was probably was paid to weave, and truth is not a necessary by-product.

The **third example** comes from listening to disenchanted "old school" journalists (print and broadcast). These voices seem to be a mixture of midcareer and older practitioners, from "print" (an increasingly outdated term these days) and broadcast media. Again and again, I hear various versions of the famous rant from the Oscar-winning 1976 movie *Network*, where Peter Finch, playing the part of an anchorman named Howard Beale, says, "I'm as mad as hell, and I'm not going to take this anymore!" But the words I repeatedly hear from these modern versions of Howard Beale are more nuanced and revolve around being unwilling to comply with corporate demands to mislead the public in various ways. These journalists all tell stories of widespread soul-destroying corporate media censorship and propaganda that they just cannot tolerate anymore. The censorship and propaganda are found from small local outlets all the way to the top stars of major networks. Their stories mimic versions of my own story: they just could no longer tolerate the ethical erosion of their chosen profession. So, they take an income hit and go independent. Some succeed, others not so much. For some, they seem to never be able to completely leave their old reality behind. "You can take the journalist out of the *New York Times*, but you can't take the *New York Times* out of the journalist" is one saying describing the latter.

What has changed? Is the present reality any different from what has always existed, going back to the "yellow journalism" days of William Randolph Hearst that emphasized sensationalism over facts? Trying to make sense out of the world, I began asking the "old school" journalists whom I encountered what they think about all of this. And what I have discovered is that there is yet another insidious form of attack on our educational system, driven by the corporate interests of large "nongovernmental organizations" (including the Bill and Melinda Gates Foundation) that have used grants to journalism schools as a way to drive changes in how their graduates have been trained. To be blunt, this is yet another story of the gradual erosion of integrity due to the pernicious influence of massive accumulations of wealth by a few who weaponize that wealth to advance both their own power and various social agendas.

Under the influence of large "grants" (I think they would more appropriately be called "strategic investments"), many journalism schools have taken to teaching "advocacy journalism." Which is basically a fancy term for propaganda. These news media hire journalists specifically to report with skewed biases on topics of interest to these corporations (or governments), often with corporate sponsorship. Let that sink in for a moment. The advocacy journalists are often paid by an outside organization with an agenda. For instance, we now know that the CDC spent a billion dollars in 2021 to place stories that promote vaccines in a positive light and to fight vaccine-related "misinformation." So, let's figure this out what exactly is going on, starting by defining terms. The definition of advocacy journalism from *Merriam-Webster* is:

"Journalism that advocates a cause or expresses a viewpoint."

To me, that sounds like how one might define propaganda. So, am I wrong? The definition of "propaganda" (*Merriam-Webster*) is:

"the spreading of ideas, information, or rumor for the purpose of helping or injuring an institution, a cause, or a person."
"ideas, facts, or allegations spread deliberately to further one's cause or to damage an opposing cause."

The definitions for "advocacy journalism" and propaganda are essentially the same. A whole lot of doublespeak. It is truly a Brave New World.

Now, why is this important? Because increasingly journalism is taught by those who believe that "classic journalism"—which required that both sides of an issue be presented (you know—"fair and balanced")—is outdated and deserves to die a quiet death.

This is exemplified by a Wiki definition of advocacy journalism that is frankly astounding:

> *Classic tenets of journalism call for objectivity and neutrality. These are antiquated principles no longer universally observed....* We must absolutely not feel bound by them. If we are ever to create meaningful change, advocacy journalism will be the single most crucial element to enable the necessary organizing. It is therefore very important that we learn how to be successful advocacy journalists. For many, this will require a different way of identifying and pursuing goals.

So, who teaches "advocacy journalism," and who funds such teachings?

Well, for starters—let's go to one of the premier journalism schools in the USA—Columbia University. How do they view "advocacy journalism"? At Columbia University, one of their programs proudly announces the following:

> **"CALLING FOR COALITIONS: BUILDING PARTNERSHIPS BETWEEN JOURNALISTS AND ADVOCATES"**
> Journalism is being hit hard globally, and some even predict the end of independent journalism in the global south, especially in Africa. It's time to look at what may survive. Philanthropic funding will become more essential, and donors will be eager to expand partnerships between journalism and advocacy groups. Through this project, the Bill & Melinda Gates Foundation's Media Partnerships team explored the dynamics of such collaborations. Drawing from multiple case studies, the project provided recommendations for foundations, nonprofits and media organizations that maximize impact, respecting a shared covenant" [351].

Their partner in developing advocacy journalism training programs, with the expansion of such funding is... the Bill & Melinda Gates Foundation. This all seems truly unbelievable, except that this is reality.

But now there is another new "style" of journalism that has become quite the fad. This subset of advocacy journalism is called "solutions journalism," and it is the term that the Bill & Melinda Gates Foundation likes to use for their funding mechanisms to influence governments, citizens, and leaders. They have funneled hundreds of millions of dollars to media to skew the news to promote their causes and points of view [352]. Of course, advocacy journalism is basically a "nicer, kinder" form of propaganda... Right? You know, like when people call censorship—"cancel culture." Because cancel culture sounds so much "nicer" than censorship. After all, what Twitter, LinkedIn, and You Tube are doing by banning people and content is for all "our" benefit, right?

Speaking plainly, what these modern media companies are doing is really a form of book burning. For an example of how that might work out, read Ray Bradbury's masterpiece *Fahrenheit 451*. Large donors or sponsors are giving money to media corporations to bias reporting via "solutions journalism." As we know now, many governments, including our own, are also influencing what is allowed to be discussed and in what ways. The sponsors can be nongovernmental organizations, governments, or global nongovernmental organizations such as the Zuckerberg-Chan initiative, Bill and Melinda Gates Foundation, United Nations, World Health Organization, or World Economic Forum. These groups seek out "private-public" partnerships (which, as previously noted, is basically another euphemism for what Benito Mussolini defined as fascism). The organizations codify these relationships by hiring staff that use advocacy or solutions journalism, that is, propaganda to sway public opinion. Sometimes they even fund specific investigations. When this happens, who is compromised? Clearly, truth and integrity are immediate casualties. All for the greater good of the greatest number of people, of course... As defined by the organization giving the money. The article titled "Conflict of Interest? Bill Gates Gave $319 Million to Major Media Outlets, Documents Reveal" asserts:

> According to MintPress News, the Bill & Melinda Gates Foundation donated at least $319 million to fund media projects at hundreds of organizations including CNN, NBC, NPR, PBS and The Atlantic, raising questions about those news outlets' ability to report objectively on Gates and his work [352].

It is important to realize that biases and opinions have always been a part of journalism. Generally, we previously called these pieces "editorials." When editorials are grouped together, they formerly were presented as the "opinion page." An antiquated term, I know. Of course, we all know that some newspapers are "liberal" and some "conservative." Of course, biases do creep into reporting, and in fact, every newspaper's reputation is built on those biases.

But this is different. This is allowing a nonprofit governmental organization (at best), a corporation or government (at worst) to control the content of a newspaper or magazine through secret handshakes, grants, and contracts. It is allowing psyops operations a front-row seat into influencing the minds of readers. This is a whole other ballgame, and it needs to stop or at the very least, be called out and recognized for what it is: corporate and state-sponsored propaganda.

Advocacy journalists can and are influenced by governmental policies. For instance, the *NY Times* recently described a relatively new hire as "joining the *New York Times* as a technology reporter covering disinformation and all of its tentacles." The pejorative use of the word "tentacles" pretty much shows what biases this reporter is expected to have. The implication being that information not disseminated by the US government is disinformation, whether the topic be on climate change, diversity, elections, physician's right to try, or infectious disease. By the way: anyone else notice how the disinformation list keeps growing longer?

How does the Trusted News Initiative (TNI) or global information control fit into the campaign against "disinformation" [353]? The TNI is basically a treaty management organization managed by the British Broadcasting Corporation, and the corporate media consortium bound together by the TNI agreements uses advocacy journalism to control content of news media throughout the "free" world. This treaty among the largest media organizations in the world initially evolved from efforts to quell election misinformation by foreign agents. It has now become the driving force to stop anything different from the national public health policies and (for some countries) the World Health Organization. This means that only that "news" or PR spin that a government or world body wishes to be advanced can be allowed to be published or electronically distributed in mainstream media (which many now label as "state sponsored media"). Advocacy journalism that promotes a certain viewpoint fits right in with the TNI model.

Some of the known signers of the treaty include AFP, BBC, CBC/Radio-Canada, European Broadcasting Union (EBU), Facebook, *Financial Times*, First Draft, Google/YouTube, *The Hindu*, Microsoft, Reuters, Reuters Institute for the Study of Journalism, Twitter, and the *Wall Street Journal*.

The long, strange evolution of the Trusted News Initiative, from election interference to COVID-19 total information management shows the extent to which power corrupts, and that those being corrupted often have no idea that they are being corrupted. "Journalists" are trained or co-opted into buying into the idea that there is "one truth," one right answer, and that governments are honest brokers in the assessment of that truth. Such journalists are not "fair and balanced." These journalists are naive and dangerous. Governments do lie, regulatory capture is real, and what they offer as truth is often better termed mis- dis- and malinformation. Which is precisely why advocacy journalism (ergo, propaganda) is dangerous. In a democracy, if an electorate is to be able to make appropriately informed choices, the news must be free from government (and corporate interest group) interference, reported from all angles, from all points of view—not just one narrow reading of the "truth," as presented by big brother.

The problem is that this truly is a slippery slope. How does a newspaper or content provider determine what propaganda is "good" or "bad"? How is disinformation determined? Does the government get to decide? Does the "Trusted News Initiative" leadership decide for the entire Western world? What about the Bill and Melinda Gates Foundation: are they to be the arbiters of truth?

In an article titled "Donor-Driven Journalism," *National Affairs* magazine recently delved into one aspect of what has really been driving the wave of "advocacy journalism" observed in modern corporate media [354]. The article cites numerous examples of how political organizations are using advocacy journalism to skew election results. They cite the recent instance of Courier Newsroom, "which was founded in 2019 to "restore trust in media by building local reporting infrastructure in states across the U.S." It turned out to be a "clandestine political operation," in the words of Washington-based reporter Gabby Deutch. "Paid for with an initial investment of $25 million, a group called Acronym managed to fund what appears to be a local news website but in fact offers information to promote Democratic candidates for office."

The authors of this National Affairs article state that "Many large media institutions take money directly from foundations to fund particular areas of coverage. Others publish works from organizations like ProPublica, which are entirely funded by philanthropic dollars. As the news business has hemorrhaged subscribers and advertisers in recent years, many newspapers, magazines, and websites have looked to such organizations for financial support. Some of that support is helping struggling publications survive in areas where there might otherwise be a 'news desert.' But much of it is linked to a particular set of ideological goals.

"Other donors rely on the unspoken expectations that are inevitably attached to large donations. When the Ford Foundation makes large grants to the *New York Times* to fund a disability-journalism fellowship, for example, the message may not be explicit, but it is nevertheless clear: The paper is expected to make the case that government should provide more accommodations to people with disabilities. And in fact, the reporter funded by that grant published articles describing the efforts of disability advocates to keep Covid-19 restrictions in place to protect vulnerable populations."

Most troubling were their observations tying together modern trends in journalism with parallel changes that have swept through academia, and that may help explain why so many academic physicians have failed to meet both public and patient expectations during the COVIDcrisis. Under the heading "Setting Elite Priorities," they drop a series of bombshell observations: "In a sense, this is precisely what happened at universities. Initially, they expanded in new directions—often into obscure fields of interest to some faculty. Then they began expanding into areas that generate grants and donations, but they did so at the expense of the university's original mission. Higher education institutions care less today than in the past about basic education for undergraduates, just as newspapers today have less interest in reporting basic news to their readers. Large foundations like Ford and MacArthur have been able to steer American higher education by putting money into new academic centers and curricula, conferences, awards, and publications. The university of today bears but a faint resemblance to the institutions of a generation or two ago. A parallel process is well underway in the news business, transforming newspapers into vehicles for promoting causes of interest to large foundations but not to ordinary readers. In the process, the news business, like higher education, is turning itself into another source of ideological conflict in American life—as if that's what the country really needs today."

Should a newspaper, magazine, or broadcaster try to make a determination about the impact on objectivity before accepting funds or making commitments to the government? Once upon a time (note the fairy-tale prelude), most "established" legacy media tried to maintain a firewall between their "news" and "op-ed" operations. How antiquated that now seems. Does the organization being paid to present one point of view have an obligation to be transparent? To disclose conflicts of interest? Do they need to provide the public with the contract, the information on how they are being paid to bias the news they are reporting? For example, when a news organization takes money to hire journalists with CDC or Gates Foundation monies? What happens when the information control comes in the form of stopping certain types of mis- dis- or malinformation that the government doesn't want reporters to write on? Or threatens to label those who communicate such as domestic terrorists [355]? What happens when the sponsor wants the advocacy journalism to include marketing campaigns that basically target individuals viewed as opposition? Does the newspaper have an obligation to inform the public that they are being nudged? The ethical morass that this type of journalism creates is huge. All we can hope is that institutions teaching journalism begin to recognize the dangers of promoting advocacy or solutions journalism and return back to the classic tenets of journalism, those being objectivity and neutrality. And restore integrity to the discipline.

The Omicron Variant was a disruptive event that expanded the Overton window

The Overton window of political possibility is the concept that there are a limited range of ideas the public is willing to consider and accept. That politicians can only be effective by advocating policy that fits within the Overton window and that "disruptive" groups are the primary ones who can expand or contract the Overton window. In the case of COVID-19, the use of the noble lie has been used by government and world health officials as well as legacy media sycophants to restrict and even close the Overton window regarding issues such as mask use, vaccination strategies, and lockdowns. To be explicitly clear, the Overton window is a political concept that is now being actively manipulated to constrain scientific discourse.

Government officials are intentionally acting as those disruptive groups that limit the window by co-opting media outlets, social media, and big tech through NGOs, The Trusted News Initiative, and three-letter agencies. Through NIH media campaigns, Dr. Anthony Fauci, Dr. Francis Collins, and many others have acted to limit or close the Overton window. They have also effectively used smear campaigns both against groups and individuals to limit the Overton window and keep their policy positions from being questioned by the public and to expand governmental authority [294]. There is now a deep understanding of the political science behind the Overton window, and the idea is being used to market ideas to the public in ways never thought of before by governmental organizations and agencies.

The most common misconception is that lawmakers themselves are in the business of shifting the Overton window. That is absolutely false. Lawmakers are actually in the business of detecting where the window is, and then moving to be in accordance with it.
—Joseph Lehman

So, according to Joseph Lehman, president of the Mackinac Center for Public Policy, politicians can only detect the Overton window but are not typically effective in directly influencing it. An exception to this rule may be a populist, charismatic, and radicalized leader, such as President Trump. In fact, Trump's expansion of the Overton window was key to his success. Both Trump and his original campaign manager, Steve Bannon, are masterful marketers and media manipulators. Their use of the Overton window as a tool in social media marketing campaigns was most likely intentional.

As a general rule, it is believed that think tanks and social movements are the political actors who shift the Overton window one way or another. That it is these groups who act to convince voters and the public that policies outside the window should be included, or to limit what should be allowed to fit within the "approved" Overton window. One can argue that NIH has become one such advocacy group working to benefit the US government pandemic response policies (and the bureaucrat/politician careers of key HHS leaders)—by both limiting and expanding the Overton window as needed. The use of propaganda, by way of contracting news organizations and influencers, as already discussed, has fundamentally changed the role of governments and the news. The government is now actively trying to influence the Overton window.

The Great Barrington Declaration is an example of an "outsider" group of scientists that has managed to expand the COVID-19 Overton window by use of a successful social media campaign. This declaration, now with almost one million signatures, reiterates sound, scientific principles regarding pandemic response policies that basically state the following:

> The most compassionate approach that balances the risks and benefits of reaching herd immunity is to allow those who are at minimal risk of death to live their lives normally to build up immunity to the virus through natural infection, while better protecting those who are at highest risk. We call this Focused Protection.
>
> —The Great Barrington Declaration [356].

On the other side of this issue, The US Government Health and Human Services (HHS) through NIH and legacy media, via the Trusted News Initiative [357], have acted as the radical influencers to close the Overton window regarding universal vaccination, mandates, and forced vaccination of children.

US government officials have sought to limit activist scientists and physicians by libeling and setting up mainstream media in opposition[358]. The email below, which was obtained by FOIA, lists Francis Collins, Anthony Fauci, Cliff Lane, and Lawrence Tabak as involved in administering the "takedown" and character assassination of the primary authors. These senior USG HHS officials are important names to remember, as they have been

---

From: Collins, Francis (NIH/OD) [E]   (b) (6)
Sent: Thursday, October 8, 2020 2:31 PM .
To: Fauci, Anthony (NIH/NIAID) [E]   (b) (6); Lane, Cliff (NIH/NIAID) [E]
       (b) (6)
Cc: Tabak, Lawrence (NIH/OD) [E]   (b) (6)
Subject: Great Barrington Declaration

Hi Tony and Cliff,

See https://gbdeclaration.org/ This proposal from the three fringe epidemiologists who met with the Secretary seems to be getting a lot of attention – and even a co-signature from Nobel Prize winner Mike Leavitt at Stanford. There needs to be a quick and devastating published take down of its premises. I don't see anything like that on line yet – is it underway?

Francis

leading this and prior pandemic responses. This type of behavior from governmental officials cannot be tolerated in a free, open, and democratic society.

These researchers weren't fringe, and neither was their opposition to quarantining society. But in the panic over the virus, Dr. Fauci and Dr. Collins—two voices of officialdom—used their authority to stigmatize dissenters and crush debate. A week after his email, Dr. Collins spoke to the *Washington Post* about the Great Barrington Declaration [359]. "This is a fringe component of epidemiology," he said. "This is not mainstream science. It's dangerous." His message spread, and the alternative strategy was dismissed.

Dr. Fauci replied to Dr. Collins that the takedown was underway. An article in *Wired*, a tech-news site, denied there was any scientific divide and argued lockdowns were a straw man—they weren't coming back [359]. If only it were true. The next month case counts increased and restrictions returned. Dr. Fauci also emailed an article from the *Nation*, a left-wing magazine, and his staff sent him several more.

The emails suggest a feedback loop with Dr. Fauci and the media. The media cited Dr. Fauci as an unquestionable authority, and Dr. Fauci got his talking points from the media. Facebook censored mentions of the Great Barrington Declaration during this period. When the emails were discovered via FOIA, almost a year later, the *Wall Street Journal* wrote about the email content: "This is how groupthink works" [360].

Dr. Fauci has been particularly egregious in his attempts to limit the Overton window by his repeated use of the "Noble Lie." A "Noble Lie" is knowingly propagated by politicians, often to advance an agenda. A Noble Lie is intrinsically paternalistic. The deceiver assumes that lying serves the best interest of society. Fauci's misinformation and lies have been targeted at influencing behavior of public and government officials, including politicians. He has effectively used the Noble Lie to expand and limit the Overton window and in effect, control the narrative that politicians can act on and stay within the safety limits of their constituency.

The Noble Lie

In Plato's *Republic*, the Noble Lie is supposed to make citizens care more for their city-state. The Noble Lie is supposed to engender in them devotion for their city (country) and instill in them the belief that they should "invest their best energies into promoting what they judge to be the city's best interests." Writing as Socrates, Plato's introduction of the *Republic*'s notorious "Noble Lie" comes near the end of Book 3 (414b-c). "We want one single, grand lie," he says, "which will be believed by everybody—including the rulers, ideally, but failing that the rest of the city." In essence, the Noble Lie is a falsehood that, if people can be made to believe it, will cause the citizens to be strongly motivated to care for the city and for one another. This is essentially Plato's justification for state-based propaganda.

One of Fauci's first of many "Noble Lies" in this pandemic was in regard to mask use. He asserted early in 2020 that masks were not effective against preventing COVID-19. He stated this as fact, rather than admit that the USA had a mask shortage. After which he recommended people wear one or even two masks. At some point, Dr. Fauci admitted that the data were not there—one way or the other—but he was basing these decisions on the availability of masks for healthcare personnel. The CDC backed up this mandate—long after data showed that the effectiveness of most masks was limited to 10% at best. Dr. Fauci then stated the spread of the virus was unlike anything he's seen before. While data at that time showed that the severity of the illness is much like that of the flu virus, except in the elderly and high-risk populations.

Other examples of Fauci's Noble Lies are his public statements regarding gain-of-function research. The change in Fauci's position over gain-of-function research in the last eighteen months went from "it never happened" to "well, it depends on your definition of gain-of-function" to "it would have been negligent not to fund the lab." Of course, as this lie happened to be about protecting himself, it actually doesn't fit into the category of a Noble Lie—but just a lie to cover up his own involvement. Dr. Fauci agreed to fund this research. Of course, he knew about this research, and now we have independent confirmation that the so-called pandemic originated at the Wuhan Institute of Virology from precisely this research that was previously widely characterized as a conspiracy theory [361, 362].

Fauci began the vaccine campaign stating that it would require 60% of the population to be vaccinated to achieve herd immunity. He consistently nudged up the percentage of people required to be vaccinated to achieve what he calls herd immunity [363]. He started at 60%, then went to 70%, 85%, 90%, and finally 100% vaccinated with boosters would be needed to achieve herd immunity. Long after it was clear that the virus would be endemic, he was still advocating for herd immunity to push for vaccines, vaccines for children, and vaccines mandates. At one point, in a moment of honesty, he admitted that he nudged up the ratios needed with the implication being that he wanted people to take the vaccine and was trying to guilt-trip them into the jab [364].

As the concept of the Overton window has become embedded in our political sphere and public discourse, the ability of governments to manipulate the Overton window has become normalized. The use of big tech, social media, and MSM by governments to control the Overton window is now a standard pattern of behavior. The Trusted News Initiative is just one example of this. The use of the social media, MSM, government officials, and think tanks by NIH to smear epidemiologists who do not share the same set of public policy viewpoints is another.

By use of limiting what is acceptable speech on COVID-19, authoritarianism in the form of vaccine mandates, lockdowns, and mask use has come to dominate governmental responses throughout the world. And the USG public health has normalized propaganda as a tool for enforcing their policy decisions on America and the world.

The new variant of COVID-19, Omicron, exploded onto the scene in December 2021. This variant is different from earlier SARS-CoV-2 strains in five essential ways:

- More infectious and will soon be the dominant variant in the USA
- Less pathogenic
- Poorly matched to currently available vaccines
- Natural immunity is providing good protection against Omicron
- Disease symptoms are more similar to the common cold

As some governments are now touting a 3rd, 4th, or even 5th booster as mandatory, various groups, influencers, and even the WHO to some extent, including mainstream media outlets, are beginning to question the whole public policy response in the face of the emerging data about Omicron.

Suddenly, some of the data exposing that these vaccines are not as safe and effective as once thought are showing up on more and more alternative media. Even some mainstream media outlets, such as Real Clear Politics, are promoting articles that are outside of the BBC-led "Trusted News Initiative" narrative.

In the meantime, President Biden has worked hard to limit the what the public believes by promoting the meme that "For the unvaccinated, you're looking at a winter of severe illness and death for yourselves, your families, and the hospitals you may soon overwhelm." Unfortunately for him, he forgot to follow the "Overton Golden Rule"—politicians can only detect the Overton window and cannot act to contract or expand it. So, in this case, public response to his pronouncement about a winter of severe illness and death was met with widespread derision.

In December 2021, the Israeli Ministry of Health overtly tried to limit the Overton window regarding their childhood vaccine mandates by running a smear campaign against me. My sin being that I questioned the use of these vaccines for children and was against the mandates. Unfortunately for them, they also forgot the "Overton Golden Rule," that politicians can only act within the confines of the Overton window and that window has now been expanded due to Omicron. The majority of parents were already against vaccinating children in the USA, and that point of view is expanding, due to the Omicron variant having very mild symptoms.

Of course, for those caught in the mass formation psychosis (hypnosis), Omicron still has not been enough to shake their obsession with the vaccines and mandates. However, for the persuadable third—there is a shifting of perspective. With this shift, this expansion of the Overton window, politicians will be able to expand what is politically acceptable speech and maybe will bring some sanity back into this pandemic response by the US Government and the world.

CHAPTER 25

Mendacious *New York Times* Warning about Censorship

Once upon a time, long, long ago, there were a few daily publications in the USA that at least tried to separate editorial opinion from reporting. There was a "social contract" of sorts between the American electorate and what were believed to be leading members of the Fourth Estate, otherwise known as the press (stemming from the First Amendment), in which—in exchange for some semblance of objectivity and independent "watchdog" service to the public—the press would enjoy broad legal protections. This was codified in the common practice of segregating editorial opinion (which was clearly advocacy) from "hard news" reporting. And when I was younger (or just more naive), this seemed fairly clear. But over time, the *New York Times*, the paper of record in the United States, whose lead so many other newspapers follow, has gradually eroded the separation between news analysis and opinion, to the point where they now routinely inject government propaganda, mis- and disinformation, and political opinion into their daily reporting.

Welcome to the world of "advocacy journalism."

Advocacy journalism includes both propaganda and censorship. And censorship takes many forms. One of the most egregious is a tactic known as "a hit piece," a standard part of the deplatforming playbook. According to the Oxford Advanced Learning Dictionary, a hit piece is "an article that deliberately tries to make somebody/something look bad by presenting information about them that appears to be true and accurate but actually is not." One of the strategies used to create the illusion of accuracy is to have an in-person interview, but then misrepresent the context or content.

A critical analysis of this process, written for the *New Yorker Magazine* in 1989 by acclaimed journalist Janet Malcolm, gets right to the heart of my own experiences with "advocacy journalism."

Ms. Malcolm writes:

> Every journalist who is not too stupid or too full of himself to notice what is going on knows that what he does is morally indefensible. He is a kind of confidence man, preying on people's vanity, ignorance, or loneliness, gaining their trust and betraying them without remorse. Like the credulous widow who wakes up one day to find the charming young man and all her savings gone, so the consenting subject of a piece of nonfiction writing learns—when the article or book appears—his hard lesson. Journalists justify their treachery in various ways according to their temperaments. The more pompous talk about freedom of speech and "the public's right to know"; the least talented talk about Art; the seemliest murmur about earning a living.
>
> The catastrophe suffered by the subject is no simple matter of an unflattering likeness or a misrepresentation of his views; what pains him, what rankles and sometimes drives him to extremes of vengefulness, is the deception that has been practiced on him. On reading the article or book in question, he has to face the fact that the journalist—who seemed so friendly and sympathetic, so keen to understand him fully, so remarkably attuned to his vision of things—never had the slightest intention of collaborating with him on his story but always intended to write a story of his own.

So, has anything changed in the last thirty years? Sadly, things have gotten much worse. What started as a relatively rare occurrence has become common practice.

During the last year, I learned about this firsthand when the *New York Times* published a hit piece clearly aimed at discrediting me. Subsequently, various reporters and podcasters have asked me how the *New York Times* was able to gain my trust and get me to agree to an interview. Was I "like the credulous widow," described above, suffering from "vanity, ignorance, or loneliness"? Was I hoodwinked? Or did I go into this with eyes wide open?

In the interest of helping to "vaccinate" readers with the sad truth of how journalism is practiced at the *New York Times* (and most of corporate media), I offer the following anecdote.

Davey Alba emailed me identifying herself as a *New York Times* reporter. "We are interested in profiling you and I wanted to reach out and see if this is something that you'd be up for. My beat is traditionally misinformation, but *I wanted to understand and potentially correct the record on some other mainstream publications' quick write-ups of what your views about Covid-19 have been in the past year.* I've listened to your podcast with Joe Rogan, the five video interviews with the *Epoch Times*, and your podcast with Bret Weinstein. I've heard you say you aren't anti-vaxx—you have some concerns about how quickly the treatment has been developed and pushed out to millions of people around the world, but that it has helped, especially in the case of older adults. And it seems like as a pioneer of mRNA, you had some concerns about the stability of the technology and had pushed for other treatments like pills and drugs, before going the route of a vaccine. I also understand that it's not like you haven't lived through this before, since you were involved in a company trying to develop drug treatments during the Zika virus epidemic of 2015–2016. *I wanted to make sure that this understanding of your views is correct, and to give you a chance to respond to your detractors.*"

At first, I was reluctant. On the one hand, I had had many personal experiences where reporters said similar things and used similar tactics to seduce me into believing that they would not write "hit pieces." I knew enough to be highly skeptical. On the other hand, Mattias Desmet had spoken to me of the moral imperative and importance of trying to speak to all sides in order to help minimize the risk of the overall society falling even deeper into the mass formation process described in his seminal tome *The Psychology of Totalitarianism* [310]. And I genuinely wanted to engage with both sides of the political spectrum. And I liked the idea of setting the record straight. Finally, I knew a *New York Times* reporter who I believed was trustworthy. Many years before, I had been a "Whistleblower." Upon the advice of my medical ethics mentor at the time, I had gone to the press concerning what I knew about the death of young Jesse Gelsinger at the hands of University of Pennsylvania gene therapy guru James Wilson. I had worked closely with *New York Times* reporter Sheryl Gaye Stolberg. She wrote a summary of the incident titled "The Biotech Death of Jesse

Gelsinger" [365]. Over time, I had come to trust Ms. Stolberg and to respect her reporting. When I received an email from her in support of Ms. Alba, it had some real weight. She wrote: "I would also direct you to Davey's initial email, in which she says she would like *to understand and potentially correct the record on what other publications have written.* I've worked for The Times for 25 years and I know that we demand accuracy and fairness. I understand why you are media-shy. But I would argue that if you are going to pick one and only one outlet to work with, it should be *The Times.* So, that is my two cents!" Ms. Alba followed up with an email stating that the *Times* "is committed to fairness and accuracy, and I assure you I am approaching this story from as objective a standpoint as I can. We do not republish lies in our paper." So that is how *New York Times* reporter Davey Alba gained my confidence and my agreement to the interview. I allowed her to come into my home, at our farm, for two days.

In the end, the *New York Times* article was a pure hit piece filled with inaccuracies and outright lies. Ms. Alba used tactics that are morally indefensible—treachery and deceit—to attack me personally and misrepresent my views, much like the tactics described by Ms. Malcolm more than thirty years earlier. I had not seen this at the time, but the *New York Times* made the following statement when they hired Ms. Alba: "Davey Alba is joining the *New York Times* as a technology reporter covering disinformation and all of its tentacles." This clearly was an advocacy journalism hire. I do have to wonder if and how much the CDC might have paid the *Times* to support Ms. Alba's employment! By way of full disclosure, our attorney has filed a formal complaint for defamation against both the paper and the (former) *New York Times* reporter.

I hope that my story has been helpful for all of you who may be contacted by the legacy media at some point in the future. Do not agree to interviews with journalists who seek to advance an agenda, particularly ones who are specifically hired to do so!

It's stories like mine that led my publisher, Tony Lyons, to write the article that is reproduced below. Tony is a dedicated free-speech advocate and the president and publisher of Skyhorse Publishing, which has published more than 10,000 books in print and 58 *New York Times* bestsellers.

* * *

The Paper of Record Has Clearly Become the Ministry of Truth
by Tony Lyons

The *New York Times*, a bastion of censorship and corruption, warns the world that "America Has a Free Speech Problem."

In a bold but clearly disingenuous statement from its famed Editorial Board, "a group of opinion journalists whose views are informed by expertise, research, debate, and certain longstanding values," the *New York Times* issued a cautionary statement:

> For all the tolerance and enlightenment that modern society claims, Americans are losing hold of a fundamental right as citizens of a free country: the right to speak their minds and voice their opinions in public without fear of being shamed or shunned.

The Editorial Board pounded the point home:

> People should be able to put forward viewpoints, ask questions and make mistakes, and take unpopular but good-faith positions on issues that society is still working through—all without fearing cancellation . . . Freedom of speech requires not just a commitment to openness and tolerance in the abstract. It demands conscientiousness. . . . We believe it isn't enough for Americans to just believe in the rights of others to speak freely; they should also find ways to actively support and protect those rights.

Of course, the *New York Times* should be teaching by example. In fact, it has not supported free speech, protected the First Amendment, or allowed honest debate. It has not allowed competing perspectives about the most important issues of the day. It has been a mouthpiece for greedy corporations and corrupt government officials.

In support of their interests, and at the expense of those of American citizens, it censored *The Real Anthony Fauci* [189] by Robert F. Kennedy Jr. in every conceivable way. It ranked the book #7 on its nonfiction bestseller list even though Kennedy's book outsold any other book in America that week by thousands of copies. Then it refused to allow Skyhorse Publishing to place an advertisement for the book because its censorship division, ironically called "Standards Management," decided that the book itself constituted misinformation, despite their stated policy that "Standards" only looks into whether an ad itself is "non-defamatory and accurate."

The *New York Times* followed up with a scathing hit piece targeting Kennedy as "a leading voice in the campaign to discredit coronavirus vaccines and other measures being advanced by the Biden White House to battle a pandemic that was…killing close to 1,900 people a day." It accused him of circulating "false information," without indicating what that information is or explaining why it's false, and of comparing the government pandemic response to the Holocaust, even though he clearly didn't do that.

Finally, they refused to review *The Real Anthony Fauci* or so much as comment on its historic grassroots success, even though it's become a cult classic, selling over 1,000,000 copies and launching a worldwide movement against government corruption and corporate greed.

> "Despite all the lying, or maybe in reaction to it," Tucker Carlson wrote, "Robert F. Kennedy Jr. is becoming a legitimate folk hero."

He is a folk hero because he stood up, grabbed a bullhorn, and spoke truth to power. He's risked everything and lost a lot. He's realized that you either care about justice or you care about personal consequences. And for him there have been many.

After suppressing freedom of speech for two years, after defending a specific, myopic, and harmful narrative, the Editorial Board of the *New York Times* decided it was the perfect time to take a strong stance against censorship and cancel culture.

The irony of the most powerful and impactful violator of First Amendment rights lamenting the lack of free speech and offering up ideas to protect the rights of Americans was palpable, inescapable, and despicable.

Like Captain Renault in the movie *Casablanca*, when he closes Rick's Café Américain and proclaims, "I'm shocked, shocked to find that gambling is going on in here," the *New York Times* gladly accepted its winnings. Their profitability has soared during the worst and most pervasive period of censorship in recent American history. They have done absolutely nothing to protect the free speech rights of hundreds, if not thousands, of doctors, nurses, scientists, and concerned citizens who have tried to discuss views, make arguments, and analyze scientific studies that challenge the prevailing COVID narrative. They have silenced debate, worked tirelessly to chastise, vilify, and discredit those whose positions they disagree with, and failed to investigate serious claims of government corruption.

Nevertheless, they claim to lament that "when public discourse in America is narrowed, it becomes harder to answer...the urgent questions we face as a society."

What could be more important, more urgent, than the truth about corruption at the highest levels of government, about a pandemic response that led to more serious illness and death than was necessary, about the most powerful public health official in the country being more concerned with helping Big Pharma maximize return on investment and mitigate risk than protecting people?

As the *NYT* wrote, the worst kind of censorship is cancel culture, and the worst kind of cancel culture is the "piling on" kind. Why then, one might ask, did the *NYT* run a hit piece about Robert F. Kennedy Jr. that covered essentially the same subject matter as a dozen other hit pieces against Kennedy? Why now? Why this target? His family thinks he's wrong about vaccines, the *Times* noted. His friends think he's wrong about vaccines. Dr. Fauci thinks he's wrong about vaccines. Ever heard that before? Any analysis about vaccine safety? Any facts? Any citations? Any discussion of Dr. Fauci's despicable corruption as described in *The Real Anthony Fauci*, Kennedy's recent and epic takedown of Fauci? No, no, no, no, and no. What was the *New York Times* doing when the whole world was attacking Robert F. Kennedy Jr.?

Where was the *New York Times* when Robert F. Kennedy Jr., Dr. Robert Malone, Dr. Judy Mikovits, Dr. Pierre Kory, Dr. Paul Marik, Dr. Ryan Cole . . . and so many other impressive voices were being stifled? Here's an easy answer: they were "piling on."

The *NYT* has stated that it won't "publish ad hominem attacks," but it does publish hit pieces that any rational person understands are meant to discredit a book that they don't mention and obviously haven't read. They protect corrupt government officials against the unsuspecting public by forwarding policy statements or official memos that they have not thoroughly vetted, investigated, or corroborated. They are the worst kind of coconspirators: the kind that claim to be protecting their victims.

The *New York Times* writes that:

> At the individual level, human beings cannot flourish without the confidence to take risks, to pursue ideas and express thoughts that others might reject. . . . When speech is stifled or when dissenters are shut out of the public discourse, a society also loses its ability to resolve conflict, and it faces the risk of political violence.

That's where we are in America today. There is no debate, no public discourse, and we have lost the ability to resolve conflict. We have separated the country into two Americas, at least partially because of the policies and practices of the *New York Times*.

The *New York Post* has pointed out that the *New York Times* "published lies to serve a biased narrative." They accused the *Times* of "malicious misreporting" and cite a book called *The Grey Lady Winked* by Ashley Rindsberg.

Rindsberg is quoted as calling the *New York Times* "a truth-producing machine." He believes that the "fabrications and distortions" they've peddled since the 1920s were a system of twisting facts to manipulate public opinion about everything from "Hitler's Germany and Stalin's Russia to Vietnam and the Iraq War." The "reporting" is designed to "support a narrative aligned with the corporate whims, economic needs and political preferences" of the *New York Times*. He believes that they have consistently created "false narratives."

The *New York Post* says the *Times* has the resources to do it: "With close to $2 billion in annual revenue, the *Times* has the money, prestige, experience and stature to set the narratives that other news outlets invariably follow."

Rindsberg alleges that a former *Times* bureau chief in Berlin was a Nazi collaborator and that another star reporter for the *New York Times* parroted Soviet propaganda to defend Stalin.

The *NYT* coverage in the lead-up to the Vietnam and Iraq Wars seemed like government disinformation designed to support going to war. More recently, Rindsberg points to the stories that the *New York Times* published about Russia putting a bounty on US soldiers in Afghanistan, which the Biden administration later conceded was misinformation, and the story about Capitol Police Officer Brian Sicknick being "murdered by rampaging Trump supporters," though it was later proven that he had died of a stroke.

Similarly, Glenn Greenwald accused the *New York Times* of participating in "one of the most successful disinformation campaigns in modern electoral history." The *Times*, which before the 2020 election dismissed the Hunter Biden laptop as Russian Disinformation, later conceded that it was authentic.

It seems likely the *New York Times* coverage of the COVID pandemic isn't any different from its coverage of Hitler, Stalin, Vietnam, the Iraq War, January 6th, the Russian bounty on American soldiers, or the Hunter Biden laptop. Like most of the major big-tech platforms, they appear to have worked closely with Dr. Fauci and others, as representatives of the US Government, to control and propagate a specific narrative and to do what the government can't legally do itself—censor ideas that it disagrees with or narratives that might be harmful to its corporate partners.

As discussed above, the *New York Times* actively suppressed Robert F. Kennedy Jr.'s book and his allegations of corruption against Dr. Anthony Fauci. It defended Dr. Fauci without any investigation, without a full, free, and fair discussion of what is clearly the most important book of the decade. By ignoring Kennedy's book, by refusing to review it, by not allowing advertisements, by misrepresenting its success on its bestseller list, it clearly did everything in its power to avoid any debate whatsoever about the real science behind the origins of COVID or the best practices for controlling the virus and protecting the public. The *New York Times* has shown a total disregard for the scientific process, individual due process rights, or for any real search for truth.

And, once again, it did all this while lecturing us about the importance of free speech.

We have arrived at Orwell's 1984. Doublespeak is the universal language. The paper of record floods the world with disinformation, claims to be working tirelessly to protect the American people, and has clearly become The Ministry of Truth.

Reading Robert F. Kennedy Jr.'s book, *The Real Anthony Fauci*—the book Big Pharma, Dr. Fauci, the US Government and the *New York Times* will do absolutely anything to prevent you from reading—has become an act of rebellion, a blow to fascism, and a clear message that censorship in America just doesn't work.

CHAPTER 26

Lockdown Harms and the Silence of Economists

By Mikko Packalen and Jayanta Bhattacharya

Mikko Packalen is an Associate Professor of Economics at the University of Waterloo.

Jay Bhattacharya is a Senior Scholar of Brownstone Institute and a Professor of Medicine at Stanford University. He also serves as a research associate at the National Bureau of Economics Research and a senior fellow at the Stanford Institute for Economic Policy Research and the Stanford Freeman Spogli Institute.

* * * * *

As professional economists, we have watched the response of much of the economics profession to COVID-era lockdowns with considerable surprise. Given the evident and predictable harms of lockdowns to health and economic well-being, we expected economists to raise the alarm when lockdowns were first imposed. If there is any special knowledge that economists possess, it is that for every good thing, there is a cost. This fact is burned into economists' minds in the form of the unofficial motto of the economics profession that "there ain't no such thing as a free lunch."

From the depths of our souls, economists believe that the law of unintended consequences applies to every social policy, especially a social policy as all-encompassing and intrusive as lockdown. We economists believe that there are trade-offs in everything, and it is our particular job to point them out even when the whole world is yelling at the top of its voice to be quiet about them. It may still be a good idea to adopt some policy because the benefits are worth the cost, but we should go in with our eyes open about both.

That lockdown would, in principle, impose overwhelming costs on the population at large is not surprising. The scope of human activity touched by lockdown is overwhelming. Lockdowns closed schools and playgrounds, shuttered businesses, and barred international travel. Lockdowns told children they could not visit their friends, put masks on toddlers, and dismissed university students from campus. They forced elderly people to die alone and prevented families from gathering to honor their elders' passing. Lockdowns canceled screening and even treatment for cancer patients and made sure that diabetics skipped their checkups and regular exercise. For the world's poor, lockdown ended the ability of many to feed their families.

Economists, who study and write about these phenomena for a living, had a special responsibility to raise the alarm. And though some did speak [366], most either stayed silent or actively promoted lockdown. Economists had one job—notice costs. On COVID, the profession failed.

There are personal reasons for this docility that are easy to understand. First, when public health officials first imposed lockdowns, the intellectual zeitgeist was actively hostile to any suggestion that there might be costs to pay. The lazy formulation that lockdowns pitted lives versus dollars took hold of the public mind. This provided lockdown proponents with an easy way to dismiss economists whose inclination was to point out costs. Given the catastrophic toll in human life that epidemiological modelers projected, any mention whatsoever about pecuniary harm from lockdown was morally crass. The moral zeal with which lockdown proponents pushed this idea undoubtedly played an important role in sidelining economists. No one wants to be cast as a heartless Scrooge, and economists have a particular aversion to the part. The charge was unfair given the costs in lives that the lockdowns have imposed, but no matter.

Second, economists belong to the laptop class. We work for universities, banks, governments, consulting agencies, corporations, think tanks, and other elite institutions. Relative to much of the rest of society, the lockdowns posed much less harm on us and maybe even kept some of us safe from COVID. Narrowly speaking, lockdowns personally benefited many economists, which may have colored our views about them.

In this essay, we will leave these personal interests aside, though they are important, and focus only on the intellectual defense that some economists have put forward for their defense of lockdown. That economists have human weaknesses and interests that might render them less willing to speak taboo thoughts or against self-interest is not surprising. More interesting are the reasons (inadequate, we believe) that economists have given for their support of lockdowns, since, if correct, they would provide a rational defense against the charge we make in this essay that the economics profession, as a whole, has failed to do its job.

Spring 2020

In April 2020, the United Nations' World Food Program warned that 130 million people will starve as a result of the stumbling global economy [367]. The U.N.'s forecasts of the health impacts of this economic collapse were especially dire for children; they predicted hundreds of thousands of children in the world's poorest countries would die [368]. They would be collateral damage from the Great Lockdown, as the International Monetary Fund termed it in 2020 [369].

It was natural to expect scores of economists to refine these estimates and quantify how our response to the virus in rich countries would hurt the world's poor by disrupting global supply chains. Such work would increase awareness of the costs of our response to the virus.

Our supposition of economists' sense of duty to the world's poorest was well justified. For decades economists have fiercely defended the global economic system on the grounds that it has helped lift more than a billion people out of extreme poverty and increase life expectancy everywhere. The global economy has some significant flaws—vast inequality and climate change are often noted. But the worldwide network of trade has an essential role in facilitating economic development that brings sustained improvements to the lives of the world's poorest, economists have argued.

The expected rush to quantify the global collateral damage from rich countries' lockdowns never materialized. With few exceptions, economists most decidedly did not lean into quantifying lockdown harms either in developing countries or rich countries.

Precautionary Principle and Lockdown Love

Already in March 2020, economists considered lockdowns to be worthwhile. Their reasoning was a glorified version of the precautionary principle. Several research teams quantified how large the economic damage would have to be for lockdowns to be beneficial on net. Using epidemiologists' guesses of how many lives lockdowns might save, these analyses calculated the dollar value of the life years saved by lockdowns [370, 371].

In the early days of the epidemic, there was fundamental scientific uncertainty about the nature of the virus and the risk it posed. Faced with this uncertainty, many economists (joining other scientists less well trained in thinking about decision making under uncertainty) adopted a peculiar form of the precautionary principle. The implicit counterfactual exercise in these analyses took at face value the output from compartment models with dubious assumptions about critical parameters, such as the infection fatality rate from the model and compliance with lockdown policy [372]. Unsurprisingly, these early analyses concluded that lockdowns would be worthwhile, even if they were to cause extensive economic disruptions.

Applied to the COVID crisis, the precautionary principle says that when you have scientific uncertainty, it may make sense to assume the worst case about the biological or physical phenomenon you want to prevent. This is what the early economic analyses of lockdowns did by taking at face value the early estimates produced by epidemiological models (such as the Imperial College Model) of alarming COVID deaths in the absence of lockdowns.

The idea was that since we do not know with certainty, for instance, about the infection fatality rate, immunity after infection, and the correlates of disease severity, it is prudent to assume the worst. Therefore, we must act as if two or three out of a hundred infected people will die; there is no immunity after infection; and everyone, no matter what age, is equally at risk of hospitalization and death after infection.

Every one of these extreme suppositions turned out wrong, but of course, we could not have known that with certainty at the time, although there was already some evidence to the contrary. Scientific uncertainties are notoriously hard to resolve in advance of the time-consuming scientific work to resolve them, so maybe it was prudent to assume the worst. Unfortunately, fixating on the worst-case scenario then spurred long-lasting unfounded fears among the public and economists.

This all sounds very reasonable, but there was a curious asymmetry in the application of the precautionary principle in these analyses. With the benefit of hindsight, it should be clear that this application of the precautionary principle to the uncertainties of March 2020 was shockingly incomplete. In particular, it was not reasonable to assume the best case about the harms from the interventions you want to impose while at the same time accepting the worst case about the disease.

There are harms from the lockdown policies that any responsible economist should have considered before deciding that lockdowns were a good idea even then. A consistent application of the precautionary principle would have considered the possibility of such collateral lockdown harms, assuming the worst as the principal dictates.

In the panic of March 2020, economists assumed the best about these collateral harms. They adopted the implicit position that the lockdowns would be costless and that there was no other choice but to enforce lockdowns, at first for two weeks and then for as long as it might take to eliminate community disease spread. Under these assumptions motivated perhaps by a curiously asymmetric application of the precautionary principle, economists stayed silent while governments adopted lockdown policies wholesale.

In addition to the asymmetric treatment of scientific uncertainty about COVID epidemiology and lockdown harms, economists erred in two additional ways in applying the precautionary principle. First, when evidence arose contrary to the worst case, economists insisted on continuing to believe the worst case. One example of this rigidity is the negative reaction by many (including many economists) to studies that showed the infection fatality rate from COVID to be much lower than initially feared [373]. Motivating much of this reaction was the thought that this new evidence might lead the public and policymakers to not believe the worst about the disease's deadliness and thereby not comply with lockdown orders. A second example is economists' support (with some exceptions [374, 375]) in 2020 for continued school closures in the US in the face of ample evidence from Europe that showed that schools could be opened safely.

Second, while the precautionary principle is useful for aiding decision making (particularly, it can help avoid decision paralysis in the face of uncertainty), we must still consider alternate policies. Unfortunately, in the Spring of 2020, economists—in their rush to defend lockdowns—largely closed their eyes to any alternatives to lockdowns, such as age-targeted focused protection policies [356, 376]. These mistakes further solidified the economics profession's ill-advised support for lockdowns.

Rational Panic?

A second strand of analysis by economists in Spring 2020 was perhaps even more influential in turning economists in favor of lockdowns [377]. Economists observed that most of the decline in movement and economic activity occurred before governments imposed any formal lockdown orders. The conclusion? The decline in economic activity in Spring 2020 was driven not by lockdowns, but by voluntary changes in behavior. Fear of the virus induced people to engage in social distancing and other precautionary measures to protect themselves, economists reasoned.

Having concluded that lockdowns do not significantly impede economic activity, economists have seen little need to quantify any domestic or global collateral damage from lockdowns.

To governments, this consensus among economists provided considerable relief and arrived just in time. At around the same time in the spring of 2020, it became evident that the depth of the economic contraction was much larger than first anticipated [378]. It was essential to politicians to blame this economic damage on the virus itself rather than the lockdowns, since they were responsible for the latter but not the former. And economists obliged.

But was this conclusion about the lack of marginal lockdown harms justified? Economists were no doubt correct that movement and business activity would have changed even without any lockdowns. Vulnerable older people were wise to take some precautionary measures, the elderly in particular. The staggeringly steep age gradient in mortality risk from infection with novel coronavirus was already known by March 2020 [379].

Nevertheless, the argument that people would have voluntarily locked down anyway even in the absence of a formal lockdown is spurious. First, suppose we take the argument that people rationally and voluntarily restricted their behavior in response to the threat of COVID as correct. One implication would be that formal lockdowns are unnecessary, since people will voluntarily curtail activities *without lockdown*. If true, then why have a formal lockdown at all? A formal lockdown imposes the same restrictions on everyone, whether or not they are able to bear the harm. By contrast, public health advice to restrict activities voluntarily for a time would permit those—especially the poor and working class—to avoid the worst lockdown-related harms. That some (though not all) people did curtail their behavior in response to the disease threat is thus not a sufficient argument to support a formal lockdown.

Second, and perhaps more important, not all of the fear of COVID has been rational. Surveys conducted in Spring 2020 show that people perceived the population mortality and hospitalization risks to be much greater than they actually are [380]. These surveys also indicate that people vastly underestimate the degree to which the risk rises with age. The actual mortality risk from COVID is a thousand times higher for the elderly than it is for the young [381]. Survey evidence indicates that people mistakenly perceive age to have a far smaller influence on the mortality risk [382].

This excess fear has received little media coverage until recently. For example, studies on fear published in July and December 2020 [383] gained little traction at the time but were discussed by the *New York Times* in March 2021 [384] and by other high-profile media outlets shortly after that. These delays indicate a persistent (but now finally easing) unwillingness by the media to accept these facts that are strong evidence that the public fear of COVID has not corresponded to objective facts about the disease.

So, our indictment that economists have paid insufficient attention to the harms from lockdowns thus cannot be evaded by recourse to a rational fear of COVID in the population.

Panic as a Policy

There is an even deeper problem with the rational panic argument. In part motivated by the precautionary principle, many governments adopted a policy of inducing panic in the population to induce compliance with lockdown measures. In a sense, lockdowns themselves drove the panic and distorted the risk perceptions of economists, just as they distorted the risk perception of the public at large. Lockdowns were, after all, an unprecedented policy tool in modern times, a tool that the World Health Organization and the

Western media still in January 2020 ruled out as a reasonable policy option. It was not clear even to influential scientists like Neil Ferguson whether the West would be willing to copy Chinese-style lockdowns or comply with them if implemented [385].

Then in March 2020, lockdowns were widely adopted and became an integral part of the decision to panic the population to induce compliance [309, 386]. The earliest lockdowns fomented fear elsewhere, and each successive lockdown then further magnified it. Because lockdowns do not distinguish who is at greatest risk from the virus, they are likely also a key culprit to the public's lack of understanding about the steep link between age and COVID mortality risk.

Because economists' estimates of lockdown impacts have ignored these fear spillovers from lockdowns to other jurisdictions, the conclusion that lockdowns inflict no significant economic harm is decidedly not justified. The large voluntary decline in movement and business activity was not a purely rational response to COVID risks. Excessive COVID fears fomented by lockdowns drove the decline in mobility and economic activity. Excess COVID fears thus induced a behavioral response that was partly irrational.

The lockdowns of Spring of 2020 were thus likely responsible for much more of the decline in economic activity than the consensus among economists admits. Economists have been unwilling to examine the implications of this fact, just as economists have been unwilling to examine the implications of the broader issue that governments stoked fear among the public as a part of the anti-COVID policy.

A Conservative Evaluation

Let us leave aside the controversy over whether the reduction in human movement in Spring 2020 was a rational response to the risk posed by the virus or a panic-induced overreaction. In truth, it was likely a mix of both. Let us then take at face value a lockdown study by economists that showed that "only" 15% of the decline in economic activity can be attributed to lockdowns [387]. (We will leave aside the fact that some economic studies on lockdowns have found the share of the decline in economic activity attributable to formal lockdown orders to be considerably higher, even 60%.) If the conservative 15% estimate is correct, would that imply that lockdowns were worth the cost [388]? No.

Recall the early UN estimates that forecasted the starvation of 130 million people in poor countries due to the global economic decline [389]. Suppose that only 15% of that figure is attributable to lockdowns. Taking 15% of 130 million yields a number that represents immense human suffering attributable to lockdowns, even by this overly conservative reckoning. And we have not begun to count the other harms of lockdown, which include hundreds of thousands of additional children in South Asia dead from starvation or inadequate medical care, the collapse of treatment networks for tuberculosis and HIV patients, delayed cancer treatment and screening, and much else [390].

In other words, if lockdowns are indeed responsible for only a tiny share of the decline in economic activity—as many economists have claimed—the total size of the local and global collateral costs from lockdowns is still enormous. The collateral harms to human health and life caused by lockdown are far too large to be dismissed, even under the rosy assumption that panic would have happened in the absence of lockdown.

It bears noting also that the long-run impact of lockdowns on business activity is yet uncertain. The arbitrariness of lockdown rules may chill future business confidence and entrepreneurial activity much more than voluntary movement and economic activity reductions. Economists' silence on lockdown harms also indicates a belief that *every* lockdown comes without harm. In reality, each lockdown causes its own set of unpredictable collateral consequences, since they interdict normal human and economic interactions in different ways.

The Role Economists Have Played

Economists' conclusion that lockdowns can do no marginal harm is thus unwarranted. The evidence put forth by economists does not justify abandoning attempts to quantify the global and local collateral health costs of lockdowns. Lockdowns are not a free lunch.

For economics, the failure to document the collateral damage from lockdowns is fundamental. The very purpose of economics is to provide an understanding of the pains and successes in society. Economists' role is to synthesize the facts and trade-offs and point out how policy assessments depend on our values, as well. When economists turn a blind eye to the pains in our society, as they have in the past year, governments lose crucial indicators needed to design balanced policies.

In the short term, such blindness reaffirms the elites' unwavering belief that the course is correct. As long as only the potential benefits of lockdowns are examined and discussed in the media, it is hard for the public to object to lockdowns. But slowly yet inevitably, the truth about the pains, both big and small, is revealed in the long run. Neither the reputation of economics nor the legitimacy of our political system will fare well if the divide between the elite and those who felt the collateral damage all along is too wide when this divide is finally revealed. By not documenting the pains caused by lockdowns, economists have served as apologists for draconian government responses.

To be sure, some economists have questioned the lockdown consensus throughout the pandemic, and more recently, others have started to express their doubts, as well. Also, to the profession's credit, scores of economists did respond to the pandemic with considerable vigor in an attempt to help policymakers make informed decisions. Whether these sincere efforts were directed in the best way is another matter. Nevertheless, the economics profession will be haunted for a long time for our failure to speak up for the poor, the working class, the small businessmen, and the children who have borne the brunt of the lockdown-related collateral harms.

Economists also erred in closing ranks so quickly and so vociferously to build the ill-advised consensus on lockdowns. One economist even labeled—publicly—those who questioned the consensus as "liars, grifters, and sadists." Another economist organized a boycott on Facebook of a health economics textbook (written by one of the authors of this piece long before the epidemic started) in response to the publication of the Great Barrington Declaration, which opposed lockdowns and favored a focused protection approach to the pandemic. Amidst such chilling edicts from the profession's leaders, it is not surprising that the consensus on lockdowns has been challenged so rarely. Economists and others were intimidated against pointing out lockdown costs.

The attempts to stifle scientific debate on lockdowns have been costly but have come with one silver lining. The use of such underhanded tactics to support a consensus view is always an implicit admission that the arguments supporting the consensus are themselves understood to be too feeble to withstand closer scrutiny.

Economists' rush to consensus on lockdowns has also had broader ramifications for science. Once the scientific discipline tasked with quantifying the trade-offs in life decided that the linchpin of our COVID response—lockdowns—involved no trade-offs, it became natural to expect science to give us unambiguous answers in all COVID matters. Economists' silence on lockdown costs, in essence, gave others a carte blanche to ignore not just lockdown costs, but also the costs of other COVID policies such as school closures.

Once the aversion to pointing out the costs of COVID policies took hold among scientists, science came to be widely seen and misused as an authority [391]. Politicians, civil servants, and even scientists now constantly hide behind the "follow the science" mantra rather than admit that science merely helps us make more informed decisions. We no longer dare acknowledge that—because our choices always involve trade-offs—the virtue of pursuing one course of action over another always rests not just on the knowledge we get from science, but also on our values. We have seemingly forgotten that scientists merely produce knowledge about the physical world, not moral imperatives about actions that involve trade-offs. The latter requires understanding our values.

The prevalent misuse of science as a political shield in this manner may in part reflect the fact that, as a society, we are ashamed of the value system that our COVID restrictions have implicitly revealed. This criticism applies to economics, as well. Much of what economists have done in the past year has been in the service of the rich and the ruling class at the expense of both the poor and the middle class. The profession has sought to hide its values by pretending that lockdowns have no costs and by actively stifling any criticism of the misguided lockdown consensus.

Economists Should Be Gardeners, Not Engineers

Economists' embrace of lockdowns is questionable also from a theoretical perspective. The complexity of the economy and differing tastes of individuals have generally tilted economists in favor of individual freedom and free markets over government planning. Governments lack the information needed to steer the economy efficiently through centralized planning. Yet, in the context of lockdowns, many economists suddenly appeared to expect governments to understand very well which functions of society are "essential" and most valued by citizens and who should perform them.

In a matter of mere weeks in the Spring of 2020, a great many economists were seemingly transformed into what Adam Smith had 260 years earlier derided as a "man of system" [392]. By this, he meant a person under the illusion that society is something akin to a game of chess, that it follows laws of motion that we understand well and that we can use this knowledge to wisely direct people at will. Economists suddenly forgot that our understanding of society is always very incomplete, that the citizenry will always have values and needs beyond our ken and will act in ways that we can neither fully predict nor control.

From another perspective, economists' support for lockdowns is not surprising. The lockdown consensus can be seen as the natural end result of modern economists' strong technocratic bent. While economics textbooks still emphasize the profession's liberal roots and lessons, among professional economists, there is now a widespread belief that almost any societal problem has a technocratic, top-down solution.

This shift in economics is remarkable. The attitude among economists today is very different from the days when historian Thomas Carlyle attacked the profession as "the dismal science." His complaint was that economists of his day supported individual freedom too much, rather than systems that he favored in which the wise and powerful would govern every aspect of the lives of the purportedly unsophisticated masses.

This technocratic orientation of the economics profession is evident in the ongoing debate among economists over which professional analogy best captures the work of modern economists. Engineer, scientist, dentist, surgeon, car mechanic, plumber, and general contractor are among the many analogies that economists have proposed to describe what economists today ought to do. Every one of these analogies is justified based on modern economists' supposed ability to offer technocratic solutions to nearly every societal problem.

We view economists' proper role in directing the lives of our fellow citizens as much more limited. The role of a gardener is more apt for economists than, say, the role of either an engineer or a plumber. The tools and knowledge our profession has developed are not sophisticated enough to justify thinking that we economists ought to try to fix all the ills of our society, employing technocratic solutions in the same way that engineers and plumbers do. Just as gardeners help gardens thrive, we economists too should stick to thinking of ways to assist individuals and economies prosper rather than offer all-encompassing solutions that dictate what individuals and companies ought to do.

Economists surprised the public also with their cavalier attitude toward the plight of small businesses, devastated by lockdowns. The profession's central tenets rest on the virtues of competition. Yet economists' foremost wonder about the intense duress experienced by small businesses during lockdowns seems to have been whether the closures will have a "cleansing" effect by eliminating the worst-performing firms first. To the dismay of many, the dismal science has had very little to say about how lockdowns have favored big business and what this will mean for market competition and consumer well-being in the years to come.

Economists' reluctance to challenge policies that favor big business is regrettable yet understandable. Increasingly, we economists work for big business—the digital giants in particular. We send our students to work for Amazon, Microsoft, Facebook, Twitter, and Google, and we count it a great success when they land jobs with those prestigious companies. Being on good terms with these companies is important also because of these companies' data and computational resources. Both are now crucial for successful publishing and associated career advancement in economics. Rare is the economist who is immune to the power wielded by the digital giants within the economics profession.

The Path Forward

To regain its bearings, the economics profession must rethink its values. In recent years so much has been written about the increasing emphasis on methods and big data in economics at the expense of theoretical and qualitative work [393–397]. As empirical techniques and applications have taken over the profession, economics has become a stagnant or perhaps even a receding discipline in its understanding of basic economic trade-offs that once comprised the core of economic training. How many professional economists still agree with Lionel Robbins's famous definition, "Economics is the science that studies human behavior as a relationship between ends and scarce resources which have alternative uses"? How much of the work of today's economists serves this goal well?

This dynamic is no doubt partly to blame for the profession's misguided espousal of lockdowns. Overt emphasis on quantitative methods in empirical work has made economists less familiar with the economy itself, a trend that the disconnect between the perceived and actual precision of economists' theoretical modeling has amplified. Economists have obsessed with the finer technical details of empirical analyses and the internal logic of theoretical models to a degree that has effectively blinded much of the profession from the bigger picture. Unfortunately, without understanding the bigger picture, getting the small details correct is of little use.

That economists famously are not blessed with much intellectual humility likely also played a part in the profession's rushed ascent to agreement on lockdowns. Economists demonstrated little desire to explore the many limitations and caveats inherent in the profession's lockdown analyses even though those analyses were often by people with little or no prior training or interest in epidemiology or public health, and even though those analyses served to support the most intrusive government policies in a generation. Economists did not heed epidemiologists' prior warnings about the need to be very humble when connecting insights from models to our complex reality [398].

The fact that economists' concern for the poor vanished so quickly in the spring of 2020 also speaks of a distinct lack of empathy. Because most economists are blessed with incomes that place us in the upper-middle class or higher, we (with some exceptions, of course) live lives that are often disconnected from the poor in our own country, much less in developing countries. Because of this disconnection, it is hard for economists to understand how the poor near them in rich countries and globally would experience and respond to lockdowns.

Economics should reinvigorate itself with a renewed emphasis on connecting with the lives of the poor both in rich countries and globally. Training in the profession should emphasize the value of empathy and intellectual humility over technique and even theory. The economics profession should celebrate empathy and intellectual humility as the hallmarks of a model economist.

Reforming economics will bear considerable fruit in the form of trust by the public in the recommendations that economists make about policy, but it will not be easy. Changing the profession's values requires sustained effort and the kind of patience that the profession sorely lacked when it rushed to defend lockdowns.

In terms of reassessing lockdown harms, there is reason for optimism. Economics served the world well when it defended the global economic system during the past several decades on the basis that economic progress serves a crucial role in advancing the well-being of the world's most vulnerable people. That this happened so recently gives hope that economists will soon yet regain their interest in the lives of the world's poorest.

Rather than hide behind the false belief that lockdowns are a free lunch, it is crucial that economists soon evaluate the global impacts of rich countries' lockdowns. A better understanding of our lockdowns' global effects will facilitate a more compassionate COVID response in rich countries, and also a better response to future pandemics—the kind of response that values how our response in rich countries influences the economic and health outcomes in the less prosperous parts of the world.

It is equally important that economists soon examine and assess with vigor the domestic pains caused by lockdowns, school closures, and other COVID restrictions. Documenting the highs and lows of society is, after all, the profession's foremost task. Economics can ill-afford to overlook this core mission much longer.

CHAPTER 27
Debt versus Sovereignty

The unresolved conflict between Jefferson and Hamilton enables WEF hegemony.

Policies and practices designed to drive either individuals or nation-states into debt have long been a preferred method for political coercion, co-optation, enslavement, incremental dominance, and control. A form of subtle, creeping indentured servitude. Neither individuals, communities, businesses nor nation-states can be free when they are indebted (financially or otherwise) to another. This subtle method of control of both nation-states and their citizens has been consciously, intentionally, and strategically deployed by central banks for centuries. This is the method by which the World Economic Forum, itself a guild representing the interests of the largest corporations (and their controlling owners), seeks to transform itself into a fascist totalitarian world government.

Furthermore, as so crudely and bluntly illustrated in the case of the arbitrary freezing of financial assets belonging to political opponents by Canadian Prime Minister Justin Trudeau and Deputy Prime Minister Chrystia Freeland (both WEF-trained "young leaders"), if given opportunity and technical capability to directly deprive political dissenters of access to existing owned financial assets, tyrants will act arbitrarily and capriciously to directly create financial dependency by weaponizing the system that enables global coordination of central bank transactions [399]. This Trudeau/Freeland tactic was not a novel innovation; rather, it represented a crude, transparent, explicit deployment (at the level of individual dissenters) of a financial weapon that has been enabling surreptitious political control of both individuals and nation-states throughout recorded history.

If you occasionally experience a vague sense that you are being inten-tionally controlled via indebtedness, you should probably listen to that inter-nal voice. To illustrate with examples from the present, the most common explanation for why physicians have not spoken up about the weaponiza-tion and manipulation of public health information and policies during the COVIDcrisis is that they are deeply indebted due to the loans taken out to enable their extended and expensive education and have no practical choice other than to comply with the mandates imposed on them by government, insurance agencies, and their host institutions (academic or private hospi-tal chains). They have a profound financial conflict of interest—comply or go bankrupt. In large part, the physicians and medical scientists who have spoken up about the compromised medical ethics, regulatory standards, mis- and disinformation propagated by governments and WHO (including intentionally withheld or manipulated medical and epidemiological infor-mation) have been financially independent, often senior with high status or established independent medical practices, or otherwise have been decou-pled from mechanisms or institutions that have been weaponized to force compliance with centralized edicts. In other words, the majority of those who have spoken out have freedom to speak because they are (relatively) financially independent.

To provide a broader historic example that helps illustrate the point, a case has been made that the Stock Market crash of 1929 was engineered by the central banks. This hypothesis is grounded in the observation that Goldman Sachs, Lehman Bank, and others profited from the Crash of 1929. In a quote that many will find oddly relevant to the COVIDcrisis, House Banking Committee Chairman Louis McFadden (D-NY) said of the Great Depression:

> It was no accident. It was a carefully contrived occurrence. . . .
> The international bankers sought to bring about a condition of
> despair here so they might emerge as rulers of us all.

U.S. Congressman Louis McFadden, speaking about the 1929 Stock Market Crash.

At the macro scale, the same holds true for nation-states and those who seek to function as political leaders of either Nations or States. The leaders functionally must sell their autonomy to the highest bidders in order to gain office. They have very little operational autonomy, even if they have good intentions to implement constructive and adaptive changes that will advance the interests of their state or country. The central banks and those who control them actively promote political forces and agendas (including war!) to drive nation-states into indebtedness so that they can functionally extract a form of rent and control the politics of these captured organizations so that the banks and their owners can control global affairs to benefit their commercial interests. Please see *The Federal Reserve Cartel: The Rothschild, Rockefeller and Morgan Families* by John Morse for a detailed historical analysis of these strategies and behaviors [400]. Both Thomas Jefferson and Alexander Hamilton foresaw this. Jefferson fought to preserve the autonomy of United States citizens, while Alexander Hamilton fought to enable and empower what were essentially the financial oligarchy of his day—which has persisted remarkably intact through time!

Yet another example. The Bank of International Settlements (BIS) is the most powerful bank in the world, a global central bank for the Eight Families who control the private central banks of almost all Western and developing nations. The first president of BIS was Rockefeller banker Gates McGarrah—an official at Chase Manhattan and the Federal Reserve. McGarrah was the grandfather of former CIA director Richard Helms. The Rockefellers—like the Morgans—had close ties to London. A case has been made that the Rockefellers and Morgans were just "gofers" for the European Rothschilds. BIS is owned by the Federal Reserve, Bank of England, Bank of Italy, Bank of Canada, Swiss National Bank, Nederlandsche Bank, Bundesbank, and Bank of France.

Historian Carroll Quigley wrote in his epic book *Tragedy and Hope* that BIS was part of a plan:

> To create a world system of financial control in private hands able to dominate the political system of each country and the economy of the world as a whole . . . to be controlled in a feudalistic fashion by the central banks of the world acting in concert by secret agreements [401].

Sound familiar? Certainly familiar to those who have read the Klaus Schwab and Thierry Malleret books *COVID 19: The Great Reset* and *The Great*

Narrative for a Better Future, whereby these two WEF leaders lay out plans to further refine a world system of financial controlled private interests. A "new world order" [402, 403].

As summarized by John Morse [400]:

> The US government had a historical distrust of BIS, lobbying unsuccessfully for its demise at the 1944 post-WWII Bretton Woods Conference. Instead, the Eight Families' power was exacerbated, with the Bretton Woods creation of the IMF and the World Bank. The US Federal Reserve only took shares in BIS in September 1994.
>
> BIS holds at least 10% of monetary reserves for at least eighty of the world's central banks, the IMF, and other multilateral institutions. It serves as financial agent for international agreements, collects information on the global economy, and serves as lender of last resort to prevent global financial collapse.
>
> BIS promotes an agenda of monopolistic capitalist fascism. It gave a bridge loan to Hungary in the 1990s to ensure privatization of that country's economy.
>
> It served as conduit for Eight Families funding of Adolf Hitler, led by the Warburg's J. Henry Schroeder and Mendelsohn Bank of Amsterdam. Many researchers assert that BIS is at the nadir of global drug money laundering. It is no coincidence that BIS is headquartered in Switzerland, favorite hiding place for the wealth of the global aristocracy and headquarters for the P-2 Italian Freemason's Alpina Lodge and Nazi International. Other institutions that the Eight Families control include the World Economic Forum, the International Monetary Conference, and the World Trade Organization. Bretton Woods was a boon to the Eight Families. The IMF and World Bank were central to this "new world order."

Thomas Jefferson was obsessed with the importance of establishing sound monetary policies while enabling and initially guiding the political experiment in self-governance embodied in the United States Government and was horrified by policies advocating printing of paper fiat currency. Jefferson foresaw what we have repeatedly observed over the past few decades: a flood of increasingly worthless paper (fiat currency decoupled from any commodity), which has caused:

> a general demoralization of the nation, a filching from industry of its honest earnings, wherewith to build up palaces, and raise gambling stock for swindlers and shavers, who are to close their career of piracies by fraudulent bankruptcies.
>
> *Thomas Jefferson to Nathaniel Macon, 12 January 1819*

A case can be made that one of the most consistent predictors of emergence of periods dominated by the politicized madness of crowds (e.g., mass formation psychosis), which Mattias Desmet dissects in his seminal book *The Psychology of Totalitarianism,* is not the fragmentation of society, but rather loss of faith in the economic transactional infrastructure that is required for the very existence of macro scale social organization and cohesion [310]. This may be one of the central drivers of the social fragmentation that is the immediate predecessor of the Mass Formation Psychosis/Hypnosis phenomenon.

As summarized by John McClaughry [404]:

> For Jefferson, deficit spending was simply and unarguably immoral. Government debt would, he believed, lead us into an "English career of debt, corruption, and rottenness, closing with revolution." His prescription was straightforward: hold government expenditures to a minimum, raise the funds to meet those expenditures by taxation, and plan to collect a surplus to extinguish the public debt. He even advocated a constitutional amendment prohibiting the federal government from incurring debt at all and denied that government had the power to make paper money legal tender for private debts.
>
> When our present-day leaders—of both parties—accept the idea that they can run enormous federal deficits year after year, when they accept the idea that printing new money is less painful than cutting spending or raising taxes, they accept ideas that Jefferson would have branded not only as economically disastrous, but as morally repugnant.

To further illustrate this point, it is helpful to compare and contrast the awareness of the dangers inherent in incremental loss of autonomy and commitment to individual autonomy of Thomas Jefferson to the deeply embedded sense of entitlement and elitism of Alexander Hamilton.

> "Single acts of tyranny may be ascribed to the accidental opinion of the day; but a series of oppressions, begun at a distinguished period, and pursued unalterably through every change of ministers (administrators) too plainly proves a deliberate, systematic plan of reducing us to slavery."
>
> *Thomas Jefferson, A Summary View of the Rights of British America (ed. 1774)*

> "If a nation expects to be ignorant & free, in a state of civilization, it expects what never was & never will be. The functionaries of every government have propensities to command at will the liberty & property of their constituents. There is no safe deposit for these but with the people themselves; nor can they be safe with them without information. Where the press is free and every man able to read, all is safe."
>
> *Thomas Jefferson, The Papers of Thomas Jefferson, Retirement Series, Volume 9: 1 September 1815 to 30 April 1816*

> "All communities divide themselves into the few and the many. The first are the rich and well born, the other the mass of the people. The voice of the people has been said to be the voice of God; and however generally this maxim has been quoted and believed, it is not true in fact. The people are turbulent and changing; they seldom judge or determine right. Give therefore to the first class a distinct, permanent share in the government."
>
> *Alexander Hamilton. Said on June 19, 1787. The Records Of The Federal Convention Of 1787, book edited by Max Farrand. Volume I, p. 299, 1937.*

In many ways, Jefferson and Hamilton represent the great tension at the heart of the American experiment in self-governance, and the battle between the two political forces that these individuals represent is the central conflict that has dominated American politics since even before the Declaration of Independence, Constitution, and Bill of Rights were drafted. Thomas Jefferson, together with John Adams, James Madison, and Thomas Paine, were the leaders of the populist block during the time of the American Enlightenment and the creation of the United States.

In contrast, Alexander Hamilton was closely allied with the Rothschild banking and finance family. For example, with Rothschild financing Alexander Hamilton founded two New York banks, including the Bank of New York. The Rothschild family owns the Bank of England and leads the European Freemason movement. All US Masonic lodges to this day are warranted by the British Crown, whom they serve as a global intelligence and counterrevolutionary subversion network. Hamilton was one of many founding fathers who were Freemasons. George Washington, Benjamin Franklin, John Jay, Ethan Allen, Samuel Adams, Patrick Henry, John Brown, and Roger Sherman were also Masons. Roger Livingston helped Sherman and Franklin write the Declaration of Independence. He gave George Washington his oaths of office while he was grand master of the New York Grand Lodge of Freemasons. Washington himself was grand master of the Virginia Lodge. Of the general officers in the Revolutionary Army, thirty-three were Masons. The First Continental Congress convened in Philadelphia in 1774 under the presidency of Peyton Randolph, who succeeded Washington as grand master of the Virginia Lodge. The Second Continental Congress convened in 1775 under the presidency of Freemason John Hancock. In 1779 Benjamin Franklin became grand master of the French Neuf Soeurs (Nine Sisters) Masonic Lodge, to which John Paul Jones and Voltaire belonged. Franklin was also a member of the more secretive Royal Lodge of Commanders of the Temple West of Carcassonne, whose members included Frederick Prince of Wales.

During the creation of the nation, populist Thomas Jefferson argued that the United States needed a publicly owned central bank so that European monarchs and aristocrats could not use the printing of money to control the affairs of the new nation. However, larger forces were set in motion that favored a privately owned central bank for the new nation. In 1789 Alexander Hamilton became the first treasury secretary of the United States (under President George Washington). Thomas Jefferson was appointed secretary of state. William Randolph became the nation's first attorney general and secretary of state under George Washington, but his family returned to England loyal to the Crown. John Marshall, the nation's first Supreme Court justice, was also a Mason. The Rothschilds sponsored Hamilton's arguments for a private US central bank and in the end carried the day. In 1791 the Bank of the United States (BUS) was founded, with the Rothschilds as main owners.

What we have seen play out over the last few years of the COVIDcrisis is only the most recent skirmish and effort by these massively capitalized central banking families, acting under the guise of their latest surrogates BlackRock, State Street, and, most important, Vanguard via the World Economic Forum and the many acolyte/surrogates that have been trained via the five yearlong "young leaders program" and placed into positions of power and influence throughout the world [405]. This is who owns "Big Pharma" as well as the old media [406], and this is how the massive lies, defamation campaigns, mis- and disinformation have been propagated globally.

What can you do about this at a personal level? Simple. *Get out of debt.* That is the starting point. These organizations and their masters use debt to control you, to control the regional government where you live, and to control your Nation State.

It is really hard to wean yourself off of their addictive financial products, particularly at first. You get used to the convenience of the "Credit Card," to the nice, financed car, to the big house. And they constantly manipulate you to make you think that you need these things, using the same tools that they have been using to manipulate the public during this COVIDcrisis I know it is hard. We basically went abruptly stone cold sober, in terms of loans and financial instruments. Jill and I lost hundreds of thousands in real estate equity we had built up over decades when the "Great Recession" of 2008 hit. We never declared bankruptcy, but we were cleaned out. We sold everything we could, relocated from Northern Georgia up to Virginia, rented or leased a series of rough farms that we partially rebuilt while leasing, and then Jill found an unimproved parcel (which is now our little gem of a farm) about six years ago. No water, no electric, no well, no septic, no fences. Negotiated with the current owner for direct purchase from her, bought an office trailer and moved it onto the land (illegally…), contracted for a portable toilet service, and got to work. The first few years were rough. Winters were cold.

We used a local gym to get showers. But gradually, gradually, we cleared all remaining debt while we built out the farm to what it is now. I do not think I would have been able to speak so freely about my concerns regarding what has gone on in public health if I were still fully dependent on "the system" and in debt to some company (or employer) that was able to hold a metaphorical "Sword of Damocles" over my neck.

It was rough. Jill and I got through it. And you can also. Freedom is worth it.

CHAPTER 28
The World Economic Forum, a Trade Organization on Steroids

Words and their meanings are tricky things. Clever people, con artists, liars, and CIA agents (which are all of the above, in my experience) are often very skilled at using language and manipulating both meaning and emotions to hide their true intent.

Wikipedia, an organization that is generally very friendly to the World Economic Forum and its agenda, defines the WEF as follows:

> The World Economic Forum (WEF) is an international non-governmental and lobbying organization based in Cologny, Canton of Geneva, Switzerland. It was founded on 24 January 1971 by German engineer and economist Klaus Schwab. The foundation, which is mostly funded by its 1,000-member companies—typically global enterprises with more than five trillion US dollars in turnover—as well as public subsidies, views its own mission as "improving the state of the world by engaging business, political, academic, and other leaders of society to shape global, regional, and industry agendas.

Simplifying that for sake of discussion, the WEF is a trade organization that is designed to advance the business interests of extremely wealthy companies and their owners (generally referred to as "Davos Man"). Global enterprises with more than five trillion US dollars in turnover is a very small and elite group. Five trillion dollars (five thousand billion) in annual revenue makes for a very exclusive club, as illustrated by the latest Forbes 2000 international ranking (from 2021).

What is a trade organization? Back to Wikipedia.

A trade association, also known as an industry trade group, business association, sector association or industry body, is an organization founded and funded by businesses that operate in a specific industry. An industry trade association participates in public relations activities such as advertising, education, publishing, lobbying, and political donations, but its focus is collaboration between companies. Associations may offer other services, such as producing conferences, holding networking or charitable events, or offering classes or educational materials.

Yup. If the shoe fits, wear it.

Not surprisingly, in its own mission statement, the WEF defines itself as follows: "The World Economic Forum is the International Organization for Public-Private Cooperation." Which is really a carefully wordsmithed way of saying that the WEF is a centralized trade organization for promoting international corporatism.

Public-private cooperation as a political and economic structure is known by two other terms: corporatism and fascism. Benito Mussolini is often credited with a very succinct definition of corporatism in the disputed quote "Fascism should more properly be called corporatism because it is the merger of state and corporate power." Whether or not this represents an accurate English translation of his Italian words, the statement reflects a fundamental political truth.

Merriam-Webster defines corporatism as:

The organization of a society into industrial and professional corporations serving as organs of political representation and exercising control over persons and activities within their jurisdiction.

Andrew Stuttaford, writing in the *National Review* during October 2020 (here is his personally archived version), provides an alternative definition that I think really gets to the root of the issue. In his opinion, corporatism as advocated by the WEF consists of:

[A] hydra-headed ideology with origins in the premodern, and a very mixed past—sometimes benignly (it influenced the formation of West Germany's social market economy) and sometimes not (it was an important element in pre-war fascist theory). The different forms corporatism has taken make it tricky to define with precision, but they share a common core: the conviction that society should be organized by and for its principal interest groups—let's call them "stakeholders"—intermediated by, and ultimately subordinate to, the state. The individual does not get a look in.

The context for this remarkable statement is a prescient article titled "A Useful Pandemic: Davos Launches New 'Reset,' this Time on the Back of COVID" [407]. In my opinion, in some future listing of COVID heroes, Mr. Stuttaford surely deserves to be in the top 10. Here are some other profoundly prescient articles from the same author:

"The Great Reset: If Only It Were Just a Conspiracy," November 27, 2020 [408]. This one has another notable quote relevant to this topic: "The 'Great Reset' masterminded by the World Economic Forum is just corporatism by another name."

And then there is this, which cuts right to the bone of the matter.

"Larry Fink, 'Emperor'?" February 19, 2022 [408]. And yet another key money quote: "As BlackRock and other large index-fund managers continue pushing stakeholder capitalism, America slouches toward corporatism."

"Stakeholder capitalism." There is another benign-sounding term that requires definition and understanding. Turns out that this is a phrase largely pioneered and championed by Klaus Schwab, leader of the WEF. It is at the very heart of the self-concept of the WEF. Schwab's definition is as follows:

Stakeholder capitalism is a form of capitalism in which companies seek long-term value creation by taking into account the needs of all their stakeholders, and society at large.

On the following page is Mr. Schwab's graphical representation of how he defines this term.

This is called a "flower diagram," for those who want to know, and

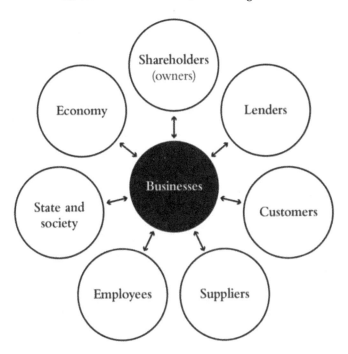

this type of diagram is often attributed to the consulting company Deloitte. Notice that business is at the center of this worldview, and State and society are lumped together and relegated to the position of one of many "stakeholders" that business needs to take into account. This is the political objective that is at the center of the Uniparty globalist vision.

Under this concept, we all exist to serve and enable businesses and their economic growth objectives. Good to know. Puts things in their proper perspective. No role here for faith-based organizations!

Now, let's see what the prophetic Andrew Stuttaford has to say about stakeholder capitalism [409]:

> Recently, one expression of corporatism, 'stakeholder capitalism,' has won strong support on both sides of the Atlantic. This might be expected in Europe, but that it has been taken up by the Business Roundtable and many leading firms in the U.S.—allegedly a bastion of both free enterprise and democracy—is depressing. Looked at optimistically, the BRT and its C-suite cheerleaders are useful idiots. Looked at realistically, they are part of a managerial class grubbing for the power that flows from other people's money.

Stakeholder capitalism rests on the notion that a company's management owes a duty to more than its shareholders. It's something that Klaus Schwab, the WEF's founder and executive chairman, has been advocating for a long time. A key feature of the Great Reset is the idea that stakeholder capitalism should, one way or another, be adopted.

That would reduce a company's shareholders to just another category of 'stakeholder,' effectively transferring the power that capital should confer away from its owners and into the hands of those who administer it. They are then accountable to, well, it's not quite clear whom. It's not difficult to grasp why so many corporate bosses are enthused by stakeholder capitalism.

But stakeholder capitalism is a betrayal of democracy as well as of shareholders. The power it gives to managers is increasingly being used to support an agenda influenced by a cabal of activists, NGOs, representatives of the "international community," and politicians too arrogant to go through the usual legislative process.

And there we have it. The logic of "stakeholder capitalism," as developed by Klaus Schwab, is at the root of the whole shitshow that we can now see in the gross mismanagement of the global public health response to COVID-19 and the COVIDcrisis. This is where the rot took hold.

Turning back to Andrew Stuttaford, here is more from his August 2021 analysis of "stakeholder capitalism" [410]:

Stakeholder capitalism is an expression of corporatism. Some of these rogues, cynics, if you like, of the "wrong" sort, have their eyes on an even bigger prize, securing for themselves an important—and, one way or another, well-rewarded—role in the corporatist society that is now under construction in this country. Such a society is not, regardless of the sound of that adjective, one dominated by big business, but one, run, in theory anyway, by and for various interest groups, players in an orchestra, with the state acting as a conductor. Corporatism can be relatively benign—its traces are visible in, say, post-war West Germany—it is also the socioeconomic model (again, in theory) underlying fascist and fascist-adjacent regimes in mid-century Europe and Argentina. The U.S. is not headed the whole way down that path, but our current iteration of corporatism will end up as considerably more assertive than anything seen during the years of the Wirtschafts-wunder—and it is more likely to lead to economic decline than an economic miracle. It won't be great for democracy either.

Cynics of the right sort, on the other hand, were presumably gambling that a bit of fancy talk—that they were fully in favor of the BRT's "transformative statement" and so on, might be enough to keep the enemies of shareholder primacy at bay.

And seen through this lens, what transpired between Elon Musk and the Twitter Board of Directors is a huge blow to the logic of stakeholder capitalism as implemented over at corporate Twitter, prompting cries of anguish concerning the need to "protect" the Twitterati by the BBC [411] (a bastion of stakeholder capitalism logic including "nudging" and the Trusted News Initiative), as well as from Thierry Breton, the EU's "commissioner for the internal market."

This is the same logic that has led Barack Obama to promote censorship [412]. But in reality, the true primary agenda is to protect the interests of corporate elites ("Davos Man") who apparently often have a child-like need to be venerated for their social contributions. Or maybe it is all just a convenient smoke shield to obscure their real agenda—to own everything.

It is time to wake up and recognize that the levers of global power are being taken over by a commercial trade organization that represents the interests of the 1,000-member companies—typically global enterprises with more than five trillion US dollars in turnover, which are its primary contributors.

All of the other smokescreens, wordplay/lobbying, coordinated censorship and propaganda, trappings, and training programs that the WEF has implemented are merely tools designed to achieve the business objectives of those 1,000 companies and their wealthy owners. And it is being hidden behind a curtain called "stakeholder capitalism." This is corporatism or fascism (choose your favorite term) deployed on a global scale, financed by the global titans of industry, "Davos Man."

PART THREE

TREATMENT PLAN – SYNTHESIS AND CONCLUSIONS

Everywhere I travel, during almost every interview or podcast, I get asked some form of the question "How do we move forward from this?" Indeed. Most were like I was, living life in sort of a reality bubble, consuming corporate "mainstream" media. Most of us believing for the most part that the United States government was far from perfect, but also far better than most others. That the NIH, FDA, and CDC had some issues with pharmaceutical industry capture, some issues with corrupted leadership, but for the most part were committed to the mission of protecting citizens' health.

Then during the COVIDcrisis, it became as if I, and most of us, had actually been living in a darkened room. Living in Plato's cave. And we backed into a light switch, or somehow stepped out into the light, and we could never see things the same way again. I had forgotten about fellow Salk Institute trainee Michael Crichton's "Gell-Mann Amnesia effect" (discussed in Chapter 32). Dr. Paul Marik once lamented to me that he once had a quiet life, would come home from work and read the *New York Times* thinking nothing of his acceptance of all that was written, but now he could not even stand to look at that paper.

The subtitle of this book bravely and optimistically asserts that there are better times ahead. Frankly, trying to imagine how we get from here to better times for ourselves and our children has been the biggest challenge for me in writing this volume. What better future, when oligarchs, monopolists, and totalitarians invested in the logic of transhumanism, the next industrial revolution that will lead to a fusion of man and machine, and a utilitarian belief system that if only they could collect enough data on all of us then they could force socialism to work. Their kind of globalized socialism, based on deployment and enforcement of social credit scores, central bank digital currency, personal identification numbers for all, and medical tyranny coupled with a command economy. Socialism for us, not for them. All for our own good, of course. A world in which you will own nothing and be happy. One based on a new economic model in which all assets and resources are held and "managed" by small numbers of individuals and megacorporations, who will allocate those resources (using a rent-based model, of course) in the best possible way, to allow us all to live in the best of all possible worlds. Yes, they acknowledge that socialism has repeatedly failed, but that was only because there was not enough data, enough processing power, enough control of thought, speech, and behavior. After all, they tell us, the biggest threat to Democracy is free speech. And independence. And both national and individual sovereignty. And too much decentralization. The independent nation-state is an anachronism. The time has come for the bankers and economists, and the megarich whom they service, to rule the world properly. For our own good.

I guess I am old-fashioned. Please just leave me out of this workers' paradise, thank you very much. Like Voltaire's *Candide*, I prefer to go work in the garden.

As I have traveled all over the USA, Canada, Mexico, and Europe over the last three years, and spent so, so many hours on countless Zoom calls and podcast recordings with people from all over the world, my sense is that there could be a better future lurking under cover of the "Great Reset." Many refer instead to the "Great Awakening." Of a more decentralized future, in which intentional communities share a commitment to integrity, the sanctity of human dignity, and a commitment to community—to one another as human beings. A world where servant leadership is nurtured and valued, rather than the current worship of self-aggrandizing narcissistic sociopathic monopolists. A world in which children are recognized and respected as the only true form of immortality on this Earth.

How do we get there? If only 10% of people really want to be free, as Dr. Mattias Desmet teaches, and the rest merely want to be told what to do, must that ten percent just retreat into small refuges? Become preppers, hoarding stocks of food, ammunition, and firearms while awaiting the black helicopters and United Nations stormtroopers? Do we dive headlong into a new dark ages? Or is there another way? Could there be a new renaissance, a new age of enlightenment?

Yet once again I say, do not allow me or anyone else tell you what to think. Think for yourself. We need to work together to find new solutions, new models, new philosophies. Utilitarianism, Marxism, Malthusianism. These ways of imagining the nature of man and his environment are yesterday's visions. Let's walk together for a while, and think a bit about what a better tomorrow might look like. What I want to know is, *"Are you kind?"*

The following is an outline and overview of my suggestions for a treatment plan. It will not be easy. It never is. Now, let's get to work and see if we can together find a path toward the better times ahead.

CHAPTER 29
What to Do with a Problem Like HHS?

Defining the Problem: HHS and the Administrative State

Many had come to believe that when Dr. Anthony Fauci either resigned or was removed from his position as Director of the National Institute of Allergy and Infectious Diseases (NIAID), then the whole COVIDcrisis problem of chronic, strategic, and tactical administrative overreach, dishonesty, mismanagement, and ethical breaches within the US Department of Health and Human Services (HHS) would be resolved. Under this theory, Dr. Fauci was responsible for policies that were developed during the AIDScrisis and then flourished during the COVIDcrisis, and once the tumor is removed, the patient will recover. I strongly disagree with this magical thinking; I believe that Dr. Fauci represents a symptom, not the cause of the current problems within HHS. Dr. Fauci, who joined the HHS bureaucracy as a way to avoid the Viet Nam draft and personifies many of the administrative problems that have accelerated since that period, would merely be replaced by another NIAID director who might even become worse. The underlying problem is a perverted bureaucratic system of governance that is completely insulated from functional oversight by elected officials.

The "administrative state" is a general term used to describe the entrenched form of government that currently controls almost all levers of federal power in the United States, with the possible exception of the Supreme Court of the United States (SCOTUS). The premature leaking of the SCOTUS majority decision concerning *Roe v. Wade* to corporate press allies was essentially a preemptive strike by the administrative state in response to an action that threatened its power. The threat being mitigated was the constitutionalist logic upon that the legal argument was based, that authority to define rights not specifically defined in the US Constitution as being federally granted vests with individual states. Played out under the political cover of one of the most contentious political topics in modern US history, this was merely another skirmish demonstrating that the entrenched bureaucracy and its allies in the corporate media will continue to resist any constitutional or statutory restrictions on its power and privilege. Resistance to any form of control or oversight has been a consistent bureaucratic behavior throughout the history of the United States government, and this trend has accelerated since the end of the Second World War. More recently, this somewhat existential Constitutionalist threat to the Administrative State was validated in the case of West Virginia vs The Environmental Protection Agency, in which the court determined that when federal agencies issue regulations with sweeping economic and political consequences, the regulations are presumptively invalid unless Congress has specifically authorized the action. With this decision, for the first time in modern history boundaries have started to be imposed on the expansion of the power of unelected senior administrators within the federal bureaucracy.

Legal Underpinning for the Administrative State.
"Nondelegation Doctrine"

Administrative law rests on two fictions. The first, the nondelegation doctrine, imagines that Congress does not delegate legislative power to agencies. The second, which flows from the first, is that the administrative state thus exercises only executive power, even if that power sometimes looks legislative or judicial. These fictions are required by a formalist reading of the Constitution, whose Vesting Clauses permit only Congress to make law and the president only to execute the law. This formalist reading requires us to accept as a matter of practice unconstitutional delegation and the resulting violation of the separation of powers, while pretending as a matter of doctrine that no violation occurs.

The nondelegation doctrine is a principle in administrative law that Congress cannot delegate its legislative powers to other entities. This prohibition typically involves Congress delegating its powers to administrative agencies or to private organizations.

In *J.W. Hampton v. United States*, 276 U.S. 394 (1928), the Supreme

Court clarified that when Congress does give an agency the ability to regulate, Congress must give the agencies an "intelligible principle" on which to base their regulations. This standard is viewed as quite lenient and has rarely, if ever, been used to strike down legislation.

In *A.L.A. Schechter Poultry Corp. v. United States*, 295 U.S. 495 (1935), the Supreme Court held that "Congress is not permitted to abdicate or to transfer to others the essential legislative functions with which it is thus vested."

"Chevron deference"

One of the most important principles in administrative law, which is the branch of law governing the creation and operation of administrative agencies, is the "Chevron deference." This is a term coined after a landmark case, *Chevron U.S.A., Inc. v. Natural Resources Defense Council, Inc.*, 468 U.S. 837 (1984), referring to the doctrine of judicial deference given to administrative actions.

In essence, the Chevron deference doctrine is that when a legislative delegation to an administrative agency on a particular issue or question is not explicit but rather implicit, a court may not substitute its own interpretation of the statute for a reasonable interpretation made by the administrative agency. In other words, when the statute is silent or ambiguous with respect to the specific issue, the question for the court is whether the agency's action was based on a permissible construction of the statute.

Generally, to be accorded Chevron deference, the agency's interpretation of an ambiguous statute must be permissible, which the court has defined to mean "rational" or "reasonable." In determining the reasonableness of a particular construction of a statute by the agency, the age of that administrative interpretation as well as the congressional action or inaction in response to that interpretation at issue can be a useful guide.

Judicial Threats to the Administrative State

None of the issues involved in current debates over these two core doctrines of administrative law has the power to fully deconstruct the administrative state. But current debates and decisions could contribute some constitutionally informed limits on the power, discretion, and independence of unelected administrators. Together, recent and pending Supreme Court decisions might help reconstruct a constitutional state which is more closely aligned with the original intent and vision of the founders.

Very few appreciate that these issues underlie recent decisions concerning whom to appoint to the Supreme Court. Trump's first two appointments to the high court—Neil Gorsuch and Brett Kavanaugh—were two of the nation's leading judicial minds on administrative law, and White House Counsel Don McGahn made clear that this was no coincidence. So too with Trump's appointments to the lower courts, which included administrative-law experts such as the D.C. Circuit's Neomi Rao and Greg Katsas, and the Fifth Circuit's Andrew Oldham.

COVIDcrisis and the Administrative State

The arc of the history of the COVIDcrisis encompasses collusive planning between a wide range of corporate interests, globalists, and the administrative state. This includes pandemic event planning with all of the above actors for a coronavirus outbreak, such as Event 201, which was held at Johns Hopkins University in partnership with the WEF and the Bill and Melinda Gates Foundation [413]; subsequent efforts to cover up administrative state culpability in creating the crisis; followed by gross mismanagement of public health policies, decision making, and communication all acting in lockstep with the preceding planning sessions [414]. This dysfunctional planning-response coupling revealed for all to see that the US Department of Health and Human Services has become a leading example illustrating the practical consequences of this degenerate, corrupt, and unaccountable system of government.

Across two administrations led by presidents who have championed very different worldviews, HHS COVIDcrisis policies have continued with little or no change; one administration seemingly flowing directly into the next with hardly a hiccup. If anything, under Biden the HHS arm of the US administrative state became more authoritarian, more unaccountable, and more decoupled from any need to consider the general social and economic consequences of their actions. As this has progressed, the HHS bureaucracy has become increasingly obsequious and deferential to the economic interests of the medical-pharmaceutical-industrial complex. This is most clearly evident in the maintenance of a state of medical emergency, which provides HHS bureaucrats with almost unlimited powers to bypass constitutional restrictions, despite the clear evidence that there is no longer any medical emergency. Maintaining the ruse of an official public health emergency has been necessary both to maintain power as well as US Government contract revenue for those corporations who have been making obscene profits from selling the "Emergency Use Authorized" medical countermeasures that have been allowed to bypass long-established regulatory, bioethical, and legal liability norms. A public-private partnership like nothing the US had ever seen before, making the War Profiteering against which Harry Truman had campaigned look like child's play [415].

There is an organizational paradox that enables immense power to be amassed by those who have risen to the top of the civilian scientific corps within HHS. These bureaucrats have almost unprecedented access to the public purse, are technically employed by the executive, but are also almost completely protected from accountability by the executive branch of government that is tasked with managing them—and therefore these bureaucrats are unaccountable to those who actually pay the bills for their activities (taxpayers). To the extent these administrators are able to be held to task, this accountability flows indirectly from Congress. Their organizational budgets can be either enhanced or cut during following fiscal years, but otherwise they are largely protected from corrective action including termination of employment absent some major moral transgression. In a Machiavellian sense, these senior administrators function as The Prince, each federal health institute functions as a semiautonomous city-state, and the administrators and their respective courtiers act accordingly. As validation for this analogy, we have the theater observed on C-SPAN each time a minority congressperson or senator queries an indignant scientific administrator, such as has been repeatedly observed with Anthony Fauci's haughty exchanges during congressional testimony.

In his masterpiece *The Best and the Brightest: Kennedy-Johnson Administrations*, David Halberstam cites a quote from *New York Times* reporter Neil Sheehan to illustrate the role of the administrative state on the series of horrifically poor decisions that resulted in one of the greatest US public policy failures of the 20th century—the Vietnam War [416]. In retrospect, the parallels between the mismanagement, propaganda, willingness to suspend prior ethical norms, and chronic lies that define that deadly fiasco are remarkably similar to those that characterize the COVIDcrisis response.

And as in the present, the surreptitious hand of the US intelligence community was often in the background, always pushing the boundaries of acceptable behavior. Quoting from Halberstam and Sheehan:

> Since covert operations were part of the game, over a period of time there was in the high levels of the bureaucracy, particularly as the CIA became more powerful, a gradual acceptance of covert operations and dirty tricks as part of normal diplomatic-political maneuvering; higher and higher government officials became co-opted (as the President's personal assistant, McGeorge Bundy would oversee the covert operations for both Kennedy and Johnson, thus bringing, in a sense, presidential approval). It was a reflection of the frustration which the national security people, private men all, felt in matching the foreign policy of a totalitarian society, which gave so much more freedom to its officials and seemingly provided so few checks on its own leaders. To be on the inside and oppose or question covert operations was considered a sign of weakness. (In 1964 a well-bred young CIA official, wondering whether we had the right to try some of the black activities on the North, was told by Desmond FitzGerald, the number-three man in the Agency, "Don't be so wet"—the classic old-school putdown of someone who knows the real rules of the game to someone softer, questioning the rectitude of the rules.) It was this acceptance of covert operations by the Kennedy Administration which had brought Adlai Stevenson to the lowest moment of his career during the Bay of Pigs, a special shame as he had stood and lied at the UN about things that he did not know, but which, of course, the Cubans knew. Covert operations often got ahead of the Administration itself and pulled the Administration along with them, as the Bay of Pigs had shown—since the planning and training were all done, we couldn't tell those freedom-loving Cubans that it was all off, could we, argued Allen Dulles. He had pulled public men like the President with him into that particular disaster. At the time, Fulbright had argued against it, had not only argued that it would fail, which was easy enough to say, but he had gone beyond this, and being a

public man, entered the rarest of arguments, an argument against it on moral grounds, that it was precisely our reluctance to do things like this which differentiated us from the Soviet Union and made us special, made it worth being a democracy. One further point must be made about even covert support of a Castro overthrow; it is in violation of the spirit and probably the letter as well, of treaties to which the United States is a party and of U.S. domestic legislation. . . . To give this activity even covert support is of a piece with the hypocrisy and cynicism for which the United States is constantly denouncing the Soviet Union in the United Nations and elsewhere. This point will not be lost on the rest of the world—nor on our own consciences for that matter,' he wrote Kennedy. But arguments like this found little acceptance in those days; instead, the Kennedy Administration had been particularly aggressive in wanting to match the Communists at new modern guerrilla and covert activities, and the lines between what a democracy could and could not do were more blurred in those years than others.

These men, largely private, were functioning on a level different from the public policy of the United States, and years later when *New York Times* reporter Neil Sheehan read through the entire documentary history of the war, that history known as the Pentagon Papers, he would come away with one impression above all, which was that the government of the United States was not what he had thought it was; it was as if there were an inner U.S. government, what he called "a centralized state, far more powerful than anything else, *for whom the enemy is not simply the Communists but everything else, its own press, its own judiciary, its own Congress, foreign and friendly governments—all these are potentially antagonistic.* It had survived and perpetuated itself," Sheehan continued, "often using the issue of anti-Communism as a weapon against the other branches of government and the press, and finally, *it does not function necessarily for the benefit of the Republic but rather for its own ends, its own perpetuation; it has its own codes which are quite different from public codes. Secrecy was a way of protecting itself, not so much from threats by foreign governments, but from detection from its own population on charges of its own competence and wisdom.*" Each succeeding Administration, Sheehan noted, was careful, once in office, not to expose the weaknesses of its predecessor. After all, essentially the same people were running the governments, they had continuity to each other, and each succeeding Administration found itself faced with virtually the same enemies. Thus the national security apparatus kept its continuity, and every outgoing President tended to rally to the side of each incumbent President.

The parallels of organizational culture are uncanny and, as previously discussed, have flourished under the guise of the need to manage the national biodefense enterprise. Since the 2001 *Amerithrax* Anthrax spore "attacks," HHS has increasingly been horizontally integrated with the intelligence community (see Chapter 21) as well as with the Department of Homeland Security to form a health security state with enormous ability to shape and enforce "consensus" through widespread propaganda, censorship, "nudge" technology, and intentional manipulation of the "Mass Formation" hypnosis process using modern adaptations of methods originally developed by Dr. Joseph Goebbels [417].

The Administrative State and Inverted Totalitarianism

The term "inverted totalitarianism" was first coined in 2003 by the political theorist and writer Dr. Sheldon Wolin, and then his analysis was extended by Chris Hedges and Joe Sacco in their 2012 book *Days of Destruction, Days of Revolt* [418]. Wolin used the term "inverted totalitarianism" to illuminate totalitarian aspects of the American political system, and to highlight his opinion that the modern American federal government has similarities to the historic German Nazi government. Hedges and Sacco built upon Wolin's insights to extend the definition of inverted totalitarianism to describe *a system where corporations have corrupted and subverted democracy,* and where macroeconomics has become the primary force driving political decisions (rather than ethics, Maslow's hierarchy of needs, or vox populi). Under inverted totalitarianism, every natural resource and living being becomes commodified and exploited by large corporations to the point of collapse, as excess consumerism and sensationalism lull and manipulate the citizenry into surrendering their liberties and their participation in government. *Inverted totalitarianism* is now what the government of the United States has devolved into, as Wolin had warned might happen many years ago in his book *Democracy Incorporated* [418]. The administrative state has turned the USA into a "managed democracy." This monster is also referred to as the "*deep state*," the civil service, the centralized state, or the *administrative state*.

Political systems that have devolved into inverted totalitarianism do not have an authoritarian leader, but instead are run by a nontransparent group of bureaucrats. The "leader" basically serves the interests of the true bureaucratic administrative leaders. In other words, an unelected, invisible ruling class of bureaucrat-administrators runs the country from within.

Corporatist (Fascist) Partnering with the Administrative State

As noted earlier, because science, medicine, and politics are three threads woven into the same cloth of public policy, we will have to work to fix all three simultaneously. The corruption of political systems by global corporatists has filtered down to our science, medicine, and healthcare systems. The perversion of science and medicine by corporate interests is expanding its reach; it is pernicious and intractable. Regulatory capture by corporate interests runs rampant throughout our politics, governmental agencies, and institutes. The corporatists have infiltrated all three branches of government. Corporate-public partnerships that have become so trendy have another name; that name is *fascism*—the political science term for the fusion of the interests of corporations and the state. Basically, the tension between the interest of the republic and its citizens (which Jefferson felt should be primary) and the financial interests of business and corporations (Hamilton's ideal) has swung far too far to the interests of corporations and their billionaire owners at the expense of the general population.

Development of inverted totalitarianism is often driven by the personal financial interests of individual bureaucrats, and many Western democracies have succumbed to this process. Bureaucrats are easily influenced and co-opted by corporate interests due to both the lure of powerful jobs after federal employment ("revolving door") and the capture of legislative bodies by the lobbyists serving concealed corporate interests.

In the case of both the Centers for Disease Control and Prevention (CDC) and the National Institutes of Health (NIH), there are also direct financial ties that bind corporations, philanthropic capitalist nongovernmental organizations (such as the Bill and Melinda Gates Foundation), and the administrative state. The likes of you and I cannot "give" to the federal government, as under the Federal Acquisition Regulations this is considered to be a risk for exerting undue influence. But the CDC has established a nonprofit "CDC Foundation." According to the CDC's own website [419]:

> Established by Congress as an independent, nonprofit organization, the CDC Foundation is the sole entity authorized by Congress to mobilize philanthropic partners and private-sector resources to support CDC's critical health protection mission.

Likewise, the NIH has established the "Foundation for the National Institutes of Health," currently headed by CEO Dr. Julie Gerberding (formerly CDC director, then president of Merck Vaccines, then chief patient officer and executive vice president, Population Health & Sustainability at Merck and Company—where she had responsibility for Merck's ESG score compliance). Dr. Gerberding's career provides a case history illustrating the ties between the administrative state and corporate America.

These congressionally chartered nonprofit organizations provide a vehicle whereby the medical-pharmaceutical complex can funnel money into the NIH and CDC to influence both research agendas and policies.

And then we have the strongest ties that bind the for-profit medical-pharmaceutical complex to CDC and NIH employees and administrators, the Bayh-Dole act. Wikipedia provides a succinct summary:

> The Bayh–Dole Act or Patent and Trademark Law Amendments Act (Pub. L. 96-517, December 12, 1980) is United States legislation permitting ownership by contractors of inventions arising from federal government-funded research. Sponsored by two senators, Birch Bayh of Indiana and Bob Dole of Kansas, the Act was adopted in 1980, is codified at 94 Stat. 3015, and in 35 U.S.C. § 200–212, and is implemented by 37 C.F.R. 401 for

federal funding agreements with contractors and 37 C.F.R 404 for licensing of inventions owned by the federal government.

A key change made by Bayh–Dole was in the procedures by which federal contractors that acquired ownership of inventions made with federal funding could retain that ownership. Before the Bayh–Dole Act, the Federal Procurement Regulation required the use of a patent rights clause that in some cases required federal contractors or their inventors to assign inventions made under contract to the federal government unless the funding agency determined that the public interest was better served by allowing the contractor or inventor to retain principal or exclusive rights. The National Institutes of Health, National Science Foundation, and the Department of Commerce had implemented programs that permitted nonprofit organizations to retain rights to inventions upon notice without requesting an agency determination. By contrast, Bayh–Dole uniformly permits nonprofit organizations and small business firm contractors to retain ownership of inventions made under contract and that they have acquired, provided that each invention is timely disclosed and the contractor elects to retain ownership in that invention.

A second key change with Bayh-Dole was to authorize federal agencies to grant exclusive licenses to inventions owned by the federal government.

While originally intended to create incentives for federally funded academia, nonprofit organizations, and federal contractors to protect inventions and other intellectual property so that the intellectual products of taxpayer investments could help drive commercialization, the terms of Bayh-Dole have now also been applied to federal employees, resulting in massive personal payments to specific employees as well as the agencies, branches, and divisions for which they work. This creates perverse incentives for federal employees to favor specific companies and specific technologies that they have contributed relative to competing companies and technologies. This policy is particularly insidious in the case of federal employees who have a role in determining the direction of research funding allocation, such as is the case with Dr. Anthony Fauci [420–422].

Treating the Disease: HHS, The Administrative State, and Inverted Totalitarianism

To help understand and prioritize the stack of possible responses to the advanced state of corruption within the US HHS, it is useful to think of a pyramid-shaped hierarchy of problems and issues. The origin of these issues and the overall Administrative State can be traced to the Pendleton Act of 1883, which was established to end the patronage system that had preceded it. Of necessity, the brief analysis below will only highlight a few of the issues with a particular focus on the COVIDcrisis, as a comprehensive summary and action plan would require hundreds if not thousands of pages of texts, graphs, and figures.

To provide context concerning the size of the HHS Administrative State, the president's FY 2022 HHS budget proposes $131.8 billion in discretionary budget authority and $1.5 trillion in mandatory funding [423]. In contrast, president's FY 2022 budget request for DoD is $715 billion [424]. According to Federal News Network, the president's Budget Request included approximately $62.5 billion for NIH, compared to $42.9 billion the agency received in the 2022 continuing resolution, and $42.8 billion in the final 2021 budget [425]. The request represents a 7.2% increase for research project grants, a 50% increase in the buildings and facilities appropriation, and a 5% increase for training. The 2023 proposal includes $12.1 billion more for pandemic preparedness, and an additional $5 billion to stand up the new Advanced Research Project Agency for Health (ARPA-H). Based on 2022 numbers, the NIH budget (alone, not including ASPR/BARDA) represents 8.7% of the entire DoD budget [426].

Stopping Administrative State COVIDcrisis Overreach

The foundation of the HHS COVIDcrisis mismanagement is built upon the authorization that has allowed the HHS arm of the Administrative State to suspend a wide range of federal statutes and functionally bypass various aspects of the Bill of Rights of the US Constitution: the "Determination that a Public Health Emergency Exists." First signed by HHS Secretary Alex Azar on January 31, 2020, it was then renewed by Azar/Trump effective April 26, 2020, and again on July 23 (Azar/Trump), again on October 2, 2020 (Azar/Trump), January 7, 2021 (Azar/Trump), and then we switch presidential administrations. The Biden administration did not miss a beat

[427]. On January 22, 2021, Acting HHS Secretary Norris Cochran notified governors across the country of details concerning the ongoing public health emergency declaration for COVID-19 [428]. Among other things, Acting Secretary Cochran indicated that HHS will provide states with sixty days' notice prior to the termination of the public health emergency declaration for COVID-19. HHS Secretary Xavier Becerra then began renewing the Determination that a Public Health Emergency Exists on April 15, 2021, renewed July 19, 2021; October 15, 2021; January 14, 2022; and April 12, 2022. Based on this schedule, another renewal was authorized in July 2022. All of this is based upon the authority granted to the HHS arm of the Administrative State by Congress when it passed the Pandemic and All Hazards Preparedness Reauthorization Act (PAHPRA) in 2013.

According to the Office of the Assistant Secretary for Preparedness and Response, the Pandemic and All Hazards Preparedness Reauthorization Act (PAHPRA) amended section 564 of the Federal Food, Drug and Cosmetic (FD&C) Act, 21 U.S.C. 360bbb-3, is intended to provide more flexibility to the Health and Human Services secretary to authorize the US Food and Drug Administration (FDA) to issue an Emergency Use Authorization (EUA) [429]. The secretary is no longer required to make a formal determination of a public health emergency under section 319 of the Public Health Service Act, 42 U.S.C. 247d before declaring that circumstances justify issuing an EUA. Under section 564 of the FFD&C Act, as amended, the secretary now may determine that there is a public health emergency or significant potential for a public health emergency that affects, or has significant potential to affect, national security or the health and security of US citizens living abroad and involves a biological, chemical, radiological, or nuclear agent or disease or condition that may be attributable to such agent(s). The secretary may then declare that the circumstances justify emergency authorization of a product, enabling the FDA to issue an EUA before the emergency occurs.

Based on my understanding of Federal Administrative Law, the PAHPRA is unconstitutional and should be immediately rescinded by the courts due to the nondelegation doctrine. In my opinion, this is the first action that should be taken to dismantle the HHS overreach that has yielded the COVIDcrisis public health fiasco and will not require a major electoral turnover before proceeding. As previously discussed, the "nondelegation doctrine" is arguably the most significant Administrative State issue being actively considered within the current Supreme Court. The theory is predicated on the Constitution's Article I, which provides that all legislative powers herein granted shall be vested in Congress. This grant of power, the argument goes, cannot be redelegated to the executive branch. If Congress grants an agency effectively unlimited discretion (as it has with PAHPRA), then it violates the constitutional "nondelegation" rule. If the PAHPRA is overturned, then the whole cascade of HHS Administrative State actions that have enabled bypassing of normal bioethical (see the "Common Rule" 48 CFR § 1352.235-70 - Protection of human subjects) and both normal drug and vaccine regulatory procedures would collapse. Furthermore, the PAHPRA is what enables the Emergency Use Authorization (EUA) of drugs and vaccines, and if PAHPRA is overruled, the regulatory authorization for these unlicensed EUA-allowed would be jeopardized [430]. In addition to challenging the legitimacy of the PAHPRA based on the nondelegation doctrine, similar challenges should be raised with the 21st Century Cures Act (HR 34; PL: 114-255) and Public Law 115-92 (HR 4374).

Dismantling the HHS Administrative State

The leadership hierarchy of the US Federal Administrative State is structured along the same lines as the military, with a progressive series of general service ranks (GS-1 through GS-15, with 15 being the most senior) that are led by a separate leadership group called the Senior Executive Service (SES V through I, with SES I being most senior), which oversees civilian government operations. According to the Office of Personnel Management [431]:

> The Senior Executive Service (SES) lead America's workforce. As the keystone of the Civil Service Reform Act of 1978, the SES was established to ". . . ensure that the executive management of the Government of the United States is responsive to the needs, policies, and goals of the Nation and otherwise is of the highest quality." These leaders possess well-honed executive skills and share a broad perspective on government and *a public service commitment that is grounded in the Constitution.*
>
> Members of the SES serve in the key positions just below the top Presidential appointees. SES members are the major

link between these appointees and the rest of the Federal work-force. *They operate and oversee nearly every government activity in approximately 75 Federal agencies.*

The U.S. Office of Personnel Management (OPM) manages the overall Federal executive personnel program, providing the day-to-day oversight and assistance to agencies as they develop, select, and manage their Federal executives.

In general, the SES is the leadership of the administrative state, but it is not the only category of employment that has amassed power. Dr. Anthony Fauci, one of the highest paid federal employees ($434,312 base salary), is exempt from being a member of the SES but rather serves taxpayers as a medical officer at the National Institutes of Health in Bethesda, Maryland. Medical officer was the 10th most popular job in the US Government during 2020, with 33,865 employed under this category. Anthony S. Fauci is employed at the highest medical officer rank of RF-00 under the employees appointed and compensated as special consultants under 42 U.S.C. § 209(f).

Despite the fact that Dr. Fauci is a consultant, he is still subject to 42-160 Conduct Laws and Regulations, which states that Title 42 employees must comply with all ethical and conduct-related laws and regulations applicable to other Executive Branch employees. These include laws concerning financial interests, financial disclosure, and conduct regulations promulgated by the Department, the Office of Government Ethics, and other agencies. Discharge of Title 42 employees under the ethical and conduct-related laws and regulations applicable to Executive Branch employees, or to 42-140 Performance Management and Conduct breaches (for example, lying in sworn congressional testimony), often requires up to two years of legal processes, which gives rise to the common practice of assigning such personnel to a proverbial "broom closet" office without windows, telephone, or assigned tasks.

Jeffrey Tucker of the Brownstone Institute has summarized one set of strategies developed to dismantle the Administrative State. President Trump tried to break the power of the SES using a series of executive orders (E.O. 13837, E.O. 13836, and E.O.13839) that would have diminished the access of federal employees (including the SES) to labor-union protection when being pressed on the terms of their employment [432]. All three of these were struck down with a decision by a DC District Court. The presiding judge was Ketanji Brown Jackson, who was later rewarded for her decision with a nomination to the Supreme Court, which was affirmed by the US Senate. Jackson's judgment was later reversed, but Trump's actions were embroiled in a juridical tangle that rendered them moot. However, in light of the recent Supreme Court decisions, it is possible that the structure of these executive orders may withstand future judicial action. Two weeks before the 2020 general election, on October 21, 2020, Donald Trump issued an executive order (E.O. 13957) on "Creating Schedule F in the Excepted Service." that was designed to overcome the prior objections and involved creation of a new category of federal employment called Schedule F. Employees of the federal government classified as Schedule F would have been subject to control by the elected president and other representatives, and these employees would have included:

> Positions of a confidential, policy-determining, policy-making, or policy-advocating character not normally subject to change as a result of a Presidential transition shall be listed in Schedule F. In appointing an individual to a position in Schedule F, each agency shall follow the principle of veteran preference as far as administratively feasible.

The order demanded a thorough governmental review of what is essentially a reclassification of the SES.

> Each head of an executive agency (as defined in section 105 of title 5, United States Code, but excluding the Government Accountability Office) shall conduct, within 90 days of the date of this order, a preliminary review of agency positions covered by subchapter II of chapter 75 of title 5, United States Code, and shall conduct a complete review of such positions within 210 days of the date of this order.

The *Washington Post*, which often functions as the official organ of the Administrative State, certainly appreciated the power of this approach when it was proposed, breathlessly posting an op-ed titled "Trump's newest executive order could prove one of his most insidious" [433]:

The directive from the White House, issued late Wednesday, sounds technical: creating a new "Schedule F" within the "excepted service" of the federal government for employees in policymaking roles, and directing agencies to determine who qualifies. Its implications, however, are profound and alarming. It gives those in power the authority to fire more or less at will as many as tens of thousands of workers currently in the competitive civil service, from managers to lawyers to economists to, yes, scientists. This week's order is a major salvo in the president's onslaught against the cadre of dedicated civil servants whom he calls the "deep state"—and who are really the greatest strength of the U.S. government.

Jeffrey Tucker summarizes the subsequent cascade of events:

Ninety days after October 21, 2020 would have been January 19, 2021, the day before the new president was to be inaugurated. The *Washington Post* commented ominously: "Mr. Trump will try to realize his sad vision in his second term, unless voters are wise enough to stop him."

Biden was declared the winner due mostly to mail-in ballots.

On January 21, 2021, the day after inauguration, Biden reversed the order. It was one of his first actions as president. No wonder, because, as The Hill reported, this executive order would have been "the biggest change to federal workforce protections in a century, converting many federal workers to 'at will' employment."

How many federal workers in agencies would have been newly classified at Schedule F? We do not know because only one completed the review before their jobs were saved by the election result. The one that did was the Congressional Budget Office. Its conclusion: fully 88% of employees would have been newly classified as Schedule F, thus allowing the president to terminate their employment.

This would have been a revolutionary change, a complete remake of Washington, DC, and all politics as usual.

If the HHS Administrative State is to be dismantled, so that it will become possible to manage the various Executive Branch agencies once again, Schedule F provides an excellent strategy and template to achieve the objective. If this most important of all tasks is not achieved, then we will remain at risk that HHS will once again attempt to trade our national sovereignty for additional power by aligning with the WHO, as was recently attempted in the case of the surreptitious January 28, 2022, proposed modifications to the International Health Regulations [434]. These actions, which were not made public until April 12, 2022, clearly demonstrate that the HHS Administrative State represents a clear and present danger to the US Constitution and national sovereignty and must be dismantled as soon as possible.

Stopping Corporate-Administrative Collusion and Corruption

The third core problem that must be addressed involves the various laws, administrative policies, and surreptitious practices that have empowered the symbiotic (or parasitic?) alliance that has formed between the medical-pharmaceutical complex and the HHS Administrative State. Once again, it is important to recognize the fundamental political structure that has been created: a Fascist Inverted Totalitarianism. The face of modern fascism is often stereotyped by the corporate press as a group of tiki torch-waving Proud Boys in uniforms marching in Charlottesville and committing acts of violence in person with bats or via automobile. But this is not modern fascism; it is a group of mostly young men aping superficial features of the German Third Reich while wearing outdated uniforms and chanting repugnant slogans designed to provoke outrage. Fascism is a political system that is otherwise known as Corporatism, that being the fusion of corporate and state power. And as previously discussed, currently the real power of the US Government lies in the Administrative State. To break up these "public-private partnerships" that compromise the ability of HHS to perform essential oversight duties and truly protect the health of American Citizens from the rapacious practices and disgusting ethics of the medical-pharmaceutical complex (in which they behave as predators, and we have become the prey), we must sever the financial and organizational ties that bind the medical-pharmaceutical industrial complex to the HHS Administrative State, and that have been incrementally developed and deployed over many decades.

To return balance and congressionally intended function to the HHS, the following steps must be accomplished, none of which can be accomplished until the power of the HHS Administrative State has been broken and the SES has been brought to heel through combined efforts of the Supreme Court, and both a new Congress and a new Executive branch.

1. **The Bayh-Dole act must be modified**, administratively or legislatively, so that it no longer applies to federal employees. HHS scientists and administrators must not be receiving royalties from intellectual property licensed to the medical-pharmaceutical complex, as this creates multiple layers of both explicit and occult financial conflicts of interest.

2. **The congressional charters for the "Foundation for the National Institutes of Health" and the "CDC Foundation" must be revoked**. These public-private partnership organizations have created unaccountable slush funds that are exploited by the HHS Administrative State and SES to circumvent the will of Congress (by enabling activities neither funded nor authorized by Congress) and embody the fusion of interests between the medical-pharmaceutical complex and the HHS Administrative State.

3. **The regulator-industry revolving door.** The revolving door between HHS employees and medical-pharmaceutical complex must somehow be jammed shut. Mere awareness of the probability of lucrative employment by Pharma upon retirement or departure from HHS oversight roles already biases almost every action of FDA and CDC senior and junior staff. I do not know how to accomplish this from a legal standpoint, I just know that the task must be accomplished if the public interest is to be better served.

4. **Industry Fees.** The idea of forcing the medical-pharmaceutical complex to pay for the cost of regulation was naive, and this practice must also be halted. If the taxpaying citizens of the USA want safe and effective vaccines and drugs, then they need to pay for the cost to insure that Pharma is forced to play by the rules. And when it does not, the resulting actions and fines must be so powerful that they cannot just be written off as a cost of doing business.

5. **Vaccine liability indemnification** is another legislative strategy that has clearly failed to meet its intended purpose. The vaccine industry has become an unaccountable monster that is consuming both adults and

children. The National Childhood Vaccine Injury Act (NCVIA) of 1986 (42 U.S.C. §§ 300aa-1 to 300aa-34) was signed into law by United States President Ronald Reagan as part of a larger health bill on November 14, 1986, and has created an incentive structure with the familiar problem of coupling private profit to public risk and has resulted in widespread corruption of both FDA/CBER and CDC.

6. **Speedy approvals**. Yet another "innovation" developed by Congress with wide latitude for implementation by the Administrative State, the Prescription Drug User Fee Act (PDUFA) was a law passed by the United States Congress in 1992 that allowed the Food and Drug Administration (FDA) to collect fees from drug manufacturers to fund the new drug approval process. The inefficiency of the FDA regulatory process has led (largely via administrative fiat) to a series of "expedited approval" pathways, which in turn have been amplified and exploited by Pharma to advance its own objectives, often at the expense of the public. Another case of unintended blowback in which the best laid plans have been twisted by the Administrative State to the point of no longer serving the original intent of Congress. This is another situation that deserves legal scrutiny in light of the revisitation of the nondelegation doctrine.

7. **External Advisors**. External advisors are often used to provide cover for bureaucrats, and particularly for SES staff, so that a carefully handpicked external committee can be relied upon to produce the intended result while allowing the administrator to avoid responsibility and maintain plausible deniability for decisions that may be unpopular with the citizenry but lucrative or otherwise beneficial for the medical-industrial complex. Once again, while the original intent may have been noble, in practice this has become just another tool that the Administrative State has bent to do its bidding as well as that of its corporate partners.

8. **Transparency, conflicts of interest, and data.** If we have learned anything from the COVIDcrisis, it is that the HHS Administrative State is quite willing to withhold data from both outside scientists and the general public. Clearly this must stop, and once again recent district court decisions kindle hope that forcing the SES and Administrative State to become more open and transparent is an achievable objective.

9. **Too big to fail.** Many of the subdivisions of HHS have become too large and unwieldy, and a rigorous assessment of mission, priorities, productivity,

and value provided must be performed followed by breaking up the large power centers (NIAID being one example), refocusing the overall enterprise on health and wellness, and eliminating nonessential functions.

Conclusions

Many voices have been raised that advocate some combination of pitchforks and torches for what the COVIDcrisis has clearly revealed to be a politicized and corrupted HHS and its associated subsidiary agencies and institutes. It may be that it will be necessary to create a parallel organization, mature it to the point that it can assume the essential functions of the current HHS, and then demolish the (at that point) obsolete HHS structure. But in the interim, I am convinced that the reforms proposed above could certainly advance the ball downfield toward an HHS that would provide greater value to US taxpayers and citizens, and that could be more effectively controlled by Congress and the Executive rather than operating largely autonomously to serve the interests of the Administrative State itself.

CHAPTER 30
Groupthink and the Administrative State

This book is really a journey of discovery, an attempt to perform a root cause analysis of the tragedy of what we have all experienced, trying to make sense of the horrid public policy decision making that has resulted in the COVID-crisis. Key milestones along the way have included recognizing the role of the Trusted News Initiative in censoring and slandering counter mainstream narratives and narrators, the rise and global penetration of advocacy journalism, developing a deeper understanding of the role of the World Economic Forum/Great Reset and its young leader training/indoctrination program, the process of Mass Formation or Mass Psychosis (the psychological basis of totalitarianism), widespread regulatory and other forms of government capture, the role of the administrative state, exploitation of the "crisis" by central banks and massive investment funds, the weaponization of infectious disease Fearporn as both a media business model and a political tool, "Nudge" technology and governmental behavior control, and so many other factors that have contributed to the emergent global "COVIDcrisis" phenomenon that has destroyed millions of lives, businesses, children's education, faith in the integrity of science and medicine, and triggered an economic crisis that threatens to bring down the pillars that support Western economies and banking systems.

The tendency of many is to focus on one of these as the root cause, and to overlook the complex global interplay of all factors, a very human bias to seek a single factor or individual that should be held accountable. Favorites are often the WEF/Klaus Schwab (who somehow created himself as a caricature of evil), Bill Gates (likewise), Larry Fink/Blackrock, Bank of America, as well as the Vanguard and State Street investment funds, the small number of banking families that control most of the central banks, the

United Nations, the World Health Organization/Veterinarian Dr. Tedros, the rise of the Administrative-corporate state/inverted totalitarianism, Anthony Fauci, Deborah Birx, the list goes on and on. All have played a role in fostering this global disaster that began with introduction of a new variant of an existing RNA respiratory virus into the global human population and seems to be winding down as one of the greatest global policy failures in human history.

Borrowing a term from economics, most of these interacting factors are more macroscopic in scope. But what about the more microscopic phenomena? Are there systemwide or general organizational behaviors or processes that have contributed to the resulting catastrophic public policy? Have any of these types of effects contributed to the decision making? Are there widespread organizational practices that have enabled something like an emergent fractal process such as that described by Harrison Koehli in his recent essays that critically compare and contrast the work of both Mattias Desmet and Andrew Lobaczewski (see *Political Ponerology: The Science of Evil, Psychopathy, and the Origins of Totalitarianism* [435])? Is there a component of this mess that is a consequence of how our governments and large businesses are organized, some seemingly benign fundamental organizational behaviors that could be clearly identified and therefore are amenable to being altered so that we could reduce the risk of future overreaction and collective global madness?

Many believe that the many tentacles of postmodern relativism and arrival at the logical endpoint of trends in modern liberal individualism provide one explanation, a cultural bias-based explanation for the COVIDcrisis [436, 437]. But to the extent that is true, it will require deep political "generational"-level changes to fix that problem.

What about some of the simple organizational assumptions that are widely taught and dearly held by corporations and governmental policy makers?

With this prelude, let's consider something that has been on my mind a lot lately. There is a style of group decision making that leads to poor group decisions, ergo: groupthink, that has come to dominate management practices in both government and large businesses. Why is that happening?

Groupthink is something distinct from the idea of mass formation that Professor Dr. Mattias Desmet has developed and expanded on. Let's take a moment to examine the theories behind modern management practices in the context of the entire COVIDcrisis. Management practices that have resulted in the public policy responses of the US government as well as the global response by the World Health Organization (WHO) and other nations.

Groupthink has now permeated the administrative state, to such an extent that the methods used to combat groupthink have all but disappeared from the mind-set of group managers.

So, just to recap, what is groupthink?

> *Groupthink*
> Groupthink is a psychological phenomenon that occurs within a group of people in which the desire for harmony or conformity in the group results in an irrational or dysfunctional decision-making outcome. Cohesiveness, or the desire for cohesiveness, in a group may produce a tendency among its members to agree at all costs (Wiki).

Groupthink was popularized by author Irving Janis, first in an article in 1971 and then expanded on in his famous book *Victims of Groupthink*. The focus of Janis's work was on small-group dynamics or theory. Dr. Janus researched and wrote about the behavior of small groups and how groups work in a positive way and also in a negative way. As he was proceeding with his academic work, he realized that he was seeing examples of group behavior gone awry, gone bad, particularly in the context of various federal decisions that occurred during the '60s and then up to his present (originally the Viet Nam War mismanagement, then Watergate).

Dr. Janis analyzed the behaviors of small groups under pressure making key decisions and then also came up with some really clear guidance on how to combat groupthink. Through careful examination of a series of case studies focused on US government foreign policy successes and major policy failures, Janis described how closed insular self-reinforcing groups tend to behave and how they can go wrong.

Friedrich Nietzsche (1844–1900) alluded to the same phenomena by writing: "Madness is the exception in individuals, but the rule in groups." That statement informs Irving Janis's whole analysis and his personal theory, which is captured in the book *Victims of Groupthink: A Psychological Study of Foreign Policy* [438], and then these ideas were later expanded in his other works.

However, the cautionary tales of Janis's groupthink case studies seem to have been left out of modern training, and a primary emphasis in most MBA, corporate, and governmental training programs has shifted to how to develop and reinforce group cohesion and consensus—the precise opposite of what is needed for good group decision making and avoidance of groupthink. The style of Tuckman's "stages of group development" (also known as "norming, storming, forming and performing") is now typically considered the "best practice" for small group dynamics by governments and large corporations. This is what was taught to me when I participated in the yearlong postgraduate Harvard Medical School Global Clinical Scholars Program. This process was first developed in 1965 and has continued to guide most managers in large corporations.

Tuckman's stages of group development
The forming–storming–norming–performing model of group development was first proposed by Bruce Tuckman in 1965, who said that these phases are all necessary and inevitable in order for a team to grow, face up to challenges, tackle problems, find solutions, plan work, and deliver results (Wiki).

This group development process is considered so important to human resource officers that even the American worldwide employment website for job listings, *Indeed*, has a webpage to describe it:

Forming, storming, and norming are stages of psychological development a team encounters while working on a project. Teams pass through each stage to become acquainted, face challenges, tackle problems, find solutions, and eventually focus on achieving a common goal. There are definitive stages teams pass through to develop and complete projects successfully. The stages indicate the common steps many teams follow during development and establishment. Team development is the process of learning to work together effectively as a team.

The problem is that the methods behind Tuckman's "stages of group development" do not address how to avoid groupthink. In fact, Tuckman's stages of group development encourage groupthink.

Here are the five stages:

1. Forming (Group Formation): include displaying "eagerness, socializing, generally polite tone, sticking to safe topics, being unclear about how one fits in, and some anxiety and questioning."

2. Storming (Initial Group Meetings): "resistance, lack of participation, conflict related to differences of feelings and opinions, competition, high emotions, and starting to move towards group norms."

3. Norming (Group grope): *"improved sense of purpose and understanding of goals, higher confidence, improved commitment, team members are engaged and supportive, relief—lowered anxiety, and starting to develop cohesion."*

4. Performing: "Characteristics of Performing include higher motivation, *elevated trust and empathy, individuals typically deferring to the team's needs,* effective production, consistent performance, and demonstrations of interdependence and self-management (also referred to as self-organization)."

5. Adjourning

This is literally the backbone of the processes that lead to groupthink! Yet it is the managerial technique throughout industry and government for group decision making. No wonder the administrative state is failing!

Through the COVIDcrisis, we have discovered that our leaders in different agencies (not just in the US, but also in Canada and around the world) have been making some pretty ill-advised public health management decisions. Of course, as previously discussed, other issues are also at play here: regulatory capture, inverted totalitarianism, top-down (White House, Birx or Fauci) decision making, etc. But at the core, groupthink has come to dominate group decision making in the 21st century. I believe it is partially because the widespread teaching of Tuckman's "stages of group development" has dominated governmental and corporate organizational training programs during the late 20th and early 21st centuries.

As described and defined by Dr. Janis, the core to understanding the development of groupthink psychopathology is the decision-making empowerment of a cohesive group, a small group, that wants to come together, avoid controversy, and believes that they are elite, they are the best of the best. It is fascinating to look back at this core textbook that really first introduced and defined the term "groupthink," which brought it into the public consciousness, and then to evaluate the COVIDcrisis from that perspective. Have we actually seen the characteristics of groupthink in the people that have been leading our response to the COVIDcrisis in the USA, New Zealand, Australia, Canada, and of course the World Health Organization? The answer has become self-evident, a firm—yes.

I'm not in any way saying that the WEF didn't play a role, and the WHO, and the UN, and Pfizer, and BioNTech, and Moderna, and all of these other factors. Many things can be happening at the same time.

One of the things that I found most striking right off the top in Dr. Janis's work is what he describes as *hard-headed actions by soft-headed groups*. He refers to soft-headed thinking as the product of these kinds of cohesive groups where everybody wants to agree with one another. That it is more important to agree with one another than it is to be right. That's one of the characteristics.

Quoting Janis:

> Adhere to group norms and pressures towards uniformity. Just as with ingroups of ordinary citizens, a dominant characteristic appears to be remaining loyal to the group. It's important to be loyal by sticking with the decisions to which the group has committed itself even when the policy is working badly and has unintended consequences that disturb the consciousness of the members. So it's bothering people on a deep level about ethics and things like that, but it's more important that they stay cohesive. In a sense, *the members consider loyalty to the group the highest form of morality.*

That's a crucial statement, a "mic drop" moment. Dr. Janis goes further:

> That loyalty requires each member to avoid raising controversial issues, questioning weak arguments, or calling a halt to soft-headed thinking.

What is a better description of what we've observed during the COVIDcrisis years?

I also want to highlight another of Dr. Janis's insights. Paradoxically, soft-headed groups are likely to be extremely hardhearted toward out-groups and perceived enemies. So, when I think of the behavior that has emerged from the interactions between governmental propagandists, social media, or mainstream media, the behaviors appear to demonstrate classic groupthink features.

One of the other key points that Janis emphasizes all the way through his book is that to avoid groupthink you have to bring in outsiders, or you have to have some mechanism to cause the group to continually reevaluate the decisions it has made. The group must continually test their assumptions regarding reality.

That's clearly something that didn't happen. Quoting again:

> The more amiability and *esprit de corps* among the members of the policy making in-group, the more likely the group will fall into groupthink.

So, think back to Operation Warp Speed, Big Tech, Twitter, TNI, WHO, national governments. Think how they behaved. The leaders of these groups clearly shared a lot of esprit de corps during COVID-19 (they believed that they were saving the world)! They clearly knew one another, felt warm toward one another. The greater the amiability and esprit de corps among the members of a policy-making in-group, the greater is the danger that independent, critical thinking will be replaced by groupthink. Which is then likely to result in irrational and dehumanizing actions directed against out-groups, exactly as we have seen. With the propaganda, the attacks, the manipulation of the media, the defamation, the gaslighting, all of this.

The attacks on the Great Barrington Declaration authors was classical groupthink-driven group behavior. It is clear to all in retrospect that the Great Barrington authors offered solid fact-based insight about lockdowns and presented their findings in a responsible and professional manner to the White House. The ideas were offered in good faith. What did the White House Coronavirus task force do? They didn't just argue with them, they didn't just disagree with them. They tried to destroy them, and the people who authored them. Just exactly what Dr. Janis proposes is the typical psychology of these in-groups that form, and that has historically resulted in some of the greatest public policy failures in the United States history. I suspect that history will record the COVIDcrisis as another one of those greatest policy failures.

If we follow the advice and insights of Dr. Janis, we may be able to actually develop, teach, and apply some very clear structures and processes to combat groupthink. Examining the COVIDcrisis response from this perspective, as another more modern case study, provides a very interesting opportunity to actually understand what happened, and provide prescriptions based on a true, honest analysis, and to develop plans and education designed to avoid it in the future.

I suggest that it is really most productive and adaptive to frame this in a nonpartisan fashion. We're all interested in government and effective governmental decision making, whether you're left, right, center, up, or down. We all want good government. We want value for our money. And unfortunately, there is a historic tendency all too often repeated that those who don't remember history are bound to repeat it. And what we've had in the COVIDcrisis is a great case study in the failure to learn the lessons of American history regarding American foreign policy failures, and to apply those lessons in the context of a public health response.

I don't think this has to be a Democrat or Republican issue, a liberal or conservative issue. I think we can all agree that good government is something we all want, and that we should put in place good government policies and best practices, even if we still have the administrative state, even if the incoming administration is not able to break up the senior executive service and the inverted totalitarian structures as effectively as they may wish to do. Even if we still have those administrative state structures in place, we need to be able to learn from this. Lessons learned based on open-minded, rigorous root cause analysis.

Clearly, one of the core lessons has to be that we need to avoid small cohesive in-group-based decision making that doesn't allow the in-group and its consensus to be challenged. Dr. Janis talks about this as a pretzel problem. You have to have enough cohesion in the leadership group but also be able to routinely test and challenge the intellectual products that the leadership group produces. One alternative model is that a unitary executive just makes unilateral decisions. Under such a system, the populace places all decision making in the hands of a king or dictator, and that's that. Here in the United States of America, we rejected that model during our founding, and have a tendency to want to use groups and employ group decision making. We think that model provides more diversity of opinion, and that this intellectual diversity will result in better decision making and eventual outcomes. Here in the United States, we generally agree on the idea that diversity of opinions is good. I think we can generally agree on that. We want diversity of opinion, but the benefits of intellectual diversity are lost if there is too much cohesion within a decision-making group. If the group is just an elite buddy network, what we will end up with is a group that primarily acts to reinforce and protect one another.

To address this problem, Dr. Janis has provided nine suggestions. Nine clear, tangible recommendations that we can use and implement in our public policy for how these decision-making groups should operate. Many of these recommendations are actually employed by leaders who are trained by the US military, because they have to be able to respond to a changing tactical and strategic landscape.

But I don't think it's been part of the training of Health and Human Service leadership, or in many of our other agencies (and large corporations). The big lesson here is that we can learn from COVIDcrisis leadership failures and implement policies designed to ensure that we don't have these kinds of cohesive in-group processes and the types of failures that they are prone to. Bureaucratic in-groups that just focus on themselves and protecting one another, rather than recognizing the policy failures that they are propagating and responding appropriately in service of the citizenry that pays their salaries.

Quoting again from Irving Janis, summarizing the two main conclusions from his analysis:

> Along with other sources of error in decision-making, groupthink is most likely to occur within cohesive small groups of decision makers. And that the most corrosive effects of groupthink can be counteracted by eliminating group insulation, overly directive leadership practices, and other conditions that foster premature conclusions.

As a consequence of these premature conclusions, we have all suffered. We are victims of groupthink, all of us, because of those premature conclusions. Again, quoting Dr. Janis:

> Those who take these conclusions seriously will probably find that the little knowledge they have about groupthink increases their understanding of the causes of erroneous group decisions.

We have a prescription provided by Dr. Janis. Nine different points derived from a clear cohesive analysis of multiple prior US policy failures and successes. His analysis and prescription predicted the behaviors that we've seen; it predicted the dysfunctionalism. And we could have avoided it, if we'd had a little less hubris and a little more thinking and willingness to tolerate dissent.

As for myself, when managing groups, I crave dissent. In order to have good, clear thinking, scientific thinking, you must be challenged. And yet the US Government leadership for the COVIDcrisis response has done everything it can to railroad and shut down any communication that would challenge their consensus.

Here are Dr. Janis's nine recommendations for preventing groupthink:

1. The leader of a policy-forming group should assign the role of critical evaluator to each member, encouraging the group to give high priority to airing objections and doubts. This practice needs to be reinforced by the leader's acceptance of criticism of his own judgements in order to discourage the members from soft-pedaling their disagreements.

2. The leaders in an organizations hierarchy, when assigning a policy planning mission to a group, should be impartial instead of stating preferences and expectations at the outset. This practice requires each leader to limit his briefings to unbiased statements about the scope of the problem and the limitations of available resources, without advocating specific proposals he would like to see adopted. This allows the conferees the opportunity to develop and atmosphere of open inquiry and to explore impartially a wide range of policy alternatives.

3. The organization should routinely follow the administrative practice of setting up several independent policy-planning and evaluation groups to work on the same policy question, each carrying out its deliberations under a different leader.

4. Throughout the period when the feasibility and effectiveness of policy alternatives are being surveyed, the policy-making group should from time to time divide into two or more subgroups to meet separately, under different chairmen, and then come together to hammer out their differences.

5. Each member of the policy-making group should discuss periodically the group's deliberations with trusted associates in his own unit of the organization and report back their reactions.

6. One or more outside experts or qualified colleagues within the organization who are not core members of the policy-making group should be invited to each meeting on a staggered basis and should be encouraged to challenge the views of the core members.

7. At every meeting devoted to evaluating policy alternatives, at least one member should be assigned the role of devil's advocate.

8. Whenever the policy issue involves relations with a rival nation or organization, a sizable bloc of time (perhaps an entire session) should be spent surveying all warning signals from the rivals and constructing alternative scenarios of the rivals' intentions.

9. After reaching a preliminary consensus about what seems to be the best policy alternative, the policy-making group should hold a "second chance" meeting at which every member is expected to express as vividly as he can all his residual doubts and to rethink the entire issue before making a definitive choice.

It isn't just governments that have fallen into the patterns of groupthink. It is clear that social media, big tech, and mainstream corporate media have all forgotten the insights of Dr. Janis and the lessons of groupthink.

CHAPTER 31
Don't Be Brain Dead (Think for Yourself)

The *"Gell-Mann Amnesia effect"* is a term that was coined by Michael Crichton, MD, to describe the experience of encountering unreliable information and the "approved narrative" in mainstream media when you are within your area of expertise and knowing by first-person experience that this narrative is wrong. And then, after this realization, you proceed to suspend your own critical thinking skills and place trust in these same types of "experts" (legacy/mainstream "approved" media) in another area outside of your expertise.

His point was that one must use critical thinking skill even when outside your core competencies. Crichton's writes:

> Briefly stated, the Gell-Mann Amnesia effect is as follows. You open the newspaper to an article on some subject you know well. In Murray's case, physics. In mine, show business. You read the article and see the journalist has absolutely no understanding of either the facts or the issues. Often, the article is so wrong it actually presents the story backward—reversing cause and effect. I call these the "wet streets cause rain" stories. Paper's full of them.
>
> In any case, you read with exasperation or amusement the multiple errors in a story, and then turn the page to national or international affairs and read as if the rest of the newspaper was somehow more accurate about Palestine than the baloney you just read. You turn the page and forget what you know.
>
> —*Michael Crichton (1942–2008)*

In other words, think for yourself.

As we transition into a new era of another war, we must think for our-selves. Government(s) will want to control the narrative. They will control how mainstream media and big tech respond to this crisis. Do not fall for the dominant paradigm, but instead do your own thinking. Dig deeper. We are so lucky to live in an era of alternative media. As much as the govern-ment hates both that reality as well as those of us who look for "truth," we still have this ability to think for ourselves and to find information that is not readily available on MSM.

This morning, while I was having breakfast with my wife, Jill, we were talking about our personal finances. As I sipped coffee, I said to her "*I don't know enough about economics to make an assessment on cryptocurrency, how the situation in Ukraine will affect world economies, and the global push to transition from fiat money to digital currency.*" In so many words, she said to me that I had actually spent my whole life studying politics, investing and thinking about global economies as a way of living in the world. So why in the world would I now think that I was incapable of making an analysis of the current economic situation? And then the punchline—I needed to get going and do more research into the subject. She is right, I was being intellectually lazy. More than that, I was letting the corporate mainstream media and their pushed propaganda take my mind. I was giving over my critical thinking skills to the MSM.

"Gell-Mann Amnesia" is exactly the trap I had fallen into. It happens all the time. Back in the day of the "founding fathers," 250+ years ago, people did not have the luxury of being intellectually lazy. You had to think for yourself. That is what is so different about a more rural life. The problems that crop up are constantly changing, and they are very much in the present. You have to solve them yourself. That fundamental reality is what gave rise to the United States of America, as a country and as a culture.

The United States must recreate an army of critical thinkers. It is how We the People take back our power. We must find candidates for elected office who think for themselves, and we need to work to elect them. Now is the time to not accept mediocracy and corruption in our elected officials. And no more WEF-trained hacks who do what they are told by non-US, nonelected third parties. They are not loyal to the US Constitution. They are loyal to a foreign power.

Some elements of the Republican party are revitalizing its core belief system to fight the heavy-handed governmental edicts coming from the glo-balized public health deep state. They recognize that Big Pharma and the World Economic Forum acolytes that back it have infiltrated every level of our government and virtually all "world leaders" from the Western "Democracies." They recognize that corporatism and totalitarian thinking guided by the WEF has become the norm for many elected officials. That these officials have been trained by the WEF to serve the interests of the very large transnational corporations, their owners, and the financial institutions that fund them. This must stop. The army of critical thinkers that is emerging with the Great Awakening has to support this effort in every way possible. Why? Because public health is just the "camel's nose"; the WEF has great plans for us in all aspects of our lives!

Think about this statement: "The war on cash is a global effort being waged on many fronts. My view is that the war on cash is dangerous in terms of lost privacy and the risk of government confiscation of wealth" [439]. Governments are very concerned with taxation. A digitized economy would put all of our financial transactions forever on the "cloud." There would be no privacy. In the future, those transactions will be leaked, and how those data will be used will not be up to us, as individuals. We will not have control if the government decides to release those data to financial institutions, foreign institutions, insurance agencies, etc. Our right to privacy is fundamental to being Americans. That includes the right to financial privacy. There has always been a tension between the IRS and the right to privacy, since the very precursor of the IRS in 1862. But it was not until 1913 that the IRS personal income tax division was created, and the rest is history. But the right to privacy has always been paramount, even within the IRS.

The truth is that the US government would love a way to track down and stop our "cash economies." That is to say, people who work for cash or other services, including small service providers, such as gardeners, house-keepers, plumbers, etc. You might say that our government has a very strong financial interest in going cashless. But a purely digital economy would take away yet another personal freedom. The ability to keep our personal finances from the overreach of government, global banking, and big tech. Because every financial transaction will be transparent to parties unknown. In the future, who knows who will get hold of that information or how it will be used? Combine this with the digital ID, digital passport, digitized health records connected to these accounts, ESG (social credit scoring), and facial recognition, and what privacy is left?

Whether it is Gell-Mann Amnesia, groupthink, the dominant paradigm as defined in the book *The Structure of Scientific Revolutions* [73], or mass formation psychosis acting alone or in synchrony, don't let your mind go there. Think for yourself. Think about how to protect your wealth, how to build income, how to create wealth that is government-proof, who knows what is in our future. The Internet has changed everything.

CHAPTER 32
Defend Your Sovereignty

Continuous vigilance and vigorous support of States' Constitutional rights is critical.

I believe there are more instances of the abridgement of freedom of the people by gradual and silent encroachments by those in power than by violent and sudden usurpations.
—James Madison

A gradual and silent encroachment of our freedom and federal, state, and individual sovereignty by a globalist financial corporate cartel continues to proceed. One key aspect of our current political reality in the US is that many of our laws at the federal level have been placed there by corporate stakeholders. Corporate lobbyists work tirelessly and relentlessly to insert legislation that benefits their industries and increases the wealth of their corporate clients and associated "stakeholders" into our federal laws and regulations, and to twist existing legislation so that it becomes a more perfect tool of their clients' corporate interests. In parallel, within the many branches of the US federal bureaucracy, regulatory capture has become the norm. Furthermore, it has become increasingly clear that it is grossly naive to expect solutions to these corrupting influences to come from either Congress or the entrenched and captured bureaucracy. And one hard truth learned by the former Trump administration is that the executive branch lacks the power and authority required to replace the entrenched bureaucrats that have become the ruling lords and ladies of the new inverted Totalitarian reality that is the entrenched federal bureaucracy.

The devolution of American democracy into an inverted totalitarian state is well documented. Chris Hedges of Truthdig writes [418]:

> It (Inverted totalitarianism) does not find its expression in a demagogue or charismatic leader but in the faceless anonymity of the corporate state. Our inverted totalitarianism pays outward fealty to the facade of electoral politics, the Constitution, civil liberties, freedom of the press, the independence of the judiciary, and the iconography, traditions, and language of American patriotism, but it has effectively seized all of the mechanisms of power to render the citizen impotent.

In this new federal reality, whnoere the culture of government institutions and agencies is damaged beyond repair, real change cannot occur by replacing one figurehead for another. Corrupted policy can't be easily replaced with good policy. The ideology, the rule of law, the subverted corporatist regulations are baked into federal law and into regulatory practices and cultures. The system is rigged. The unknown (or known) bureaucrats controlling the levers aren't going to let go of that easily. When they have done their time, most likely those employees will move into corporate leadership positions so long as they have protected corporatist interests during what has all too often become a form of internship-as-prelude prior to landing a "good job." As most federal employees at the top-tier are well aware of their future career opportunities, they have no incentives to try to change the status-quo and upset their potential future employers. Government institutions that are intrinsically controlled by outside interests cannot be reformed from within, so it is important to not waste too much energy trying. That's why many argue that a "siege strategy" or even the creation of completely new institution(s) may be the answer.

Those of us fighting for freedom have come full circle and now confront the issue of state's rights. This is a fundamental principle of our republic (often incorrectly referred to as "our democracy"). Each state has the ability to govern itself, within the confines of the Constitution, federal law, and the Bill of Rights. Those powers are vast but frequently under-utilized. States have the power to ensure that a strong federalized governance is not the norm.

In 2012, Obama lifted restrictions from the Smith-Mundt Act (passed in 1948) that disallowed domestic dissemination of government-funded media [440]. Then the Countering Disinformation and Propaganda Act, introduced by Republican Sen. Rob Portman of Ohio in 2016, established the Global Engagement Center under the State Department, with the following Mission and Vision statements.

> Core Mission: To direct, lead, synchronize, integrate, and coordinate efforts of the federal government to recognize, understand, expose, and counter foreign state and non-state propaganda and disinformation efforts aimed at undermining or influencing the policies, security, or stability of the United States, its allies, and partner nations.
>
> The Global Engagement Center replaced the Center for Strategic Counterterrorism Communications. *The new strategy seeks to be more effective in the information space and is focused on partner-driven messaging* and data analytics.
>
> Additionally, the Center strives to:
>
> - Enhance the capacities and empower third party, positive messengers, whether they are governments, NGOs, or other entities.
> - Leverage the entirety of the U.S. government to confront ISIL and other extremists in the information space and bring coordination and synchronization to those efforts.
> - Build a forward-looking entity within U.S. government that is agile, innovative, and embraces technological advancement."

This center authorizes grants to nongovernmental agencies to help "collect and store examples in print, online, and social media, disinformation, misinformation, and propaganda" directed at the US and its allies, as well as "counter efforts by foreign entities to use disinformation, misinformation, and propaganda to influence the policies and social and political stability" of the US and allied nations. These laws have worked to loosen the controls previously placed on the federal government regarding its ability to push propaganda.

When the changes to Smith-Mundt Act are combined with the Global

Engagement Center, it is interesting to speculate how the US population might be manipulated. The CDC has spent a billion dollars on propaganda and censorship to ensure vaccine compliance through media buys, and one has to wonder just what sort of propaganda will be pushed on the American people in the future [183].

The very fact that we have an ex-president of the USA, the very same one who presided over the lifting of restrictions of the Smith-Mundt Act, calling for censorship of the press under the guise of "PROTECTING OUR DEMOCRACY," and that he is supported in this call by so many on the left is chilling [412].

According to Obama, this is the set of principles of how content should be moderated, either by tech companies themselves or by *a government entity* [441]:

- Whether it strengthens or weakens the prospects for a healthy inclusive democracy.
- Whether it encourages robust debate and respect for our differences.
- Whether it reinforces rule of law and self-governance.
- Whether it helps us make collective decisions based on the best available information.
- Whether it recognizes the rights and freedoms and dignity of all of our citizens.
- *"Regulation has to be part of the answer,"* Obama said, calling for ways to start "slowing the spread of harmful content online."

The Universal Declaration of Human Rights, Article 19, states that: "Everyone has the right to freedom of opinion and expression; this right includes freedom to hold opinions without interference and to seek, receive and impart information and ideas through any media and regardless of frontiers." That an ex-president of the USA, who is one of the key players of the current president, that is the Biden administration, would consider censoring free speech on the above criteria is beyond the pale. Remember that big tech and social media have agreements with the White House and the WHO to censor free speech on behalf of the government [358, 442, 443].

As the declaration of emergency powers have yet to be rescinded by the president, over 2.5 years after that start of the pandemic, one has to wonder when the end is in sight. Are these powers of censorship on the Internet permanent? The "town hall" for America and for the world is now the Internet. So, these Orwellian actions are truly the hand of inverted totalitarianism. We are only allowed to "see" a few (highly redacted) documents unsealed by FOIA, and yet we can easily surmise the extent of the propaganda.

But what happens when a state like Florida becomes a threat to the federal government? Will the federal government continue to exceed its enumerated powers and seize control of more powers that have been traditionally assigned to the states? The back and forth between President Biden and Governor Ron DeSantis over Florida's COVID policies makes clear the power that the federal government has over states has been enabled by leveraging the distribution (and ability to withhold) federal tax dollars and, in the case of Florida, lifesaving medications. The actions of Governor DeSantis of Florida are so harshly criticized by those who control the levers of the federal government (as well as purchased mainstream corporate media) because he has refused to comply with the Biden administration's HHS bureaucratic edicts.

After the Biden administration admonished Florida for stagnant vaccination rates in Florida, the Feds then decided to federalize the distribution of monoclonal antibodies so that those who don't get the vaccine have no alternative treatment option [444]. This is a prime example of the retaliatory tools available to the federal government and illustrates that the federal government is able and willing to compromise the health of US citizens to punish a state that choses to be noncompliant. Remember, the states regulate medicine and public health policy within that state. *Biden refusing to send lifesaving medicine is a clear abuse of federal power.* The 10th Amendment of the United States Constitution authorizes states to establish laws and regulations protecting the health, safety, and general welfare of their citizens. The practice of medicine is not an inherent right of an individual, but a privilege granted by the people of a state acting through their elected representatives. For further details, please see the Federation of State Medical Boards statement [445].

It is state's rights—as imperfect as that system is—that protects us from the WHO/globalist power grab to leverage public health to compromise the sovereignty of the United States. The current strategic agenda of those seeking to advance globalism and global governance policies at the expense of the autonomy of nation-states is often referred to as the New World Order. Those advocating for the New World Order include the World Economic Forum, transnational corporations/globalized investment capital, the UN, the World Trade Organization, World Bank, and now the World Health Organization. These entities have, to a greater or lesser extent, gained control of the federal government through leveraging the regulatory capture of our federalized agencies, departments, and institutions. Money from these entities flows into the coffers of congressional political campaigns to influence law makers. Often, treaties and agreements on the global level codify these arrangements. This ensures that our federal government is co-opted by these entities.

Our federal government believes that it will be stronger when states have little or no control over the rules and regulations of the state. This is not what our founding fathers intended, and this is when the rights of all of us are trampled on. Our constitution codifies that each state maintains its own set of laws, rules, and regulations. The beauty of this system is that the diversity of cultures can be maintained within states. Living in Texas is very different from living in New Hampshire, and I believe that this is a good thing. Although states are stronger when there is a federal government appropriately assuming and defending those responsibilities enumerated under the US Constitution, they are also guaranteed the freedom to have their own cultural identity and rule of law. This is the beauty of our Constitution and Bill of Rights. This interweaving of state's rights under the limited umbrella of a federal government.

For many, the erosion of each state's right to monitor and control federal elections, as guaranteed in the Constitution, is the easy example to reach for as that line is also being repeatedly crossed. But the depth of this problem is much, much deeper than this single issue.

The use of federal tax dollars to control states may be a more persistent and pernicious issue. As an example, the Department of Transportation allocates funds for states that comply with seat belt laws or speeding limits but will withhold such funds for states that don't. The Department of Education allocates taxpayer funds based on the Elementary and Secondary Education Act (ESEA) goals or other "reforms," thereby controlling and harmonizing certain aspects of education in the USA across all states. In the example of COVID lockdown policies and masking, it is quite clear that the CDC and the NIH have overstepped federal powers again and again by leveraging federal tax dollars to create perverse financial incentives within hospitals and to promote "mis-, dis-, and malinformation" and to promote propaganda in both old and new media. It is time to return to the rule of law and return the power back to the state to regulate medicine and public health.

States must remember that federal funding is not the be-all, end-all. If a state can develop the political will to refuse such funds, sovereignty can be reestablished. This requires that plans can be made to mitigate the impact of such an action. These tax kickbacks are often weaponized in an arbitrary and capricious way by the federal executive branch to illegally control state policies, including public health policies. State leaders who are willing to walk away from federal funding, who are willing to work with other states and Congress to negotiate better terms for federal dollars, may offer the best hope for breaking this federal overreach.

States must reset the constitutionally correct relationship between the federal government and the States in a wide range of ways, including regulation of healthcare. If they are willing to plan for and mitigate the risks associated with taking back their constitutionally granted powers, States can leverage this new ruling to enable something closer to the intended balance of powers. "We the People" need to insist that they do so by electing strong and independent leaders at the state level, leaders who will fight for the right of each state to govern itself as our founding fathers intended. Governor Ron DeSantis and Attorney General Jeff Landry are two leaders leading the fight on this front. We must support them, so that they can continue to restore and protect the constitutional rights of each State in the Union. Other states will follow their example. I believe that it is critical to the future success of our nation to have these great leaders paving the way, being brave and setting an example for other state leaders to follow, so that the attempts of the Globalist cabal to empower the WHO to subvert constitutionally granted States Rights can be thwarted [446].

The federal government has been intentionally infiltrated by Globalists trained by the World Economic Forum. Make no mistake. The elites and transnational corporations have undermined and continue to undermine our institutions and our very Constitution. President Trump has taught us that we can't pull our punches; we must act aggressively and give no quarter. That means utilizing every tool available; these tools include using the judicial branch (the courts), working to elect and educate legislators, and of course educating and mobilizing the populace.

Individuals can have a huge role in working to ensure our freedoms, our sovereignty as a nation, by writing letters, phoning legislators, publishing independent articles and memes. Even just reposting on social media can help. We are all in this fight together. Individuals must organize, join groups and organizations, and create new groups and organizations.

Having an independent and free press that can report on our government and institutions is critical. It is how we can help Americans make informed decisions when voting. That means replacing the captured and outdated old media with new media outlets that are censorship free. Using block chain to create new social media outlets that cannot be subjected to propaganda and censorship is a good step in the right direction.

Organizing into groups that can work collectively to get the truth out is another way to help save the Sovereignty of our great nation, and that of all of the independent nations of the world. We must fight against the narratives that are being constantly pushed by government agencies that have been subverted by the Globalists. They do not want their corruption exposed. What we have learned over the past two years is that they will do whatever it takes to hide their dirty dealings from the American and world public.

The "Greatest Experiment of our time" is over. The great experiment that is American representative (Republic) democracy is crumbling. Americans no longer understand or accept the conceptual foundations necessary to sustain it. When the American Experiment was born, many thought it would not last as long as it has. This democracy was not brought down by war or famine, but the erosion of the very rights that made this nation great.

Now, the soothing words of a reasonable alternative, a more "mature" alternative to national governance, is on the brink of weaving its tentacles into the very heart of our democracy. The "New World Order" is to bring peace, harmony, and a renewed Earth to all of humanity. Fascism on a global scale, where "private-public partnerships" are taken to the next level. That is a vision of world dominance. Wrapped up in the pretty package of "build back better." Ask yourself "who" is profiting from this new world order? Is it the small businesses and farms of America? Is it even American corporations? That answer is no. It is the transnational corporations, global elites, and the small number of large international investment firms (Blackrock, State Street, Vanguard). The naive ideals of the New World Order are nothing more than putting a pretty face on a very ugly reality. Don't fall for the trap of believing that this will bring equality to the world and lift everyone out of poverty. The New World Order is nothing more than fascism on a scale never seen before, using the methods of inverted totalitarianism to control the levers of national and global power.

It is time to reteach the lessons of our founding fathers. It is time to protect America. The question is how. And that is in your hands, it is up to you.

CHAPTER 33

IppocrateOrg: People Helping One Another

By Mauro Rango and Irina Boutourline

Based in Italy and forged in response to the COVIDcrisis, the IppocrateOrg movement has assembled an international volunteer network of physicians, researchers, and health and social workers to help patients with nowhere else to turn. The Italian state-endorsed medical establishment was offering patients nothing other than nihilist inpatient hospital protocols with unacceptably high mortality rates. As the epidemic surged (particularly in Northern Italy), the founding organizers of IppocrateOrg recognized that all of the international political, financial, and corporate media structures were becoming remarkably aligned in messaging about both the risks of the virus and the treatment options. The initially chaotic and conflicting landscape of local, national, and international responses often resulted in governments dissembling and failing to provide clear, sensible answers and public health response guidance. In response to this barrage of dysfunctional confusion and lack of leadership, IppocrateOrg physicians and scientists developed and publicized new protocols for treating patients at different stages of disease, and affiliated physicians and other medical providers began deploying early treatment, saving lives, and keeping patients from even being hospitalized.

Despite the confusion, all of the sanctioned and promoted "official" voices opposed those who were offering concrete solutions and therapies to save lives using existing medications. The physicians of IppocrateOrg would not accept a treatment strategy that involved doing nothing more than administering oxygen to intubated patients while awaiting instructions from the top. Based on their clinical knowledge of the disease, those physicians made treatment decisions that no one in the top spheres of medical influence would even consider, let alone endorse. Yet in communities where these banned alternative treatment strategies were implemented, COVID mortality has been very low.

In the face of the undeniable fact that early interventions using a wide variety of existing licensed drugs save lives and almost completely eliminate hospitalization and mortality associated with SARS-CoV-2 infection, it is not surprising that questions would arise, particularly when those lifesaving drugs have been pushed aside in favor of patented medications that have not been particularly effective. These questions led IppocrateOrg leaders to seek a path to change the relationship between clinical and research medicine, and to gain autonomy from international, national, and regional Departments of Health, which appear to be captured by the medical-pharmaceutical industrial complex. This is their story.

* * * * *

IppocrateOrg: People helping one another. An Italian response to the COVIDcrisis

Authors: Mauro Rango, president of IppocrateOrg; Irina Boutourline, vice president of IppocrateOrg (Translated by Daniela Brassi)

Our efforts to envision an alternative medical community started from insights prompted by observing the dysfunctional response of corporate and state-sanctioned medicine. We are committed to the belief that it is the right of each person to assume personal responsibility for maintaining his or her own health, and that all have the right to access medical care according to their best interest. It is clear to us that we do not live in a society that puts the sanctity of human beings at the center of modern medical practice, Something has to change because the current social, economic, and corporatist model is putting the right to health in danger across the planet.

And we just can't let that happen.

As of early 2020, Italian radios, TVs, and newspapers all seemed to agree that an extremely dangerous global pandemic was underway, caused by a virus that most likely originated from bats sold for food at the Wuhan wet meat market in China: SARS-CoV-2. It sounded like the beginning of a classic American splatter movie, but instead it was everyday reality. While newspapers and news programs talked all day about bats and "patient zero," evening broadcasts were suddenly stuffed with movies like Wolfgang Petersen's *Outbreak*.

Italians found themselves confined to their homes, almost as if they were under house arrest, accepting anything to protect themselves and their loved ones. However, in spite of the lockdowns (a foreign word that suddenly become part of the Italian vocabulary), the virus continued to spread inexorably: in warlike tones, the government and its parade of virologists constantly communicated to citizens how they were going to confront the enemy.

Inexplicably, "paracetamol and watchful waiting" become the mantra of primary care physicians, the only acceptable medical treatment protocol for treating infected patients. Indeed, no effective treatment seemed possible. The only imaginable solution lay in the development and production of magical new vaccines that would finally stop the infection. Vaccines become the Italian dream, the only way out of the nightmare. During the long wait, hospitals became transmission centers for the infection, where people received no actual treatment, dying without the comfort of their families. More devastating than the virus itself, depression and hopelessness infected Italian homes.

The majority of Italians were so overwhelmed by fear that, acting with the noble intention of protecting the most vulnerable, they fueled their own self-destruction. The ever-present fear of death, triggered by journalists and talk-show scientists, prevented nonexperts from asking such basic questions as:

1. Why won't my family doctor help me?
2. Why are we told to wait for the evolution of the disease without making any attempt to use available medications or treatments?
3. Why are numerous associations of physicians arising and disobeying government mandates? Why are these associations and allied physicians treating unknown patients free of charge, devoting their attention to them, and providing prescriptions for drugs and treatments that the government denies and the corporate media derides?
4. Why are family members not allowed to say good-bye to dying relatives using proper precautions? Why are the families not even being permitted to talk to them on the phone?

Within this whirlwind of events, information, and fear, a minority managed to remain sufficiently grounded in reality and common sense. They began separating the distressing messages from a more rational and balanced interpretation of the reality of the situation. Some of them, while blocked by corporate media from expressing their alternative perspective, began to communicate through alternative information channels and new forms of media. Others, like those who now constitute IppocrateOrg, took direct action in the field, providing healthcare, listening, and understanding to many others.

The Birth of IppocrateOrg

From these roots, IppocrateOrg emerged as an organic movement of physicians and ordinary people who refused to be overtaken by the constant pressure of corporate media and state-promoted fear. Almost miraculously, people who did not know one another and had no profit motive met through daily long-distance calls to work together to develop treatment and relief strategies. In this way 70,000 people were treated by these physicians and medical care providers with only fourteen deaths—including those who were already suffering from advanced COVID disease.

With the publication of a handbook, IppocrateOrg provided the guidance needed to treat the disease in its early onset phase (with the use of anti-inflammatory drugs), its intermediate cytochemical storm phase (using cortisone, antibiotics, and heparin), and its final phase (providing assistance, through its legal team, to hospitalized patients who needed hyperimmune plasma). This set of treatment strategies, although not included in official protocols, led to the recovery of 90% of hospitalized cases.

The published IppocrateOrg handbook (*Healing at Home from Covid 19: Manual for Early Tailored Therapy*) lists therapeutic treatment plans designed for each stage of the disease, as well as examples of therapeutic tailoring (particularly in cases of prior disorders). It is still frequently consulted by Italian families, and many have provocatively gifted it to their family doctors. After the treatment protocols developed and published, a training exercise was set up for the 300 participating physicians, followed by weekly teleconferences to address the most pressing issues: a mutation of the virus, the need for adjustment of therapies, the impact of preexisting conditions on the course of the disease, a new drug "discovered" by international partners, and any other issues that could help us treat our patients. In addition, WhatsApp chats were set up where newly joined physicians could obtain help and support from doctors who were experienced with the treatment protocols. This constant information sharing, the training courses, and the shared philosophy of respect for each individual life have helped to create a new class of medical doctors. Physicians whose human qualities make them particularly sought after and in demand today, especially by like-minded people.

University researchers are now doing retrospective studies of IppocrateOrg's efforts during the COVIDcrisis. Early results have already confirmed what IppocrateOrg's 300 Italian physicians observed in the field. For example, a retrospective observational study conducted by the Center for Research in Medical Pharmacology at the University of Insubria published in *MedRxiv* as a preprint on April 5, 2021, reported near-zero lethality when people are treated right away. This, of course, leads to questions about the 160,000 "official" Italian deaths. Most of these people died because they were denied treatment by government and health agencies that were backed by scientific "researchers" whose objectivity was typically compromised by a conflict of interest stemming from the source of their research funding.

IppocrateOrg's highlights and difficulties

In 2021, in the midst of the Italian vaccination campaign, a number of events directly involved IppocrateOrg. These events acted like selective pressures causing the Association to evolve. **Let us consider the three most signifi-cant episodes**:

1. Active participation in organizing the *International Covid Summit* in Rome, a three-day event, which brought together physicians and associa-tions from around the world who had successfully treated their patients with COVID 19.

As was to be expected, given the sudden visibility of our work, people arrived from all over the world to disprove the effectiveness of the "emergency" healthcare directives adopted by most governments. From that moment on, Italian national press and TV stations began implementing an intense smear campaign against internationally renowned medical doctors and scientists, denigrating them as impostors and frauds and openly attacking Ippocrate-Org's physicians, calling them "*sellers of ginger, licorice* and ineffective cures."

2. The production of a video titled *Let's Protect Children*, which reached two million views in Italy.

The video highlighted the need to carefully weigh risks and benefits of an experimental vaccine for children. The physicians who took part in it are undergoing disciplinary proceedings by their professional bodies and await a final ruling about whether they will be allowed to continue treating patients.

3. The temporary closure of IppocrateOrg's Help Center, mainly caused by the Italian government's suspension of unvaccinated doctors.

In Italy, physicians, nurses, and teachers who decided not to get vaccinated have been hit with politically motivated disciplinary measures, including suspension from their jobs, even when their work is (or could be) conducted online. Such measures are not justified by the stated objective of avoiding patient infections, but rather have been implemented to punish those who do not comply with the diktats of the system. It has become clear that the government and allied medical-pharmaceutical industrial complex seek to eliminate such principles as therapeutic freedom of choice, risk/benefit assessment (informed consent), and the possibility of objections to an indi-vidual physician by infected caregivers self-suspending from work. Although the data have demonstrated that vaccinated persons continue to become infected, replicate, and transmit the virus, there have been no modifications to official policy. At the time of writing, suspension of unvaccinated physicians from the Italian medical register is scheduled to continue until 12/31/2022.

IppocrateOrg's responses and achieved results

Again, keeping to a similar order of triggering events, we will shed light on the key words that represent IppocrateOrg's modus operandi and the results we achieved.

1. Visibility and balance. The smear campaign unleashed after the International COVID Summit featured local TV channels televising live feeds of IppocrateOrg's president and other members of the Association, which was intended to humiliate them but, ironically, increased their visibility instead. The presence, dignity, preparation, and balance maintained by the president and other members during the broadcasts gradually elicited the interest and support of viewers who were previously unfamiliar with the movement. Moreover, upon seeing the blatant and unwarranted attacks on people who kept their cool, a great many decided to actively support the Association. "In cammino con IppocrateOrg" ("On the Path with IppocrateOrg"), for example, is a Facebook group of supporters who reacted to the media attack by defending IppocrateOrg.

2. Trust and communication. IppocrateOrg has provided free legal defense to the physicians in *Let's Protect the Children* who have been subjected to disciplinary proceedings by their professional orders and will continue to support them, especially from a communication perspective, through the alternative media information channels that are gaining more and more listeners in Italy. The open and transparent stance taken by IppocrateOrg physicians against vaccinating children helped create a greater climate of trust among those fighting with us. Beyond a few sporadic statements by individual physicians, the IppocrateOrg community decided to protect medical caregivers by not naming the doctors who were openly siding with and defending children. That choice also helped strengthen the original core of the "intentional community" that IppocrateOrg embodies.

Also, some parliamentarians and members of the government have quoted IppocrateOrg's video on television. These politicians cited the video to support their misgivings about implementing a vaccination campaign on minors, particularly as children were not getting serious forms of COVID. The video helped them make the case that, except in a few very rare cases, the risk-benefit ratio of such childhood vaccination heavily leans toward risk due to vaccine-associated side effects.

3. Destructuring old and implementing new models. The temporary suspension of the COVID Help Center (from early December 2021 until December 23, 2021) was widely reported in national newspapers and on TV. Of course, the mainstream media passed it off as the closure of the Association itself, but this episode also elicited widespread reactions from Italian society. Physicians, who were vaccinated and therefore able to work, contacted IppocrateOrg and volunteered to reopen the COVID Help Center. Other communities, including many patients treated by IppocrateOrg, then sprang up to protest the Italian authorities and share the results they achieved with treatments administered by IppocrateOrg physicians.

"Noi con IppocrateOrg" ("We with IppocrateOrg") is another community with a Facebook page that arose at the cry of its founder: "IppocrateOrg has done so much for us, now it is time for us to do something for IppocrateOrg." Tens of thousands of people have joined this Facebook group, testifying about their experience as patients assisted by IppocrateOrg, and today it distributes health information, education, and news in close contact with IppocrateOrg. Moreover, the Italian authorities' suspension and persecution of medical doctors for refusing to get vaccinated or simply warning parents against vaccinating their children has caused an even deeper reaction.

Now that the COVID emergency is over, IppocrateOrg's goals have shifted toward a real challenge: organize an alternative to the national healthcare system. This has resulted in the opening of outpatient clinics (Purpose Medicine Centers) distributed all over Italy, which will treat all kinds of diseases, not just COVID. With these medical centers, IppocrateOrg aims to remain a resource for citizens who are concerned about the poor quality of our healthcare system (as clearly highlighted by the pandemic emergency) and to promote a healthcare model that focuses on the true well-being of the individual.

To set "well-being" as a medical objective (rather than disease treatment) means adopting a holistic viewpoint, considering all aspects of life. It means creating relationships based on listening, trust, and respect, each of which are fundamental in the path toward healing. From this perspective, it becomes essential to promote a culture of prevention and healthy lifestyles, as well as to pursue access to integrated and personalized care for all.

As we move forward, the physicians of IppocrateOrg will continue to provide COVID care in addition to other illnesses, relying on continuing education (which will be further intensified) and on the already well-tested concepts of cooperation and collaboration among colleagues.

That's not all: Purpose Medicine Centers will also offer diagnostic services to allow faster results—available at home for those unable to move—and will provide psychologists and specialized health lawyers, as well as accommodations for people in financial difficulty.

The system has been organized as follows:

1. The "front liners," in contact with patients, will be the doctors not hit by government directives;
2. A nationwide "Medical Community" will support the work of the "front liners" by providing consultations with physicians of different specialties who may suggest diagnostic tests to be performed and therapies to be administered. This collective of physicians, in short, will practice the kind of integrated and personalized medicine that a single specialty-trained physician could hardly master alone;
3. At the same time, the IppocrateOrg Medical Community is opening a dialogue with all the so-called "alternative" medicine systems and their practitioners to evaluate whether as-yet unexplored avenues for the treatment of certain diseases can be pursued together.

The IppocrateOrg Association, which sponsors these outpatient clinics, requires the physicians in charge to adhere to a code of ethics that stipulates that medical practitioners commit to

- operating in science and conscience for the sole interest of the person requesting a medical examination;
- participation in weekly training courses taught by the physicians of IppocrateOrg;

- enrolling in the two-year "Master of De-Specialization" program for graduates in the healing professions (physicians, psychologists, nurses, pharmacists);
- abiding by the maximum benefit rates specified by IppocrateOrg, and providing free visits for people suspended from work because they and their families are not vaccinated;
- participation in the Association, currently being formed, which will raise funds for underprivileged people wanting to be treated by IppocrateOrg doctors;
- participation in the Medical Community to provide, in turn, advice to the "front line" doctors.

To support the development and application of new ethical and organizational models, aligned with our vision to place human beings instead of profit at the center of medical practice, IppocrateOrg also focused on the creation of a real medical school, which has been developed with four levels of instruction:

1. The De-Specialization program, aimed at graduates in the healing professions. We use the term de-specialization to emphasize the concept of returning from a *specialized* vision of medicine to the *special* vision of the individual—a medicine in which the physician always considers the big picture first, and then applies a personalized therapeutic approach.
2. The Naturopathy school, accessible to those with a high school diploma, which follows the same principles as the De-Specialization program.
3. Training for "Regional Representatives for Well-being," which has no particular educational qualifications. Regional Representatives are local liaisons for communities, families, private schools, associations, and businesses, on anything concerning individual and collective health and well-being. The training includes not only healthcare topics, but also environmental, economic, organizational, and social topics, as well. In addition to promoting their own initiatives for individual and community welfare, Representatives also work

as liaisons with IppocrateOrg physicians on medical issues for the benefit of the groups they represent.

4.   Launching a new "School of Hippocrates," designed to train a new kind of therapist, in the broadest sense of the term, to challenge, first and foremost, chronic-degenerative diseases, the true pandemic of the modern world. Further information on the school can be found at https://scuoladippocrate.life/en/scuola-di-ippocrate-english/.

The context in which IppocrateOrg is operating

Thinking big is what is enabling us to overcome the little challenges of today and the future challenges of tomorrow. On the other hand, from a larger historical, social, economic, and political perspective, analyzing the situation in which we find ourselves allows us to review and reevaluate our actions along the way. This is helping IppocrateOrg to distinguish between those issues caused by our own sociocultural conditioning and those belonging to our deeper intentions: communion with our neighbors and compassion for those in pain. Reflecting on our civilization means reflecting on ourselves, on the direction of our actions, and on the need to make individual changes.

Western Civilization has been shaped and steered, on the one hand, by an unbridled industrial crony-capitalism that has used every means at its disposal to condition people's minds, to homogenize them as much as possible, and to exert ever more refined control over them. On the other hand, the modern world is being shaped by the hidden control of a global finance system that promotes exploitation of technology, cultural trends, and the real economy itself (heedless of the crises that increasingly affect hundreds of millions of already deprived people) and that treats human beings as objects, a mere component of an industrial machine in the service of "progress."

How to get out of this situation? It will take more than slogans and protests. Escaping this system will require complex innovative thinking to meet the complexity of our times. We must reunite that which is separate. Not just the separation between people, but also the separation caused by the fragmentation within each individual. Social and individual fragmentation has been the main weapon of those who prey on others for the sole purpose of profit and who disregard ethics as well as opportunities for harmonious human development.

Under the globalized industrial systems of today, national governments have less say or sovereignty and are unable to make decisions opposing those global organizations that hold massive financial power. They have been forced to obey the edicts of a narrow élite who make decisions on a global scale, control the media, and dictate to governments actions that are intended to do nothing more than expand their own wealth and power.

Only the world's most powerful country, the United States of America, still seems to retain a negotiating edge, but this is gradually thinning. Once fickle financial power has completed its metamorphosis, transforming into a vertically structured and horizontally global power, it will act even without the support of the world's most powerful state.

In its final stage of metamorphosis into a global power system, manipulation of the masses absolutely remains the central issue. Indeed, in the predatory finance logic of today, there is no room for the "recovery and restoration" of the antecedent reality. The current fiction is to introduce into the mass consciences the feeling that we are heading toward a saving and appeasing direction—"The Great Reset."

As proof, it is enough to linger on a trivial reflection: those finding space in the mass media and ostensibly fighting in defense of the planet and its habitat will not have access to corporate media if they come into conflict with or harm the elite that governs global finance. Only narratives that support the interests of the financial elite are permitted.

But the real confirmation of what we claim comes from the fact that those promoted in the mass media as "fighters for a better world" never question the paradigm that led us to this critical point. Why do they never question that paradigm and the power governing it, and instead just "report" or describe its repercussions? Is it possible to not be aware that maintenance of the same paradigm and power system will only produce more of the same? Is it possible to not understand that by continuing to support the same system of development and the same power aggregate, both on the road to catastrophe, the future will never change?

Many volunteers from many nonprofit organizations, while committed to fighting pollution, social inequality, poverty, and all the ills afflicting humanity, disagree with our analysis. Maybe this is because some of them do not want to give up their social positions, their way of life, their status as approved spokespersons—consciously or unconsciously. But maybe, if they looked at things from a more sweeping perspective, they would discover the profound inconsistencies existing in the society they live in and the way it is run. If they entertained the possibility that there might be a better way forward, they might find themselves forced to revise and, perhaps, turn their point of view upside down.

Truly sweeping thinking, unfortunately or fortunately, is not compartmentalized.

Human nature and personal restructuring: steps to a better future

With the above considerations in mind, and since the process of change starts from within, a personal restructuring of perspective and thought seems to be the first step. True change necessarily begins with an individual journey: the mental structure patiently forged by the force of modern Western Civilization in recent decades first needs to be deconstructed in order to reconstruct the individual. To succeed in this, we need to go back a thousand steps and first understand—for ourselves—the true nature of the human being. We personally think that only by defining the nature of what it means to be human can we find a better way forward. We believe that a person's best qualities can successfully be preserved and nurtured over time, even in crises and emergencies, if the ideal soil is created and then kept fertile.

How do we create the ideal humus for planting the seed of a new civilization, a better world order?

As demonstrated by the phenomenon of entanglement and the most recent research in quantum physics, it is now a well-established fact that everything in the universe is interconnected. Therefore, the first necessary step is to move from the ego-system in which we grew up, sustained by dynamics of domination and control, to an eco-system. Within an eco-system, the disconnection and individualism typically found in the former do not exist: here, plants, animals, and humans each perform their functions naturally, parts of the same connected organism, and all deserve the same respect. Therefore, in an eco-system, social and business models are primarily based on principles of cooperation, ethics, mutual exchange, product quality, care, and genuine interpersonal connection.

For this reason, within the IppocrateOrg Association, the Origini Project (www.origini.life) was born. Starting from a reassessment of current social and cultural models, its aims are to explore the ideal habitat for human beings to evolve in harmony with the world around them. The Origini Project seeks to help create this eco-system by promoting a holistic lifestyle through certification of associations, communities, and businesses that support the goals of health, well-being, and human evolution in harmony with the environment—the greatest challenge of our century. We believe that creativity, ethical exchanges, and the development of a people-centered culture, capable of passing on an eco-systemic range of values, can form the foundation for the birth of a new Western Civilization.

CHAPTER 34
The Victory Garden

Don't just consume. Produce. Our parents did it, and so can we.

If people let government decide which food they eat and medicines they take, their
bodies will soon be in as sorry a state as are the souls of those who
live under tyranny.
—Thomas Jefferson

Many have become alarmed about the odd rash of fires breaking out in food processing plants during 2022 [447]. Whether by intent or coincidence, these fires are adding fuel to growing fear that the winter of 2022 may see widespread and somewhat idiosyncratic global food shortages [448]. Only time will tell if this is merely more fearporn, some nefarious conspiracy to advance some WEF/central bank control agenda, or the logical consequence of supply chain disruptions attributable to the Ukrainian conflict and subsequent compromise of grain and fertilizer distribution. But there is something that you can do for yourself and your family, now and into the future. Go old school—grow your own food.

A backyard garden can quite literally feed a whole family. People don't have to be dependent on international agribusinesses, nutritionally valueless food, grain from Russia or Ukraine, food imports from China and other countries, or even be dependent on high-priced organics to feed themselves and their families. Each of us has the power to create our food from scratch. So, let's walk through the history of the war gardens in the UK and US, which later evolved into what we know as the victory garden.

During World War I, food production fell dramatically in Europe because farm workers left for military service, and many farms were destroyed by the war. Furthermore, transport of goods became difficult due to the dangerous conditions required for shipping by boat. A wealthy US philanthropist and conservationist (Charles Lathrop Pack) conceived of the idea that food supply could be greatly increased by citizens planting small vegetable gardens that would supply local communities with food. That this could be done without the use of the land and manpower already engaged in larger scale agriculture, and without the significant use of transportation facilities that were otherwise needed for the war effort.

The US National War Garden Commission was organized in 1917 by Mr. Pack, and within that same year the War Garden Campaign was launched. This campaign promoted the use of surplus private and public lands for small vegetable gardens. This program resulted in over five million gardens, and the value of the produce from these gardens exceeded $1.2 billion by the end of the war. Even children were mobilized in the effort, and school victory gardens were also planted at educational institutions throughout the USA. The United State Garden Army was established by the United States Bureaus of Education and the Department of the Interior, and President Wilson took a special interest in the cause. By the end of WWI, more food was being produced by these home gardens than farmers had produced in years prior to the war!

The idea of the war garden was continued and expanded during World War II, as labor and transportation shortages once again made it hard to harvest crops and to move fruits and vegetables to market. As the government rationed foods like sugar, butter, milk, cheese, eggs, coffee, meat, and canned goods due to the war, shortages of foods became the norm. Therefore, the United States government encouraged citizens to plant "Victory Gardens," also known as "war gardens" or "food gardens for defense." Nearly twenty million gardens were planted in backyards, empty lots, and even city rooftops. New York City had the parks and public lawns devoted to victory gardens, as were portions of San Francisco's Golden Gate Park. In Hyde Park, London sections of lawn were publicly plowed for plots to publicize the movement. Neighbors and communities, all with the goal of winning the war, formed cooperatives to meet the local needs of fresh produce. Farm families, of course, had been planting gardens and preserving produce for generations. Now, urban gardens became the norm. The government and businesses encouraged people to can and preserve their own produce to save the commercial produce for the troops. People responded in mass. The produce harvested from these gardens was estimated to be 9–10 million tons. When the war effort ended, so did the victory gardens. But the idea has lived on.

With the advent of fertilizer, grain, petroleum, and energy shortages worldwide, it seems that the stage is set for the next wave of victory gardens.

A garden is a grand teacher. It teaches patience and careful watchfulness; it teaches industry and thrift; above all it teaches entire trust.
—Gertrude Jekyll

Fast-forward to my own farm. When I work in my garden, whether it be in our fruit orchard or merely routine weeding, I feel like I am doing something worthwhile. That I am creating. Growing a garden is a victory over the globalist agenda—a victory over those who wish to control every aspect of consumerism as well as every aspect of our lives. So, let's once again embrace the name of the Victory Garden, because in the very act of growing a garden, we are choosing to be a part of the production of life. To be producers, instead of consumers. That is a victory.

It is a victory to grow an abundance of food. To share that with others through cooking, giving, bartering, and even selling. Community forms from the small, everyday acts of life.

One of the most rewarding ways to both eat healthy and keep the passion high for healthy living is by growing your own food. By that I mean anything from having a parsley plant in a pot by the door of your apartment or on a window sill, to a tomato plant in a bit of soil in the backyard, to having a community garden plot or to having your own vegetable patch. Gardening is a spectrum of choices. It can even be as simple as sprouting alfalfa seeds.

When I cook with produce that I have harvested, I use resources as they become available. Cooking with what I grow is an immensely creative activity. It motivates me to eat healthy and be healthy.

Gardening is a "grand" endeavor that must be planned in advance. Many a winter or early spring, I have spent happy hours looking through seed catalogs or strategizing on where and how my vegetable garden will be cultivated. Spring brings preparing the soil and finally planting. Summer is hard work and yet, the most rewarding time for my garden. Fall is a closing up of the summer garden plot and readying for the winter, climate depending. Vegetable gardening is a seasonal activity. It puts the body and mind on track and in sync with the world around us.

> Pangloss, who was as inquisitive as he was argumentative, asked the old man what the name of the strangled Mufti was. "I don't know" answered the worthy man, "and I have never known the name of any Mufti, nor of any Vizier. I have no idea what you're talking about; my general view is that people who meddle with politics usually meet a miserable end, and indeed they deserve to. I never bother with what is going on in Constantinople; I only worry about sending the fruits of the garden which I cultivate off to be sold there." Having said these words, he invited the strangers into his house; his two sons and two daughters presented them with several sorts of sherbet, which they had made themselves, with kaimak enriched with the candied-peel of citrons, with oranges, lemons, pine-apples, pistachio-nuts, and Mocha coffee... – after which the two daughters of the honest Muslim perfumed the strangers' beards. "You must have a vast and magnificent estate," said Candide to the Turk. "I have only twenty acres," replied the old man; "I and my children cultivate them; and our labor preserves us from three great evils: weariness, vice, and want." Candide, on his way home, reflected deeply on what the old man had said. "This honest Turk," he said to Pangloss and Martin, "seems to be in a far better place than kings.... I also know," said Candide, "that we must cultivate our garden."
>
> Voltaire: *Candide – or Optimism* (1759)

A vegetable garden is also a political statement. To commit to breaking out of the supply-chain network, to living without store-bought food, is an act of resistance. If you don't want your produce coming from China, if you want to know what really went into that green veg on your plate, a garden is a must. It can also be a commitment to creating an intentional community. Whether sharing with friends and neighbors or eating a meal harvested from the earth, these are time-honored ways to create bonds.

But vegetable gardening is also more than a healthy, stress-relieving activity; it is a commitment to the future. I like to think of my vegetable garden as a small act of giving to the future. Growing food is a simple way to create surplus in times of shortages, a simple way to help relieve the stress of inflation. Beyond that, as Americans, if we truly value freedom, we need to again become committed to self-sufficiency both as a nation and as individuals. In my opinion, it is time to stop looking to other countries to fill the pantries of Americans. Just as in the days of the war garden, we can be productive and free ourselves from dependency on imported food. Our lives don't have to be filled with nonproductive endeavors. Nothing is better for the soul than using our time on this Earth for productive good.

I have spent many a fine day touring public gardens and learning about gardening techniques. But the gardens that give me the most inspiration for the future come from the war gardens, first conceived by Charles Lathrop Pack during World War I. Because producing life affirming food in the time of war shortages allowed so many to envision the better future that did eventually arrive, and we are the product of that effort by our parents and grandparents. So stand up straight and be proud of what our forefathers and mothers did for us. We are standing on the shoulders of giants.

Be well, friends. Build community. Be kind to one another. We will survive this.

CHAPTER 35
How Does it Feel to Be Vindicated?

Depressing and demoralizing. And I would do it again in a heartbeat.

Et tu, Brute?

Well, I suppose that it is a win that the HHS bureaucrats and their many paid enablers are not just backslapping and giving one another medals over how well they have managed COVID-19. At least not yet.

But we do have a modicum of chatty condescending acknowledgment of mistakes made by Drs. Rochelle Walensky (director, CDC), Paul Offit (notoriously smug coinventor of a rotavirus vaccine), and Maurice R. Hilleman (professor of Vaccinology, professor of Pediatrics at the University of Pennsylvania) and former members of the Centers for Disease Control (CDC) Advisory Committee on Immunization Practices scattered throughout media now.

An excellent commentary summarizing the stunning self-owning admissions of incompetence and culpability for massive unnecessary loss of life has been provided by Thomas Harrington writing for the Brownstone Institute, titled *Drs. Walensky and Offit: It's All in Good Fun* [449]. Personally, I can hardly bear to watch their breezy smugness as they casually chat with friends. I am reminded of the famous Hannah Arendt phrase *"the banality of evil."* Mr. Harrington points to a series of clips of the Walensky interview compiled by Phil Kerpen (unfortunately on Twitter), quoting from Alex Marinos on a Twitter thread:

> How was the decision made to ignore immunity from prior infection?
>
> In this clip, Paul Offit describes how he and another person advised in favor of accepting natural immunity, while two others

voted against it. A thread on why that was possibly the worst decision of the pandemic.

I recommend both of those abridged versions of the interviews for those (like me) who just cannot stomach the full interviews. Mr. Harrington's succinct summary nicely encapsulates my feelings about the situation and includes the following:

> All those moves to censor and professionally destroy those who had opinions different from the CDC, actions rooted precisely in the presumption that science is, in fact, black and white, and that those who get it wrong need to be professionally punished, well, that's all a figment of your primitive imagination.

Or, as Harold Pinter put it in his Nobel Prize speech when referring to the US penchant for wantonly destroying other cultures:

> It never happened. Nothing ever happened. Even while it was happening it wasn't happening. It didn't matter. It was of no interest.

So yes, excessive psychic detachment that turns fellow human beings into self-referential objects of our own minds can be rather problematic. Indeed, I think, though I can't be sure, that psychologists even have a term for it: *psychopathy*."

Bill Gates, Klaus Schwab, Anthony Fauci, Rochelle Walensky, Paul Offit, Janet Woodcock, Rick Bright, Jessica Cecil and her Trusted News Initiative. Don't forget these names. They should live in infamy. And they all share a common personality profile.

Lately, I have been getting the question "How does it feel to be vindicated?"

Dr. Jill (Glasspool-Malone) keeps noodling me to write a piece describing my feelings on this topic. Personally, I dislike focusing on the psychology of how these last two years have impacted me (and us). Much as I am very wary of the "cult of personality" aspect of my newfound fame, I have not spoken out because I sought attention; I have done this because it was the right thing to do, and I seemed to have a unique window of opportunity to speak for those whose voices were so actively suppressed. But I certainly have had to take hits for it. The slander, defamation, gaslighting, and globally coordinated character assassination have been nonstop. But as time has gone by, and more and more has been revealed about the hidden hands that seek to manage what we are allowed to hear, see, and think, I have been transformed.

The biomedical world that I thought I was living in has been revealed to be a sham. The legitimacy of the industry and discipline that I have committed my entire professional life to is in shambles. I am now embarrassed to call myself a vaccines and biodefense expert, because the fundamental corruption inherent in those domains has been so clearly revealed. I cannot unsee what I have seen. I cannot recapture all of those years spent in a profoundly corrupt academic system, spent supporting a deeply compromised discipline that appears primarily driven by financial interests rather than by what I had naively believed was a commitment to saving lives. I chose to not pursue the careers of my father and father-in-law, which were spent building weapons of war. Only to find that I had inadvertently played a significant role in enabling one of the most tragic medical follies in the history of man.

When first asked how it feels to be vindicated, I did not know what to say. It feels a long, long way from vindication. Those directly responsible are unlikely to face any form of reckoning. And rather than remorse, they seem to find the whole thing amusing. The unnecessary lives lost, the destruction of faith in the public health enterprise, vaccines in general, the entire medical/hospital system, the US Department of Health and Human Services, and government in general. *"Ha. Oh well, not our fault. Just the way things are."*

I looked inward, deep into my heart and soul, and asked the question. How does it feel? Demoralizing and depressing. I experience absolutely no pleasure whatsoever in seeing my worst fears come to pass, and in having accurately predicted so many things during the last two years. Jill and I have put everything on the line. Parked our lives, our farm, our family, in a sustained effort to try to save lives and help average people understand what was going on, what the actual "Science" was, and to try to help people to think through the issues. Going back to ground zero, to try to enable "informed consent" in a time where that fundamental bedrock of medical ethics was thrown into the dumpster. We have experienced extraordinary efforts to delegitimize us, to rewrite history, to deny us credit for intellectual and technical contributions, to slander and defame. They have destroyed the consulting business that we had built up together over decades. We have drained ourselves with the constant travel and stress of the speaking engagements. A constant stream of podcasts (up to nine per day) as a way to break through the wall of globally coordinated censorship and propaganda. I have been labeled a "right wing extremist" and "Nazi."

Here is a little story about Santa Barbara. I left my job as a carpenter and attended Santa Barbara City College between 1980 and 1982, graduating with straight A's as president of the Student Council. This was made possible in part by financial assistance from the Santa Barbara Foundation. When invited to come speak at a *Stand Up SB* event in Santa Barbara regarding COVID, I suggested it would be nice to do a fundraiser for the Santa Barbara Foundation. An opportunity to give back to the community and organization that had made my journey from carpenter and orchard farmhand to physician/scientist possible. When they contacted me, the foundation decided that they did not want to receive any support from a far-right-wing person such as myself. Et tu, Brute? This is an example of the price that has had to be paid.

And we would do it again in a heartbeat. Because it has been the right thing to do. And we found ourselves in a position where we had a chance to make a difference. We have made new friends all over the world. I now have a very different worldview than I had three years ago. I have no regrets. But I take no pleasure in the thing.

CHAPTER 36
Anonymous – Letter from a Coerced Mother

Teacher unions are not qualified to set public health policy, and yet that is precisely what they have been doing during the COVIDcrisis. I have been getting a lot of letters since I began to publicly question the US HHS policies on mask usage, suppression of early treatment, lockdowns, vaccine safety and efficacy, and in particular the vaccine mandates. These letters come to my email accounts, my social media accounts, and my US Postal Service mailbox. Of these, the most heartbreaking are the letters from parents, who face an impossible situation. Coerced by public school systems or university bureaucracies, parents face the awful choice of either having their child injected with an experimental medical procedure (gene therapy), one to which they desperately object, or denying their children their educational experience with their friends. To illustrate the difficult reality these parents face, I have included a letter that touched my heart and share an anonymized version with you below.

People in the USA and the world over were hurting in the worst way during the COVID-19 mandates. They did not know where to turn and were bombarded with a carpet bomb of corporate media propaganda and messaging (paid by the US Government, largely) asserting that the (experimental) gene therapy vaccine products are safe and effective (neither of which is true). The official propaganda and messaging smeared parents who refuse vaccination for their children, accusing them of being selfish and putting others (children, elders, teachers) at avoidable risk. Many parents opted out of public schools or adopted home schooling, but the vast majority who could not afford to do so had no choice but to have their children vaccinated to stay in the schools.

I remember when Jill and I were both working and had young children, we were always worried about how to just get by and pay the bills. Although we eventually homeschooled our boys when they reached teen age, this would not have been an option while still paying off our university school debt. My heart goes out to the mothers and fathers cruelly backed into this painful corner by university trustees, teachers union leaders, and local school board members, whose vaccine and mask mandates are supported by a massive taxpayer-funded government propaganda campaign… but not by rigorous science-based data and objective analysis.

February 2022

Dear Dr. Malone:

Subject: Vermont now recommending that schools can drop mask requirement if they have >80% students vaxxed.

I'm a busy working mom with small children living in Vermont, where my family has been living and farming for 8 generations. I do a little hobby farming myself, and I understand you are a man of the land and animals, too.

I was originally quite compliant with the pandemic response: we wore masks, I kept my healthy kids home from school, I worked from home even though it all felt nearly impossible and was a severe strain on our family. We cancelled vacations, stopped asking Grandma and Grandpa to help with the kids. My husband and I lined up to get vaccines when they came out.

After about the third time that I had to take a week off work to stay home with 3 healthy children who were in quarantine, this was in summer 2021, I started to get curious about whether all of this was really necessary, because it felt impossible and unfair that I would be denied access to my livelihood and my children would be denied school and childcare, simply because one of us made the mistake of breathing somewhere within the vicinity of a classmate who eventually tested positive for COVID-19.

I do have a, thankfully mild, (and unreported) vaccine injury in the form of menstrual disruptions: I now get a guaranteed migraine once a month. I was very sorry to hear about your vaccine injury and I am delighted that you survived it. This experience opened my mind to the fact these vaccines are not without risk and why wasn't anybody talking about risk and choice?

That was when I became aware of the work you and so many others are doing to bring a different perspective to the situation, and I've been a fan ever since. It has been difficult, being of a different mindset about the pandemic compared to most of my peers, my inner circle, our local government. And I've been grateful lately that it seems as though the end is in sight.

And now, I'm up at 2 in the morning writing to you over what feels like it could be the last straw for me: our state, which has been recommending mandatory masking in schools this whole time, is planning to end masking requirements BUT ONLY IF more than 80% of a school's students are vaccinated against Covid-19. So, in plain terms, we can drop not-very-effective mitigation measure if we increase uptake on not-very-effective mitigation measure which involves injecting our children with an experimental gene therapy for a disease that (a) my kids have already had and (b) doesn't, statistically speaking, cause any real harm for children.

Vermont is interesting because we were among the first states in the country to reach the 80% vaccination rate in adults that was expected to confer herd immunity, and here we are in 2022 coming off the worst case surge we've experienced yet. But apparently we somehow still think that vaccination is going to keep schools safer?

You better believe that there will be peer pressure at school and my daughters will come home once again asking to be vaccinated, because "everybody's doing it."

This feels like coercion of minors of the worst sort. It's illogical, wrongheaded and I am so upset I can barely type straight. And I find myself thinking, "What would Robert Malone do?" I was wondering if you have an answer to that.

Very respectfully, Jennifer

Here is my response:

Hi Jennifer, your situation touched me deeply. So, you asked me "What would Robert Malone do?" The first thing I would do is make sure I have very good lines of communication with my daughters. As they are old enough to be aware of peer pressure, I would start there.

Discuss the issues of "peer pressure"—why it is important to learn to resist such pressure, how it can affect a person emotionally. That peer pressure can cause depression, hurt, feelings of betrayal, etc. That finding true friends who won't pressure is important.

Explain that there are risks and benefits to the vaccines. I think it is important to have a conversation with your children's teachers. Explain that your children are feeling peer pressure and suffering the effects of that. If it is clear that the teacher(s) are unsupportive, it may be time to look for alternative schooling.

Your children may find this video that Jill and I produced with Children's Health Defense Hawai'i helpful (see https://vimeo.com/648749188/5333c4b078).

I know Vermont has a policy to allow home schooling, and there is a robust community for such. There are many solutions that are in between full on homeschooling and public schools. "Pods" (where a group of parents hire a teacher), co-ops—where parents take turns teaching or pay one parent to administer and teach, and private schools are all options. The most important element in this is making sure that your daughters feel like they have a community. Seeking community for your daughters outside of "school," may be as important as the actual work at school.

This all becomes difficult for students who are in high school and taking advanced coursework. But remember, such students are almost adults and should be able to resist peer pressure more effectively. Encourage your children to find like-minded individuals who will support them.

Vermont is stating an 80% vaccine compliance. At this point, I would be front and center saying "NO. Not my children. Not now, not ever for the mRNA vaccines." The reason to be front and center with the school, the teachers and your children is that if you make it non-negotiable, most likely—they will be less inclined to argue and try to persuade. If you shirk being straight forward, people will see you as someone to be coerced. Don't let yourself be that person.

As this comes down to mask use—it is time to protest locally and at the state level. This is blackmail of our children to take a vaccine which is not fully licensed—for your children, these vaccines have emergency use authorization only. Which is to say that they remain "experimental" from the standpoint of regulatory law. There is no license (marketing authorization).

If the federal state of emergency is dropped (as both I and the Truckers advocate be done), then they can no longer be distributed. But there is no emergency, as you indirectly point out in your letter. Not for your children, not with Omicron. Go to your state representatives, write letters, go to local school board meetings, etc. You can find a school board meeting information package here. Get on social media to find other people willing to stand up and spread the message that masks are not a good solution. Phone like-minded parents. Share this podcast, which covers the risks and harms being caused to our children from these vaccine and mask policies. Visit the school Principal. Let him or her know your displeasure with both mandatory mask use and the efforts to coerce your children to take the vaccine. Contact the teacher's union. Print out the studies that show that the masks worn are in-effective and get the word out. Basically, now is the time for all of us to get involved politically.

My prediction is the CDC is going to back off on this requirement. However, the fact that teacher unions are so dug in on mask use makes it a little harder to predict what influences they have on government.

So, you asked me what I would do. The honest truth is that there is no good answer except to continue to resist and to do so in a manner that is 1) peaceful, 2) effective and 3) benefits your children. You are setting an example for them that they will remember for the rest of their lives. So step up and help them as well as their and your community get through this.

Finally, if you can hang on until summer, I believe that come the next school year—the rules will have to be re-written. I can't see what is on the other side, but now is the time to speak out—for your children, and for all children.

Sincerely,
Robert

Acknowledgments

We wish to thank Gavin de Becker, whose friendship and support throughout the last two years has made this possible. Gavin–your friendship throughout has been amazing and one of the true bright lights for both of us over the past two years. You have seen us at our best and worst—and still love us both.

Robert F. Kennedy, Jr. – thank you for your friendship and advice. Your leadership is helping us all out of the darkness.

Dr. Joe Ladapo and Governor DeSantis – your leadership has inspired all of us.

Thank you to Craig Snyder for your work in editing. It made a huge difference when I needed help the most.

Kelly and Tad Coffin – we don't have words to express how much your support has meant.

I am particularly grateful for the kind mentorship and coaching generously provided to me by Dr. Peter K. Navarro, including the two groundbreaking Washington Times articles that we composed together. If Peter's leadership on hydroxychloroquine had been followed, the outcome of this story would have been completely different. Thank you for all that you and Steve Bannon have done and sacrificed, Peter, and for our original interview on War Room, where you first introduced me to the amazing world of pirate television broadcasting.

There are so many, many people who make our lives more complete and whose efforts have aided in this book. We wish to thank Senator Ron Johnson, Olivia Myers, Ian Evans, Brooke and Ann Miller, Justine Isernhinke, Nina and Colby May, Aiden Almed, the list goes on and on. We sincerely thank all of you.

I also wish to thank Dr. Richard Urso, Dr. Ryan Cole, Dr. John Littell, Dr. Katrina Lindley, and so many other physicians and scientists who have given their support and time along the way.

To the people who contributed chapters for this book, thank you for your patience and hard work. I also wish to acknowledge Kent Heckenlively for his work in helping me write and focus my thoughts with the introduction.

Finally, we wish to thank everyone who has waited so patiently for me (us) to finish this project. It is way longer and took much more work that I would have thought possible, but there is so much to be said about what has been the biggest public failure of the US government, of public health in the history of our country and the world.

If you wish to read more, you can always find us on proverbial "Galt's Gulch." That is "Who is Robert Malone" at rwmalonemd.substack.com.

The INDEX for *Lies My Gov't Told Me* is available on my substack noted just above.

References

1. Malone, R.W., et al., *Lipid-mediated polynucleotide administration to deliver a biologically active peptide and to induce a cellular immune response*, USPTO, Editor. 1989 Vical Inc: USA.
2. Malone, R.W., et al., *Lipid-mediated polynucleotide administration to reduce likelihood of subject's becoming infected* 1989 Vical Inc.: USA.
3. Malone, R.W., et al., *Generation of an immune response to a pathogen.* 1989 Vical Inc.: USA.
4. Malone, R.W., et al., *Expression of exogenous polynucleotide sequences in a vertebrate, mammal, fish, bird or human*, USPTO, Editor. 1989 USA.
5. Malone, R.W., et al., *Methods of delivering a physiologically active polypeptide to a mammal.* 1989 Vical Inc.: USA.
6. Malone, R.W., et al., *Induction of a protective immune response in a mammal by injecting a DNA sequence (includes mRNA).* 1989 Vical Inc.: USA.
7. Malone, R.W., et al., *Generation of antibodies through lipid mediated DNA delivery.* 1989 Vical Inc: USA.
8. Malone, R.W., et al., *Delivery of exogenous DNA sequences in a mammal (includes mRNA).* 1989 Vical Inc: USA.
9. Roltgen, K., et al., *Immune imprinting, breadth of variant recognition, and germinal center response in human SARS-CoV-2 infection and vaccination.* Cell, 2022. 185(6): p. 1025–1040 e14. https://www.ncbi.nlm.nih.gov/pubmed/35148837.
10. Greene, J., *Wait what? FDA wants 55 years to process FOIA request over vaccine data*, in *Reuters.* 2021. https://www.reuters.com/legal/government/wait-what-fda-wants-55-years-process-foia-request-over-vaccine-data-2021-11-18/.
11. So, L., Smith, G., *In four U.S. state prisons, nearly 3,300 inmates test positive for coronavirus—96% without symptoms*, in *Reuters.* 2020. https://www.reuters.com/article/us-health-coronavirus-prisons-testing-in/in-four-u-s-state-prisons-nearly-3300-inmates-test-positive-for-coronavirus-96-without-symptoms-idUSKCN2270RX.
12. Zabion, F.M., *Chloroquine- Definition, Properties, Uses, Mechanism Of Action, Side Effects*, in *The Biology Notes.* 2022. https://thebiologynotes.com/chloroquine/
13. Vincent, M.J., et al., *Chloroquine is a potent inhibitor of SARS coronavirus infection and spread.* Virol J, 2005. 2: p. 69. https://www.ncbi.nlm.nih.gov/pubmed/16115318.
14. Keyaerts, E., et al., *In vitro inhibition of severe acute respiratory syndrome coronavirus by chloroquine.* Biochem Biophys Res Commun, 2004. 323(1): p. 264–8. https://www.ncbi.nlm.nih.gov/pubmed/15351731.
15. Dyall, J., et al., *Repurposing of clinically developed drugs for treatment of Middle East respiratory syndrome coronavirus infection.* Antimicrob Agents Chemother, 2014. 58(8): p. 4885–93. https://www.ncbi.nlm.nih.gov/pubmed/24841273.
16. de Wilde, A.H., et al., *Screening of an FDA-approved compound library identifies four small-molecule inhibitors of Middle East respiratory syndrome coronavirus replication in cell culture.* Antimicrob Agents Chemother, 2014. 58(8): p. 4875–84. https://www.ncbi.nlm.nih.gov/pubmed/24841269.
17. Nass, M., *WHO and UK trials use potentially lethal hydroxychloroquine dose — according to WHO consultant.* 2020. https://merylnassmd.com/who-trial-using-potentially-fatal/.
18. Nass, M., *Even worse than 'Recovery,' potentially lethal hydroxychloroquine study in patients near death.* 2020. https://merylnassmd.com/even-worse-than-recovery-potentially/.
19. Nass, M., *How a false hydroxychloroquine narrative was created, and more.* 2020. https://merylnassmd.com/how-false-hydroxychloroquine-narrative/.
20. Devaux, C.A., et al., *New insights on the antiviral effects of chloroquine against coronavirus: what to expect for COVID-19?* Int J Antimicrob Agents, 2020. 55(5): p. 105938. https://www.ncbi.nlm.nih.gov/pubmed/32171740.

21. Million, M., et al., *Early treatment of COVID-19 patients with hydroxychloroquine and azithromycin: A retrospective analysis of 1061 cases in Marseille, France.* Travel Med Infect Dis, 2020. 35: p. 101738. https://www.ncbi.nlm.nih.gov/pubmed/32387409.

22. Savarino, A., et al., *New insights into the antiviral effects of chloroquine.* Lancet Infect Dis, 2006. 6(2): p. 67–9. https://www.ncbi.nlm.nih.gov/pubmed/16439323.

23. MedinCell, *The third path to fight Covid-19: Prevention.* 2020, MedinCell: Video Conference. https://www.medincell.com/en/2020/04/23/adapt-and-overcome-in-response-to-the-covid-19-crisis/.

24. FDA, *FDA Letter to Stakeholders: Do Not Use Ivermectin Intended for Animals as Treatment for COVID-19 in Humans*, F.a.D. Administration, Editor. 2020. https://www.fda.gov/animal-veterinary/product-safety-information/fda-letter-stakeholders-do-not-use-Ivermectin-intended-animals-treatment-covid-19-humans.

25. CDC, *Rapid Increase in Ivermectin Prescriptions and Reports of Severe Illness Associated with Use of Products Containing Ivermectin to Prevent or Treat COVID-19*, C.f.D. Control, Editor. 2021, CDC: Atlanta, GA. https://emergency.cdc.gov/han/2021/han00449.asp.

26. Geunot, M., *Prescriptions for Ivermectin—a deworming drug—surged to 24 times their pre-pandemic levels as people baselessly take it as a COVID-19 cure*, in *Business Insider*. 2021. https://www.businessinsider.com/Ivermectin-prescriptions-surged-24-fold-unproven-coronavirus-cure-2021-8.

27. Gonzales, C., *DOH: Doctors prescribing Ivermectin as COVID-19 cure risk losing license*, in *Inquirer.net*. 2021. https://newsinfo.inquirer.net/1411421/doh-doctors-prescribing-Ivermectin-as-covid-19-cure-risk-losing-license.

28. Luján, B.R., *Luján, Klobuchar Introduce Legislation to Hold Digital Platforms Accountable for Vaccine and Other Health-Related Misinformation.* 2021. https://www.lujan.senate.gov/newsroom/press-releases/lujan-klobuchar-introduce-legislation-to-hold-digital-platforms-accountable-for-vaccine-and-other-health-related-misinformation/.

29. Harmon, G., *Flow of damaging COVID-19 disinformation must end now.* 2021. https://www.ama-assn.org/about/leadership/flow-damaging-covid-19-disinformation-must-end-now.

30. DHS, *DHS Issues National Terrorism Advisory System (NTAS) Bulletin*, H. Security, Editor. 2022, DHS: Washington, DC. https://www.dhs.gov/news/2022/02/07/dhs-issues-national-terrorism-advisory-system-ntas-bulletin.

31. Marik, P.E., et al., *Hydrocortisone, Vitamin C, and Thiamine for the Treatment of Severe Sepsis and Septic Shock: A Retrospective Before-After Study.* Chest, 2017. 151(6): p. 1229–1238. https://www.ncbi.nlm.nih.gov/pubmed/27940189.

32. Chalifoux, R., Jr., *So what is a sham peer review?* MedGenMed, 2005. 7(4): p. 47; discussion 48. https://www.ncbi.nlm.nih.gov/pubmed/16614669.

33. Huntoon, L.R., *Tactics characteristic of sham peer review.* Journal of American Physicians and Surgeons, 2009. 14: p. 64–66.

34. Nowakowski, G., et al., *Healthcare quality improvement act: Peer review, procedure, process, and privacy.* Western Michigan University Thomas M Cooley Law Review, 2016. 33: p. 111–140.

35. Pfifferling, J.H., D.N. Meyer, and C.J. Wang, *Sham peer review: perversions of a powerful process.* Physician Exec, 2008. 34(5): p. 24–9. https://www.ncbi.nlm.nih.gov/pubmed/19456073.

36. Huntoon, L., *Sham peer review: Disaster preparedness and defense. Journal of American Physicians and Surgeons 2011; 16:2–6.* Journal of American Physicians and Surgeons, 2011. 16: p. 2–6.

37. Huntoon, L.R., *The psychology of sham peer review.* Journal of American Physicians and Surgeons, 2007. 12: p. 3–4.

38. Huntoon, L.R., *Sham peer review: the Unjust "objective test".* Journal of American Physicians and Surgeons, 2015. 2007(12): p. 12–100.

39. Huntoon, L.R., *Sham peer review: Violations of due process and fundamental fairness.* Journal of American Physicians and Surgeons, 2018. 2018: p. 66–68.

40. Huntoon, L.R., *Sham peer review: The Poliner verdict.* Journal of American Physicians and Surgeons, 2006. 11: p. 37–38.

41. Huntoon, L.R., *Sham peer review: The fifth circuit Poliner decision.* . Journal of American Physicians and Surgeons, 2008. 13: p. 98–99.

42. Times, H., *33 districts in Uttar Pradesh are now Covid-free: State govt*, in *Hindustan Times*. 2021: India. https://www.hindustantimes.com/cities/lucknow-news/33-districts-in-uttar-pradesh-are-now-covid-free-state-govt-101631267966925.html.

43. Seth, S., *Uttar Pradesh government says early use of Ivermectin helped to keep positivity, deaths low*, in *The Indian Express*. 2021. https://www.msn.com/en-in/news/localnews/uttar-pradesh-government-says-early-use-of-Ivermectin-helped-to-keep-positivity-deaths-low/ar-BB1gDp5U.

44. Kerr, L., et al., *Ivermectin Prophylaxis Used for COVID-19: A Citywide, Prospective, Observational Study of 223,128 Subjects Using Propensity Score Matching.* Cureus, 2022. 14(1): p. e21272. https://www.ncbi.nlm.nih.gov/pubmed/35070575.

45. FLCCC. *FLCCC Alliance.* 2022 April 27, 2022; Available from: https://covid19criticalcare.com/.

46. Kory, P., et al., *"MATH+" Multi-Modal Hospital Treatment Protocol for COVID-19 Infection: Clinical and Scientific Rationale.* J Clin Med Res, 2022. 14(2): p. 53–79. https://www.ncbi.nlm.nih.gov/pubmed/35317360.

47. Marik, P.E. *Doctors, Not Administrators, Should Be Treating Patients.* American Greatness, 2021. https://amgreatness.com/2021/11/22/doctors-not-administrators-should-be-treating-patients/.

48. Reis, G., et al., *Effect of early treatment with fluvoxamine on risk of emergency care and hospitalisation among patients with COVID-19: the TOGETHER randomised, platform clinical trial.* Lancet Glob Health, 2022. 10(1): p. e42-e51. https://www.ncbi.nlm.nih.gov/pubmed/34717820.

49. AP, *Oklahoma AG OKs Prescribing Ivermectin, Hydroxychloroquine,* in *Associated Press.* 2022. https://www.usnews.com/news/best-states/oklahoma/articles/2022-02-08/oklahoma-ag-oks-prescribing-Ivermectin-hydroxychloroquine.

50. Rong-Gong, L., and L. Money, *Fears of more long COVID, a 'mass disabling event' as variants rip through California,* in *Los Angeles Times.* 2022: Los Angeles. https://www.latimes.com/california/story/2022-07-26/covid-19-reinfection-worsens-long-term-risk-for-death-fatigue-heart-disorders.

51. Todd, C., *Meet the Press, MSNBC.* Fauci: Attacks On Me Are Attacks On Science And Truth. 2021; Available from: https://www.youtube.com/watch?v=Cmn3895MBkU. https://www.youtube.com/watch?v=Cmn3895MBkU.

52. Hains, T., *Fauci: Attacking Me Is Attacking Science.* Real Clear Politics, 2021. https://www.realclearpolitics.com/video/2021/06/09/fauci_attacking_me_is_attacking_science.html.

53. Laco, K., *Lawsuit filed against Biden, top officials for 'colluding' with Big Tech to censor speech on Hunter, COVID Lawsuit filed against Psaki, Fauci, Mayorkas and other top Biden administration officials.* Fox News, 2022. https://www.foxnews.com/politics/lawsuit-filed-against-biden-top-officials-colluding-big-tech-censor-speech-hunter-covid.

54. Piper, E., *Anthony Fauci, America's high priest of scientism, wears out his welcome. Scientific 'progress' unrestrained by sacred principles is fraught with dangers,* in *Washington Times.* 2021. https://www.washingtontimes.com/news/2021/apr/18/anthony-fauci-americas-high-priest-of-scientism-we/.

55. Thomas, T., *Anthony Fauci, Prophet of Scientism.* American Thinker, 2021. https://www.americanthinker.com/articles/2021/06/anthony_fauci_prophet_of_scientism.html.

56. Popper, K.R., *Objective Knowledge: An Evolutionary Approach (Revised ed.).* 1979, New York: Oxford: Clarendon Press.

57. Schiff, A., *Schiff sends letter to Google, Facebook regarding anti-vaccine misinformation.* 2019, Adam Schiff: DC. https://schiff.house.gov/news/press-releases/schiff-sends-letter-to-google-facebook-regarding-anti-vaccine-misinformation.

58. CHD, *Court Hears CHD's Arguments Against Facebook, Zuckerberg and 'Fact Checkers' Lawyers for Children's Health Defense await the ruling of Judge Susan Illston after defending CHD's lawsuit alleging government-sponsored censorship, false disparagement and wire fraud.* The Defender, 2021. https://childrenshealthdefense.org/defender/court-hears-chd-arguments-facebook-zuckerberg-fact-checkers/.

59. Stossel, J., *Here's where the 'facts' about me lie—Facebook bizarrely claims its 'fact-checks' are 'opinion',* in *New York Post.* 2021: New York. https://nypost.com/2021/12/13/facebook-bizarrely-claims-its-misquote-is-opinion/.

60. Redwood, L., *Letter from Presdient, Childrens Health Defense to Mark Zuckerberg.* 2019. https://childrenshealthdefense.org/wp-content/uploads/FACEBOOK-COMPLAINT-EXHIBIT-A-DKT-1-1-08-17-2020.pdf.

61. Ghebreyesus, T., *Vaccine Misinformation: Statement by WHO Director-General on Facebook and Instagram.* 2019, World Health Organization: Geneva. https://www.who.int/news/item/04-09-2019-vaccine-misinformation-statement-by-who-director-general-on-facebook-and-instagram.

62. Chandler, S., *Facebook's Coronavirus Misinformation Policy At Odds With Political Ads Stance*, in *Forbes*. 2020, Forbes. https://www.forbes.com/sites/simonchandler/2020/04/16/facebooks -coronavirus-misinformation-policy-at-odds-with-political-ads-stance/?sh=5e2ca1cd3610.

63. Lerman, D., *Government's National Security Arm Took Charge During the Covid Response*. Brownstone Institute, 2022. https://brownstone.org/articles/governments-national-security-arm -led-the-covid-response/.

64. Shir-Raz, Y., et al., *Censorship and Suppression of Covid-19 Heterodoxy: Tactics and Counter-Tactics*. Minerva, 2022. https://link.springer.com/article/10.1007/s11024-022-09479-4.

65. JudicialWatch, *CDC Coordinated with Facebook On COVID Messaging and 'Misinformation'; CDC Received Over $3.5 Million in Free Advertising from Social Media Companies*. Judicial Watch, 2021. https://www.judicialwatch.org/cdc-facebook-covid-messagging/.

66. Facebook, *COVID-19 and Vaccine Policy Updates & Protections*. 2022. https://www.facebook .com/help/230764881494641.

67. Schechner, S., J. Horwitz, and E. Glazer, *THE FACEBOOK FILES: How Facebook Hobbled Mark Zuckerberg's Bid to Get America Vaccinated*, in *Wall Street ournal*. 2021, WSJ. https://www .wsj.com/articles/facebook-mark-zuckerberg-vaccinated-11631880296.

68. Savage, S., *Science is a Verb*, in *Science 2.0*. 2015. https://www.science20.com/agricultural_realism /science_is_a_verb-153242.

69. Platt, J.R., *Strong Inference: Certain systematic methods of scientific thinking may produce much more rapid progress than others*. Science, 1964. 146(3642): p. 347–53. https://www.ncbi.nlm.nih.gov /pubmed/17739513.

70. Scientist, *The Method of Multiple Working Hypotheses*. Science, 1890. 15(366): p. 92–6. https: //www.ncbi.nlm.nih.gov/pubmed/17782687.

71. Chamberlin, T.C., *The Method of Multiple Working Hypotheses: With this method the dangers of parental affection for a favorite theory can be circumvented*. Science, 1965. 148(3671): p. 754–9. https://www.ncbi.nlm.nih.gov/pubmed/17748786.

72. Meerlooe, J.A.M., *The Rape of the Mind: The Psychology of Thought Control, Menticide, and Brainwashing*. 2009 ed. 1956: Progressive Press. 326.

73. Kuhn, T.S., *The Structure of Scientific Revolutions*. 1996, Chicago: University of Chicago Press.

74. TreatmentStudies. *Global Ivermectin adoption for COVID-19: 45%*. COVID-19 Treatment Studies 2022 [cited 2022 July 15, 2022]; Available from: https://ivmstatus.com/.

75. Lagier, J.C., et al., *Outcomes of 3,737 COVID-19 patients treated with hydroxychloroquine/ azithromycin and other regimens in Marseille, France: A retrospective analysis*. Travel Med Infect Dis, 2020. 36: p. 101791. https://www.ncbi.nlm.nih.gov/pubmed/32593867.

76. Malone, R.W., et al., *COVID-19: Famotidine, Histamine, Mast Cells, and Mechanisms*. Front Pharmacol, 2021. 12: p. 633680. https://www.ncbi.nlm.nih.gov/pubmed/33833683.

77. Malone, R.W., et al., *COVID-19: Famotidine, Histamine, Mast Cells, and Mechanisms*. Res Sq, 2020. https://www.ncbi.nlm.nih.gov/pubmed/32702719.

78. Hong, W., et al., *Celebrex Adjuvant Therapy on Coronavirus Disease 2019: An Experimental Study*. Front Pharmacol, 2020. 11: p. 561674. https://www.ncbi.nlm.nih.gov/pubmed/33312125.

79. Malone, R.W., *More Than Just Heartburn: Does Famotidine Effectively Treat Patients with COVID-19?* Dig Dis Sci, 2021. 66(11): p. 3672–3673. https://www.ncbi.nlm.nih.gov/pubmed/33625612.

80. Tomera, K.M., R.W. Malone, and K. Kittah, *Hospitalized COVID-19 Patients Treated With Celecoxib and High Dose Famotidine Adjuvant Therapy Show Significant Clinical Responses*. SSRN, 2020. https://papers.ssrn.com/sol3/papers.cfm?abstract_id=3646583.

81. Offord, C., *Frontiers Pulls Special COVID-19 Issue After Content Dispute*. The Scientist, 2021. https://www.the-scientist.com/news-opinion/frontiers-pulls-special-covid-19-issue-after -content-dispute-68721.

82. Lardner, R., and J. Dearen, *Pepcid as a virus remedy? Trump admin's $21M gamble fizzled*, in *Washington Post*. 2020: DC. https://www.washingtonpost.com/health/trump-admin-21m -gambit-for-pepcid-as-a-covid-remedy-fizzles/2020/07/23/84ab4aa0-cd08-11ea-99b0 -8426e26d203b_story.html.

83. Bella, T., *A vaccine scientist's discredited claims have bolstered a movement of misinformation*, in *Washington Post*. 2022: DC. https://www.washingtonpost.com/health/2022/01/24/robert-malone -vaccine-misinformation-rogan-mandates/.

84. Brennan, C.M., et al., *Oral famotidine versus placebo in non-hospitalised patients with COVID-19: a randomised, double-blind, data-intense, phase 2 clinical trial.* Gut, 2022. 71(5): p. 879–888. https://www.ncbi.nlm.nih.gov/pubmed/35144974.

85. Janowitz, T., et al., *Famotidine use and quantitative symptom tracking for COVID-19 in non-hospitalised patients: a case series.* Gut, 2020. 69(9): p. 1592–1597. https://www.ncbi.nlm.nih.gov/pubmed/32499303.

86. Mather, J.F., R.L. Seip, and R.G. McKay, *Impact of Famotidine Use on Clinical Outcomes of Hospitalized Patients With COVID-19.* Am J Gastroenterol, 2020. 115(10): p. 1617–1623. https://www.ncbi.nlm.nih.gov/pubmed/32852338.

87. Suter, F., et al., *A simple, home-therapy algorithm to prevent hospitalisation for COVID-19 patients: A retrospective observational matched-cohort study.* EClinicalMedicine, 2021. 37: p. 100941. https://www.ncbi.nlm.nih.gov/pubmed/34127959.

88. Consolaro, E., et al., *A Home-Treatment Algorithm Based on Anti-inflammatory Drugs to Prevent Hospitalization of Patients With Early COVID-19: A Matched-Cohort Study (COVER 2).* Front Med (Lausanne), 2022. 9: p. 785785. https://www.ncbi.nlm.nih.gov/pubmed/35530041.

89. Consolaro, E., et al. *A Simple Approach to Prevent Hospitalization for COVID-19 Patients (COVER2),* C.g.I. NCT04854824, Editor. 2021, Mario Negri Institute for Pharmacological Research: ClinicalTrials.gov. https://clinicaltrials.gov/ct2/show/NCT04854824.

90. Cosentino, M., et al., *Early Outpatient Treatment of COVID-9: A Retrospective Analysis of 392 Cases in Italy.* medRxiv, 2022. https://www.medrxiv.org/content/10.1101/2022.04.04.22273356v1.full.

91. Holder, J., *Tracking Coronavirus Vaccinations Around the World.* NY Times, 2022. https://www.nytimes.com/interactive/2021/world/covid-vaccinations-tracker.html.

92. Mathieu, E., et al., *Coronavirus Pandemic (COVID-19).* 2022. https://ourworldindata.org/coronavirus#explore-the-global-situation.

93. Zimnoik, B., *Parliamentary question - P-003358/2022 European Parliament Implications of statement by Pfizer executive for COVID passport.* 2022, European Parl.: Europe. https://www.europarl.europa.eu/doceo/document/P-9-2022-003358_EN.html.

94. Malone, R.W., and J.G. Malone, *DNA vaccines for eliciting a mucosal immune response (includes mRNA).* 1996: USA.

95. Friedmann, T., and R. Roblin, *Gene therapy for human genetic disease?* Science, 1972. 175(4025): p. 949–55. https://www.ncbi.nlm.nih.gov/pubmed/5061866.

96. LaFee, S., *Friedman Recognized for Pioneering Gene Therapy Research,* in *UC San Diego News Center.* 2015, UCSD: La Jolla, CA. https://ucsdnews.ucsd.edu/feature/friedmann_recognized_for_pioneering_gene_therapy_research.

97. Zhang, L., et al., *Reverse-transcribed SARS-CoV-2 RNA can integrate into the genome of cultured human cells and can be expressed in patient-derived tissues.* Proc Natl Acad Sci U S A, 2021. 118(21). https://www.ncbi.nlm.nih.gov/pubmed/33958444.

98. FDA, *Cellular & Gene Therapy Guidances,* B.B. FDA Vaccines, Editor. 2022, USG: Washington, DC. https://www.fda.gov/vaccines-blood-biologics/biologics-guidances/cellular-gene-therapy-guidances.

99. Martinez, N.M., et al., *Pseudouridine synthases modify human pre-mRNA co-transcriptionally and affect pre-mRNA processing.* Mol Cell, 2022. 82(3): p. 645–659 e9. https://www.ncbi.nlm.nih.gov/pubmed/35051350.

100. Rhea, E.M., et al., *The S1 protein of SARS-CoV-2 crosses the blood-brain barrier in mice.* Nat Neurosci, 2021. 24(3): p. 368–378. https://www.ncbi.nlm.nih.gov/pubmed/33328624.

101. Datta, G., et al., *SARS-CoV-2 S1 Protein Induces Endolysosome Dysfunction and Neuritic Dystrophy.* Front Cell Neurosci, 2021. 15: p. 777738. https://www.ncbi.nlm.nih.gov/pubmed/34776872.

102. Sarubbo, F., et al., *Neurological consequences of COVID-19 and brain related pathogenic mechanisms: A new challenge for neuroscience.* Brain Behav Immun Health, 2022. 19: p. 100399. https://www.ncbi.nlm.nih.gov/pubmed/34870247.

103. Proal, A.D., and M.B. VanElzakker, *Long COVID or Post-acute Sequelae of COVID-19 (PASC): An Overview of Biological Factors That May Contribute to Persistent Symptoms.* Front Microbiol, 2021. 12: p. 698169. https://www.ncbi.nlm.nih.gov/pubmed/34248921.

104. Kim, E.S., et al., *Spike Proteins of SARS-CoV-2 Induce Pathological Changes in Molecular Delivery and Metabolic Function in the Brain Endothelial Cells.* Viruses, 2021. 13(10). https://www.ncbi.nlm.nih.gov/pubmed/34696455.

105. Colunga Biancatelli, R.M.L., et al., *The SARS-CoV-2 spike protein subunit S1 induces COVID-19-like acute lung injury in Kappa18-hACE2 transgenic mice and barrier dysfunction in human endothelial cells.* Am J Physiol Lung Cell Mol Physiol, 2021. 321(2): p. L477-L484. https://www.ncbi.nlm.nih.gov/pubmed/34156871.

106. Grobbelaar, L.M., et al., *SARS-CoV-2 spike protein S1 induces fibrin(ogen) resistant to fibrinolysis: implications for microclot formation in COVID-19.* Biosci Rep, 2021. 41(8). https://www.ncbi.nlm.nih.gov/pubmed/34328172.

107. Shirato, K., and T. Kizaki, *SARS-CoV-2 spike protein S1 subunit induces pro-inflammatory responses via toll-like receptor 4 signaling in murine and human macrophages.* Heliyon, 2021. 7(2): p. e06187. https://www.ncbi.nlm.nih.gov/pubmed/33644468.

108. FDA, *Coronavirus (COVID-19) Update: FDA Authorizes Moderna, Pfizer-BioNTech Bivalent COVID-19 Vaccines for Use as a Booster Dose,* FDA, Editor. 2022, FDA: DC. https://www.fda.gov/news-events/press-announcements/coronavirus-covid-19-update-fda-authorizes-moderna-pfizer-biontech-bivalent-covid-19-vaccines-use.

109. Tapp, T., *Newsom Tests Positive For Covid Just 10 Days After His Second Booster Shot.* Deadline, 2022. https://deadline.com/2022/05/gavin-newsom-positive-covid-1235035122/.

110. CCCA. *Dispelling the Myth of A Pandemic of the Unvaccinated.* 2022; Available from: https://rumble.com/vtt9ge-dispelling-the-myth-of-a-pandemic-of-the-unvaccinated.html.

111. Expose, *Trudeau Panics as Fully Vaccinated account for 9 in every 10 COVID-19 Deaths in Canada over the past month; 4 in every 5 of which were Triple Jabbed.* The Exposé, 2022. https://expose-news.com/2022/06/22/trudeau-panics-9-in-10-covid-deaths-fully-vaccinated/.

112. Collie, S., et al., *Effectiveness of BNT162b2 Vaccine against Omicron Variant in South Africa.* N Engl J Med, 2022. 386(5): p. 494–496. https://www.ncbi.nlm.nih.gov/pubmed/34965358.

113. Tseng, H.F., et al., *Effectiveness of mRNA-1273 against SARS-CoV-2 Omicron and Delta variants.* Nat Med, 2022. 28(5): p. 1063–1071. https://www.ncbi.nlm.nih.gov/pubmed/35189624.

114. Andrews, N., et al., *Covid-19 Vaccine Effectiveness against the Omicron (B.1.1.529) Variant.* N Engl J Med, 2022. 386(16): p. 1532–1546. https://www.ncbi.nlm.nih.gov/pubmed/35249272.

115. Cao, Y., et al., *Omicron escapes the majority of existing SARS-CoV-2 neutralizing antibodies.* Nature, 2022. 602(7898): p. 657–663. https://www.ncbi.nlm.nih.gov/pubmed/35016194.

116. Chen, L.L., et al., *Omicron variant susceptibility to neutralizing antibodies induced in children by natural SARS-CoV-2 infection or COVID-19 vaccine.* Emerg Microbes Infect, 2022. 11(1): p. 543–547. https://www.ncbi.nlm.nih.gov/pubmed/35084295.

117. Hoffmann, M., et al., *The Omicron variant is highly resistant against antibody-mediated neutralization: Implications for control of the COVID-19 pandemic.* Cell, 2022. 185(3): p. 447–456 e11. https://www.ncbi.nlm.nih.gov/pubmed/35026151.

118. Zimmer, K., *COVID-19 Vaccine Researchers Mindful of Immune Enhancement.* The Scientist, 2020. https://www.the-scientist.com/news-opinion/covid-19-vaccine-researchers-mindful-of-immune-enhancement-67576.

119. Gostic, K.M., et al., *Childhood immune imprinting to influenza A shapes birth year-specific risk during seasonal H1N1 and H3N2 epidemics.* PLoS Pathog, 2019. 15(12): p. e1008109. https://www.ncbi.nlm.nih.gov/pubmed/31856206.

120. Lee, W.S., et al., *Antibody-dependent enhancement and SARS-CoV-2 vaccines and therapies.* Nat Microbiol, 2020. 5(10): p. 1185–1191. https://www.ncbi.nlm.nih.gov/pubmed/32908214.

121. Crawford, N., A. Harris, and G. Lewis, *Vaccine-associated enhanced diease (VAED).* Melbourne Vaccine Education Centre, 2022. https://mvec.mcri.edu.au/references/vaccine-associated-enhanced-disease-vaed/.

122. Tenbusch, M., et al., *Risk of immunodeficiency virus infection may increase with vaccine-induced immune response.* J Virol, 2012. 86(19): p. 10533–9. https://www.ncbi.nlm.nih.gov/pubmed/22811518.

123. Henry, C., et al., *From Original Antigenic Sin to the Universal Influenza Virus Vaccine.* Trends Immunol, 2018. 39(1): p. 70–79. https://www.ncbi.nlm.nih.gov/pubmed/28867526.

124. Brown, E.L., and H.T. Essigmann, *Original Antigenic Sin: the Downside of Immunological Memory and Implications for COVID-19.* mSphere, 2021. 6(2). https://www.ncbi.nlm.nih.gov/pubmed/33692194.

125. Wheatley, A.K., et al., *Immune imprinting and SARS-CoV-2 vaccine design.* Trends Immunol, 2021. 42(11): p. 956–959. https://www.ncbi.nlm.nih.gov/pubmed/34580004.

126. Hasan, A., et al., *Cellular and Humoral Immune Responses in Covid-19 and Immunotherapeutic Approaches.* Immunotargets Ther, 2021. 10: p. 63–85. https://www.ncbi.nlm.nih.gov/pubmed/33728277.

127. Aydillo, T., et al., *Immunological imprinting of the antibody response in COVID-19 patients.* Nat Commun, 2021. 12(1): p. 3781. https://www.ncbi.nlm.nih.gov/pubmed/34145263.

128. Rodda, L.B., et al., *Imprinted SARS-CoV-2-specific memory lymphocytes define hybrid immunity.* Cell, 2022. 185(9): p. 1588–1601 e14. https://www.ncbi.nlm.nih.gov/pubmed/35413241.

129. Cao, Y., et al., *BA.2.12.1, BA.4 and BA.5 escape antibodies elicited by Omicron infection.* Nature, 2022. 608(7923): p. 593–602. https://www.ncbi.nlm.nih.gov/pubmed/35714668.

130. Reynolds, C.J., et al., *Immune boosting by B.1.1.529 (Omicron) depends on previous SARS-CoV-2 exposure.* Science, 2022. 377(6603): p. eabq1841. https://www.ncbi.nlm.nih.gov/pubmed/35699621.

131. Cele, S., et al., *Omicron extensively but incompletely escapes Pfizer BNT162b2 neutralization.* Nature, 2022. 602(7898): p. 654–656. https://www.ncbi.nlm.nih.gov/pubmed/35016196.

132. Liu, L., et al., *Striking antibody evasion manifested by the Omicron variant of SARS-CoV-2.* Nature, 2022. 602(7898): p. 676–681. https://www.ncbi.nlm.nih.gov/pubmed/35016198.

133. Muik, A., et al., *Neutralization of SARS-CoV-2 Omicron by BNT162b2 mRNA vaccine-elicited human sera.* Science, 2022. 375(6581): p. 678–680. https://www.ncbi.nlm.nih.gov/pubmed/35040667.

134. Planas, D., et al., *Considerable escape of SARS-CoV-2 Omicron to antibody neutralization.* Nature, 2022. 602(7898): p. 671–675. https://www.ncbi.nlm.nih.gov/pubmed/35016199.

135. Tarke, A., et al., *SARS-CoV-2 vaccination induces immunological T cell memory able to cross-recognize variants from Alpha to Omicron.* Cell, 2022. 185(5): p. 847–859 e11. https://www.ncbi.nlm.nih.gov/pubmed/35139340.

136. Gao, Y., et al., *Ancestral SARS-CoV-2-specific T cells cross-recognize the Omicron variant.* Nat Med, 2022. 28(3): p. 472–476. https://www.ncbi.nlm.nih.gov/pubmed/35042228.

137. Naranbhai, V., et al., *T cell reactivity to the SARS-CoV-2 Omicron variant is preserved in most but not all individuals.* Cell, 2022. 185(6): p. 1041–1051 e6. https://www.ncbi.nlm.nih.gov/pubmed/35202566.

138. Goel, R.R., et al., *Efficient recall of Omicron-reactive B cell memory after a third dose of SARS-CoV-2 mRNA vaccine.* Cell, 2022. 185(11): p. 1875–1887 e8. https://www.ncbi.nlm.nih.gov/pubmed/35523182.

139. Miller, E., *EXCLUSIVE: Hospital System Faces Staff Shortages Due to COVID-19 Infections, Despite Vaccine Mandate.* The Epoch Times, 2022. https://www.theepochtimes.com/mkt_app/exclusive-large-texas-hospital-faces-staff-shortages-despite-covid-19-vaccine-mandate_4602092.html.

140. Bardosh, K., et al., *COVID-19 Vaccine Boosters for Young Adults: A Risk-Benefit Assessment and Five Ethical Arguments against Mandates at Universities.* SSRN, 2022. https://papers.ssrn.com/sol3/papers.cfm?abstract_id=4206070.

141. Brenner, H., B. Holleczek, and B. Schottker, *Vitamin D Insufficiency and Deficiency and Mortality from Respiratory Diseases in a Cohort of Older Adults: Potential for Limiting the Death Toll during and beyond the COVID-19 Pandemic?* Nutrients, 2020. 12(8). https://www.ncbi.nlm.nih.gov/pubmed/32824839.

142. Ilie, P.C., S. Stefanescu, and L. Smith, *The role of vitamin D in the prevention of coronavirus disease 2019 infection and mortality.* Aging Clin Exp Res, 2020. 32(7): p. 1195–1198. https://www.ncbi.nlm.nih.gov/pubmed/32377965.

143. Maruotti, A., F. Belloc, and A. Nicita, *Comments on: The role of vitamin D in the prevention of coronavirus disease 2019 infection and mortality.* Aging Clin Exp Res, 2020. 32(8): p. 1621–1623. https://www.ncbi.nlm.nih.gov/pubmed/32654004.

144. Cannell, J.J., et al., *Epidemic influenza and vitamin D.* Epidemiol Infect, 2006. 134(6): p. 1129–40. https://www.ncbi.nlm.nih.gov/pubmed/16959053.

145. Grant, W.B., and C.F. Garland, *The role of vitamin D3 in preventing infections.* Age Ageing, 2008. 37(1): p. 121–2. https://www.ncbi.nlm.nih.gov/pubmed/18056725.

146. Villasis-Keever, M.A., et al., *Efficacy and Safety of Vitamin D Supplementation to Prevent COVID-19 in Frontline Healthcare Workers. A Randomized Clinical Trial.* Arch Med Res, 2022. https://www.ncbi.nlm.nih.gov/pubmed/35487792.

147. Borsche, L., B. Glauner, and J. von Mendel, *COVID-19 Mortality Risk Correlates Inversely with Vitamin D3 Status, and a Mortality Rate Close to Zero Could Theoretically Be Achieved at 50 ng/mL 25(OH)D3: Results of a Systematic Review and Meta-Analysis.* Nutrients, 2021. 13(10). https://www.ncbi.nlm.nih.gov/pubmed/34684596.

148. Shin, H.S., *Reasoning processes in clinical reasoning: from the perspective of cognitive psychology.* Korean J Med Educ, 2019. 31(4): p. 299–308. https://www.ncbi.nlm.nih.gov/pubmed/31813196.

149. Tyson, B., G.Fareed, and M. Crawford, *Overcoming the COVID-19 Darkness: How Two Doctors Successfully Treated 7000 Patients.* 2022, Amazon.

150. Mandavilli, A., *The CDC isn't publishing large portions of the COVID data it collects*, in *New York Times.* 2022.

151. CDC, *COVID-19 Cases and Hospitalizations by COVID-19 Vaccination Status and Previous COVID-19 Diagnosis — California and New York, May–November 2021*, CDC, Editor. 2022, CDC: Atlanta, GA. https://www.cdc.gov/mmwr/volumes/71/wr/mm7104e1.htm.

152. Ren, J.J., et al., *A statistical analysis of vaccine-adverse event data.* BMC Med Inform Decis Mak, 2019. 19(1): p. 101. https://www.ncbi.nlm.nih.gov/pubmed/31138219.

153. Rosenblum, H.G., et al., *Safety of mRNA vaccines administered during the initial 6 months of the US COVID-19 vaccination programme: an observational study of reports to the Vaccine Adverse Event Reporting System and v-safe.* Lancet Infect Dis, 2022. https://www.ncbi.nlm.nih.gov/pubmed/35271805.

154. Demasi, M., *Covid-19: Is the US compensation scheme for vaccine injuries fit for purpose?* BMJ, 2022. 377: p. o919. https://www.ncbi.nlm.nih.gov/pubmed/35440440.

155. VAERS-Analysis, *VAERS Summary for COVID-19 Vaccines through 4/8/2022*, in *VAERS Analysis: Weekly Analysis of the VAERS data.* 2022. https://vaersanalysis.info/2022/04/15/vaers-summary-for-covid-19-vaccines-through-4-8-2022/.

156. CCCA, *Canadian Covid Care Alliance*, in *The Pfizer Inoculations Do More Harm Than Good.* 2021, Rumble: Rumble. https://rumble.com/vqx3kb-the-pfizer-inoculations-do-more-harm-than-good.html.

157. Rogan, J., *#1757 - Dr. Robert Malone, MD*, in *The Joe Rogan Experience*, J. Rogan, Editor. 2021: Spotify. https://open.spotify.com/episode/3SCsueX2bZdbEzRtKOCEyT.

158. Depue, T., *Medieval Biological Warfare*, in *Medieval Facts.* 2016. https://medievalfactsexaminer.weebly.com/articles/medieval-biological-warfare.

159. Winters, N., *Exclusive: Deleted Web Pages Show Obama Led an Effort To Build a Ukraine-Based BioLab Handling 'Especially Dangerous Pathogens'.* in *The National Pulse.* 2022: Internet. https://thenationalpulse.com/2022/03/08/obama-led-ukraine-biolab-efforts/.

160. Nightingale, H., *BREAKING: Biden official says US working with Ukraine to prevent bio research facilities from falling into Russian hands* in *The Post Millennial.* 2022. https://thepostmillennial.com/biden-official-us-ukraine-bio-research-facilities-russia.

161. Adl-Tabatabai, S. *US Embassy Quietly Deletes All Ukraine Bioweapons Lab Documents Online – Media Blackout.* News Punch, 2022. https://newspunch.com/us-embassy-quietly-deletes-all-ukraine-bioweapon-lab-documents-online-media-blackout/.

162. Xinhua, *China urges U.S. to release details of bio-labs in Ukraine*, in *Xinhua Net (English News).* 2022. https://english.news.cn/northamerica/20220308/cf0d75294eb649f098f33dcde4a01495/c.html.

163. Wei, X., *China urges US to reveal details of US-backed biological labs in Ukraine – including types of viruses stored*, in *GBN News.* 2022. https://www.gbnews.uk/news/china-urges-us-to-reveal-details-of-us-backed-biological-labs-in-ukraine-including-types-of-viruses-stored/242669.

164. Becker, K., *Russia Negotiator Charges It Now Has Evidence of 'Biological Weapons Components' in Ukraine That Show 'Good Reason' for Invasion*, in *Becker News.* 2022. https://beckernews.com/russia-negotiator-charges-it-now-has-evidence-of-biological-weapons-components-in-ukraine-that-show-good-reason-for-invasion-44309/.

165. Mark, D., *What have Fauci's friends been up to in Ukraine?*, in *Israel Unwired*. 2022. https://www.israelunwired.com/faucis-friends-bioweapons-ukraine/.

166. Rosenblum, D.G., *Biography of Deborah G. Rosenblum. Assistant Secretary of Defense for Nuclear, Chemical, and Biological Defense Programs (ASD(NCB)), and Performing the Duties of Assistant Secretary of Defense for Industrial Base Policy (ASD(IBP))*. 2022 [cited 2022]; Available from: https://www.defense.gov/About/Biographies/Biography/Article/2719352/deborah-g-rosenblum/.

167. Brest, M., *'No offensive biologic weapons' in Ukrainian biolabs US assisted, Pentagon says*, in *Washington Examiner*. 2022. https://www.washingtonexaminer.com/policy/defense-national-security/no-offensive-biologic-weapons-in-ukrainian-biolabs-us-assisted-pentagon-says.

168. Roberts, K., *'No Offensive Biologic Weapons' in Ukrainian Biolabs: Pentagon*, in *The Epoch Times*. 2022, The Epoch Times. https://www.theepochtimes.com/no-offensive-biologic-weapons-in-ukrainian-biolabs-pentagon_4380628.html.

169. Malone, R.W., *Ukraine Biolab Watchtower*. 2022. https://rwmalonemd.substack.com/p/ukraine-biolab-watchtower?r=ta0o1&s=w&utm_campaign=post&utm_medium=web.

170. Ukraine, U.S.E.i., *U.S. Embassy in Ukraine, Biological Threat Reduction Program*, U.S.E.i. Ukraine, Editor. 2022. https://ua.usembassy.gov/embassy/kyiv/sections-offices/defense-threat-reduction-office/biological-threat-reduction-program/.

171. Salazar, A., *Russian Strikes Targeting US-Run Bio-Labs in Ukraine? For years, Russia has accused the US of running bio-labs in Ukraine that could develop chemical and biological weapons.*, in *NewsWars*. 2022. https://www.newswars.com/russian-strikes-targeting-us-run-bio-labs-in-ukraine/.

172. UN, *Biological Weapons Convention*. 2022, United Nations Office for Disarmament Affairs. https://www.un.org/disarmament/biological-weapons.

173. Javad, Z., et al., *Blockchain for decentralization of internet: prospects, trends, and challenges*. Cluster Computing, 2021. 24: p. 2841–2866. https://link.springer.com/article/10.1007/s10586-021-03301-8.

174. Cross, P., *Ukraine biolabs conspiracy theory*, in *Wikipedia*. 2022. https://en.wikipedia.org/wiki/Ukraine_biolabs_conspiracy_theory.

175. Wikispooks, *"Philip Cross"*, in *Wikispooks*. 2022.

176. Fredrickson, D.S., *Asilomar and Recombinant DNA: The End of the Beginning*, in *Biomedical Politics.*, H. KE, Editor. 1991, Institute of Medicine (US) Committee to Study Decision Making: Washington (DC).

177. OTA, *Proliferation of Weapons of Mass Destruction: Assessing the Risks*. 1993, U..S. Congress: US Government Printing Office, Washington DC. http://www.princeton.edu/~ota/disk1/1993/9341/9341.PDF.

178. NTI, *Israel Biological Overview- Fact Sheet*. 2015, The Nuclear Threat Initiative. https://www.nti.org/analysis/articles/israel-biological/.

179. Wiikispooks, *Israel Biological Weapons*, in *Wiikispooks*. 2022. https://wikispooks.com/wiki/Israel/Biological_weapons.

180. ACA, *The Biological Weapons Convention (BWC) At A Glance*. 2022, Arms Control Association (ACA). https://www.armscontrol.org/factsheets/bwc.

181. Field, M., *Amid false Russian allegations of US "biolabs" in Ukraine, it's worth asking: What is a bioweapon?* Bulletin of the Atomic Scientists, 2022. https://thebulletin.org/2022/03/amid-false-russian-allegations-of-us-biolabs-in-ukraine-its-worth-asking-what-is-a-bioweapon/.

182. Wikipedia, *Biological Weapons Convention*, in *Wikipedia*. 2022. https://en.wikipedia.org/wiki/Biological_Weapons_Convention.

183. Pandolfo, C., *Exclusive: The federal government paid hundreds of media companies to advertise the COVID-19 vaccines while those same outlets provided positive coverage of the vaccines*, in *Natural News (reprinted from BlazeNews)*. 2022. https://www.naturalnews.com/2022-03-08-government-paid-media-companies-advertise-covid-vaccines.html.

184. Lajka, A., *Obama did not sign a law allowing propaganda in the U.S.*, in *AP News*. 2019, AP. https://apnews.com/article/archive-fact-checking-7064410002.

185. Goldfarb, K., *Inside Operation Mockingbird — The CIA's Plan To Infiltrate The Media*. ATI, 2021. https://allthatsinteresting.com/operation-mockingbird.

186. Bernstein, C., *The CIA and the Media*, in *Rolling Stone*. 1977. https://www.carlbernstein.com/the-cia-and-the-media-rolling-stone-10-20-1977.

187. Handout, U., *DARPA's Man in Wuhan*, in *Unlimited Handout*. 2020. https://unlimitedhangout.com/2020/07/investigative-reports/darpas-man-in-wuhan/.

188. WhatsHerFace, *Who is Robert Malone?* 2021, Youtube. https://www.youtube.com/watch?v=LwE2ZeuUyXo.

189. Kennedy, R.F., *The Real Anthony Fauci: Bill Gates, Big Pharma, and the Global War on Democracy and Public Health (Children's Health Defense)*. 2021: Skyhorse.

190. Leef, G., *How Our Federal Overseers Do Science*. National Review, 2021. https://www.nationalreview.com/corner/how-our-federal-overseers-do-science/.

191. Lamb, M., *Forbes journalist fired for revealing Fauci's income releases the emails which led to his sacking- Adam Andrzejewski wrote that the alleged corrections were minor semantic issues.*, in *Lifesite News*. 2022. https://www.lifesitenews.com/news/revealed-emails-show-how-faucis-pr-flaks-got-a-forbes-contributor-canceled/.

192. Bever, L., *Fauci criticizes Fox News for silence after pundit Lara Logan compares him to Nazi doctor*, in *Washington Post*. 2021, WP: DC. https://www.washingtonpost.com/arts-entertainment/2021/12/03/fauci-fox-nazi-comment/.

193. Weinberg, J., *The Great Recession and Its Aftermath*, in *Federal Reserve History*. 2021: Online. https://www.federalreservehistory.org/essays/great-recession-and-its-aftermath.

194. Cao, Y., A. Yismayi, and X. Sunney, *BA.2.12.1, BA.4 and BA.5 escape antibodies elicited by Omicron infection*. Nature, 2022. https://www.nature.com/articles/s41586-022-04980-y.

195. Reuters. *Jim Smith President and CEO*. 2022; Available from: https://www.thomsonreuters.com/en-us/posts/authors/jim-smith/.

196. Lapid, N., *Early Omicron infection unlikely to protect against current variants*. Reuters, 2022. https://www.reuters.com/business/healthcare-pharmaceuticals/early-omicron-infection-unlikely-protect-against-current-variants-2022-06-17/.

197. USDA, *Emergency Use Authorization of Medical Products and Related Authorities: Guidance for Industry and Other Stakeholders*. 2017, U.S. Food & Drug Administration: Wash., DC. https://www.fda.gov/regulatory-information/search-fda-guidance-documents/emergency-use-authorization-medical-products-and-related-authorities.

198. Congress, U.S., *Pandemic Response and Emergency Preparedness Act of 2020, Introduced 03/11/2020*, in *H.R. 6206*. 2020, U.S. Congress, House: Wash., DC. https://www.congress.gov/bill/116th-congress/house-bill/6206/text?r=8&s=1.

199. Santin, A.D., et al., *Ivermectin: a multifaceted drug of Nobel prize-honoured distinction with indicated efficacy against a new global scourge, COVID-19*. New Microbes New Infect, 2021. 43: p. 100924. https://www.ncbi.nlm.nih.gov/pubmed/34466270.

200. Redshaw, M., *31,470 Deaths After COVID Vaccines Reported to VAERS, Including 26 Following New Boosters*. The Defender, 2022. https://childrenshealthdefense.org/defender/vaers-deaths-covid-vaccines-boosters/.

201. Crist, C., *Delta Becomes Dominant Coronavirus Variant in U.S.* WebMD, 2022. https://www.webmd.com/lung/news/20210707/delta-dominant-us-coronavirus-variant.

202. Franklin, J., *Omicron is now the dominant COVID strain in the U.S., making up 73% of new infections*. NPR, 2021. https://www.npr.org/sections/coronavirus-live-updates/2021/12/20/1066083896/omicron-is-now-the-dominant-covid-strain-in-the-u-s-making-up-73-of-cases.

203. Holshue, M.L., et al., *First Case of 2019 Novel Coronavirus in the United States*. N Engl J Med, 2020. 382(10): p. 929–936. https://www.ncbi.nlm.nih.gov/pubmed/32004427.

204. USDA, *FDA Takes Key Action in Fight Against COVID-19 By Issuing Emergency Use Authorization for First COVID-19 Vaccine*. 2020, USDA: Wash. DC. https://www.fda.gov/news-events/press-announcements/fda-takes-key-action-fight-against-covid-19-issuing-emergency-use-authorization-first-covid-19.

205. USDA, *FDA Takes Additional Action in Fight Against COVID-19 By Issuing Emergency Use Authorization for Second COVID-19 Vaccine,*. 2020, U.S. Food & Drug Administration. https://www.fda.gov/news-events/press-announcements/fda-takes-additional-action-fight-against-covid-19-issuing-emergency-use-authorization-second-covid.

206. USDA, *FDA Issues Emergency Use Authorization for Third COVID-19 Vaccine,*. 2021, U.S. Food & Drug Administration. https://www.fda.gov/news-events/press-announcements/fda-issues-emergency-use-authorization-third-covid-19-vaccine.

207. LaFraniere, S., *Covid News: Over 200 Million Americans Are Fully Vaccinated.* NY Times, 2021. https://www.nytimes.com/live/2021/12/08/world/omicron-variant-covid.

208. CDC, *Morbidity and Mortality Weekly Report, no.71.* 2022, Centers for Disease Control and Prevention: Wash., DC. https://www.cdc.gov/mmwr/volumes/71/wr/pdfs/mm7117-h.pdf.

209. CDC, *Provisional Mortality Data—United States, 2021.* 2022, Centers for Disease Control and Prevention: Wash., DC. https://www.cdc.gov/mmwr/volumes/71/wr/mm7117e1.htm.

210. Gram, M.A., et al., *Vaccine effectiveness against SARS-CoV-2 infection or COVID-19 hospitalization with the Alpha, Delta, or Omicron SARS-CoV-2 variant: A nationwide Danish cohort study.* PLoS Med, 2022. 19(9): p. e1003992. https://www.ncbi.nlm.nih.gov/pubmed/36048766.

211. Trougakos, I.P., et al., *Adverse effects of COVID-19 mRNA vaccines: the spike hypothesis.* Trends Mol Med, 2022. 28(7): p. 542–554. https://www.ncbi.nlm.nih.gov/pubmed/35537987.

212. Alden, M., et al., *Intracellular Reverse Transcription of Pfizer BioNTech COVID-19 mRNA Vaccine BNT162b2 In Vitro in Human Liver Cell Line.* Curr Issues Mol Biol, 2022. 44(3): p. 1115–1126. https://www.ncbi.nlm.nih.gov/pubmed/35723296.

213. Seneff, S., et al., *Innate immune suppression by SARS-CoV-2 mRNA vaccinations: The role of G-quadruplexes, exosomes, and MicroRNAs.* Food Chem Toxicol, 2022. 164: p. 113008. https://www.ncbi.nlm.nih.gov/pubmed/35436552.

214. Pang, S.N.J., *Final Report on the Safety Assessment of Polyethylene Glycols (PEGs) -6, -32, -75, -150, -14M, -20M -8,.* Journal of the American College of Toxicology, 1993. 12. https://journals.sagepub.com/doi/pdf/10.3109/10915819309141598.

215. Sinopeg, *Polyethylene glycol [PEG] 2000 dimyristoyl glycerol [DMG] [mPEG2000-DMG] Cas:160743-62-4 Product Details.* 2022, SINOPEG (Manufacturer of PEG). https://www.sinopeg.com/polyethylene-glycol-peg-2000-dimyristoyl-glycerol-dmg-mpeg2000-dmg-cas-160743-62-4_p479.html.

216. Cayman, *Trade name: SM-102, Safety Data Sheet.* 2022, Cayman Chemical. https://cdn.caymanchem.com/cdn/msds/33474m.pdf.

217. Cayman, *Trade Name: ALC-0315, Safety Data Sheet.* 2022, Cayman Chemical. https://cdn.caymanchem.com/cdn/msds/34337m.pdf.

218. SAO, *Group Life COVID-19 Mortality Survey Report,"* 2022, SOA Research Institute. https://www.soa.org/4a368a/globalassets/assets/files/resources/research-report/2022/group-life-covid-19-mortality-03-2022-report.pdf.

219. Malone, R.W., *Declaration of Robert Malone, MD, MS, in Support of Plaintiffs' Malone for a Temporary Restraining Order and Preliminary Injunction,* in 2021, United States District Court Middle District of Florida. https://ia802509.us.archive.org/27/items/gov.uscourts.flmd.395057/gov.uscourts.flmd.395057.30.6.pdf.

220. Malone, R.W., P.L. Felgner, and I.M. Verma, *Cationic liposome-mediated RNA transfection.* Proc Natl Acad Sci U S A, 1989. 86(16): p. 6077–81. https://www.ncbi.nlm.nih.gov/pubmed/2762315.

221. Wolff, J.A., et al., *Direct gene transfer into mouse muscle in vivo.* Science, 1990. 247(4949 Pt 1): p. 1465–8. https://www.ncbi.nlm.nih.gov/pubmed/1690918.

222. Johnson, R.F., *Letter from U.S. Senator Ron Johnson to FDA Commissioner, Janet Woodcock, MD.* . 2021, US Senate: Wash., DC. https://www.ronjohnson.senate.gov/services/files/29E9C183-E206-49A9-8E6D-31B136B91B7D.

223. FDA, *Coronavirus (COVID-19) Update: FDA Takes Key Action by Approving Second COVID-19 Vaccine.* 2022, U.S. Food & Drug Administration: Wash., DC. https://www.fda.gov/news-events/press-announcements/coronavirus-covid-19-update-fda-takes-key-action-approving-second-covid-19-vaccine.

224. Brown, C.M., et al., *Outbreak of SARS-CoV-2 Infections, Including COVID-19 Vaccine Breakthrough Infections, Associated with Large Public Gatherings - Barnstable County, Massachusetts, July 2021.* MMWR Morb Mortal Wkly Rep, 2021. 70(31): p. 1059–1062. https://www.ncbi.nlm.nih.gov/pubmed/34351882.

225. Bhakdi, S. and A. Burkhardt, *"On COVID vaccines: why they cannot work, and irrefutable evidence of their causative role in deaths after vaccination,"* Dec 15, 2021 *(Transcript of presentation at the Doctors for COVID Ethics symposium).* Doctors for COVID Ethics symposium 2021. https://doctors4covidethics.org/on-covid-vaccines-why-they-cannot-work-and-irrefutable-evidence-of-their-causative-role-in-deaths-after-vaccination/.

226. Singanayagam, A., et al., *Community transmission and viral load kinetics of the SARS-CoV-2 delta (B.1.617.2) variant in vaccinated and unvaccinated individuals in the UK: a prospective, longitudinal, cohort study.* Lancet Infect Dis, 2022. 22(2): p. 183–195. https://www.ncbi.nlm.nih.gov /pubmed/34756186.

227. Chau, N.V.V., et al., *An observational study of breakthrough SARS-CoV-2 Delta variant infections among vaccinated healthcare workers in Vietnam.* EClinicalMedicine, 2021. 41: p. 101143. https: //www.ncbi.nlm.nih.gov/pubmed/34608454.

228. Subramanian, S.V., and A. Kumar, *Increases in COVID-19 are unrelated to levels of vaccination across 68 countries and 2947 counties in the United States.* Eur J Epidemiol, 2021. 36(12): p. 1237– 1240. https://www.ncbi.nlm.nih.gov/pubmed/34591202.

229. Smalley, J., *COVID Deaths Before and After Vaccination Programs,.* YouTube, 2021. https://www .youtube.com/watch?v=WR-pqrMWu3E.

230. FDA, *Coronavirus (COVID-19) Update: July 13, 2021.* U.S. Food & Drug Administration, 2021. https://www.fda.gov/news-events/press-announcements/coronavirus-covid-19-update-july -13-2021.

231. PublicHealthEngland, *SARS-CoV-2 variants of concern and variants under investigation in England, Technical briefing 20.* Public Health England, 2021. https://assets.publishing.service .gov.uk/government/uploads/system/uploads/attachment_data/file/1009243/Technical _Briefing_20.pdf.

232. Carbajal, E., *Nearly 60% of hospitalized COVID-19 patients in Israel fully vaccinated, data shows.* Becker's Hospital Review, 2021. https://www.beckershospitalreview.com/public-health/nearly -60-of-hospitalized-covid-19-patients-in-israel-fully-vaccinated-study-finds.html.

233. Humetrix, *Project Salus, "Effectiveness of mRNA COVID-19 Vaccines Against the Delta Variant Among 5.6M Medicare Beneficiaries 65 Years and Older.* Project Salus, 2021. https://www.justfacts .com/document/waning_effect_covid-19_vaccines_humetrix_2021.pdf.

234. PHS, S., *COVID-19 Statistical Report.* Public Health Scotland, 2021. https://publichealthscotland. scot/media/10091/21-11-10-covid19-publication_report.pdf.

235. Young-Xu, Y., et al., *Estimated Effectiveness of COVID-19 Messenger RNA Vaccination Against SARS-CoV-2 Infection Among Older Male Veterans Health Administration Enrollees, January to September 2021.* JAMA Netw Open, 2021. 4(12): p. e2138975. https://www.ncbi.nlm.nih.gov /pubmed/34910155.

236. Agency, U.H.S., *SARS-CoV-2 variants of concern and variants under investigation in England Technical briefing: Update on hospitalisation and vaccine effectiveness for Omicron VOC-21NOV-01 (B.1.1.529).* UK Health Security Agency, 2021. https://assets.publishing.service.gov.uk/ government/uploads/system/uploads/attachment_data/file/1045619/Technical-Briefing-31 -Dec-2021-Omicron_severity_update.pdf.

237. Agency, U.H.S., *COVID-19 vaccine surveillance report Week 9.* UK Health Security Agency, 2022. https://assets.publishing.service.gov.uk/government/uploads/system/uploads/attachment_data /file/1058464/Vaccine-surveillance-report-week-9.pdf.

238. Nordstrom, P., M. Ballin, and A. Nordstrom, *Risk of infection, hospitalisation, and death up to 9 months after a second dose of COVID-19 vaccine: a retrospective, total population cohort study in Sweden.* Lancet, 2022. 399(10327): p. 814–823. https://www.ncbi.nlm.nih.gov/pubmed/35131043.

239. Buchan, S.A., et al., *Estimated Effectiveness of COVID-19 Vaccines Against Omicron or Delta Symptomatic Infection and Severe Outcomes.* JAMA Netw Open, 2022. 5(9): p. e2232760. https: //www.ncbi.nlm.nih.gov/pubmed/36136332.

240. Regev-Yochay, G., et al., *4th Dose COVID mRNA Vaccines' Immunogenicity & Efficacy Against Omicron VOC.* medRxiv,, 2022. https://www.medrxiv.org/content/10.1101/2022.02.15.222709 48v1.

241. Whelan, J.P., *Comment from J. Patrick Whelan MD PhD," Vaccines and Related Biological Products Advisory Committee, Dec 8, 2020.* FDA, 2020. https://www.regulations.gov/document/FDA -2020-N-1898-0246.

242. Salk, *The Novel Coronamivus' Spike Protein Plays Additional Key Role in Illness,"* Salk News, 2021. https://www.salk.edu/news-release/the-novel-coronavirus-spike-protein-plays-additional-key-role -in-illness/.

243. Lei, Y., et al., *SARS-CoV-2 Spike Protein Impairs Endothelial Function via Downregulation of ACE2.* bioRxiv, 2020. https://www.ncbi.nlm.nih.gov/pubmed/33300001.

244. Nuovo, G.J., et al., *Endothelial cell damage is the central part of COVID-19 and a mouse model induced by injection of the S1 subunit of the spike protein.* Ann Diagn Pathol, 2021. 51: p. 151682. https://www.ncbi.nlm.nih.gov/pubmed/33360731.

245. Gundry, S.R., *Abstract 10712: Observational Findings of PULS Cardiac Test Findings for Inflammatory Markers in Patients Receiving mRNA Vaccines.* Circulation, 2021. https://www .ahajournals.org/doi/10.1161/circ.144.suppl_1.10712.

246. UK Health Security Agency, S., *COVID-19 vaccine surveillance report Week 42.* UK Health Security Agency, 2021. https://assets.publishing.service.gov.uk/government/uploads/system /uploads/attachment_data/file/1027511/Vaccine-surveillance-report-week-42.pdf.

247. Föhse, F.K., et al., *The BNT162b2 mRNA vaccine against SARS-CoV-2 reprograms both adaptive and innate immune responses.* medRxiv,, 2021. https://www.medrxiv.org/content/10.1101/2021.0 5.03.21256520v1.

248. Hertel, M., et al., *Real-world evidence from over one million COVID-19 vaccinations is consistent with reactivation of the varicella-zoster virus.* J Eur Acad Dermatol Venereol, 2022. 36(8): p. 1342–1348. https://www.ncbi.nlm.nih.gov/pubmed/35470920.

249. Risch, H., *Dr. Harvey Risch: Why Are Vaccinated People Getting COVID at Higher Rates Than the Unvaccinated?*, in *American Thought Leaders* J. Jekielek, Editor. 2022, The Epoch Times: Wash, DC. https://podcasts.apple.com/ca/podcast/dr-harvey-risch-why-are-vaccinated-people -getting-covid/id1471411980?i=1000570510124.

250. Reuters, *UPDATE 1-U.S. FDA amends J&J vaccine fact sheet to include rare bleeding risk,.* Reuters, 2022. https://www.reuters.com/article/health-coronavirus-johnsonjohnson-vaccin -idCNL4N2TR3GR.

251. Wang, H., et al., *Pathogenic Antibodies Induced by Spike Proteins of COVID-19 and SARS-CoV-2 Viruses," Research Square.* Research Square, 2020. https://europepmc.org/article/PPR/PPR355779.

252. Nagase, D., *Breaking News: Pfizer's Own Stats: 1200+/40,000 Trial Participants Dead | Interview with Dr. Nagase.*, in *The Strong and Free Truthcast.* Rumble. https://rumble.com/vqq3hw -breaking-news-pfizers-own-stats-1200-40000-trial-participants-dead-intervie.html.

253. Chen, Y., et al., *New-onset autoimmune phenomena post-COVID-19 vaccination.* Immunology, 2022. 165(4): p. 386–401. https://www.ncbi.nlm.nih.gov/pubmed/34957554.

254. McCullough, P., *Declaration of Peter McCullough, MD, MPH in Support of Plaintiffs' Petition for Preliminary Injunction.* 2021, United States District Court Middle District of Florida,. https: //ia802509.us.archive.org/27/items/gov.uscourts.flmd.395057/gov.uscourts.flmd.395057.1.8.pdf.

255. Redshaw, M., *Pfizer Skipped Critical Testing and Cut Corners on Quality Standards, Documents Reveal.* The Defender, 2021. https://childrenshealthdefense.org/defender/pfizer-skipped-critical -testing-quality-standards-covid-vaccine/.

256. LifeSite, *'Stop the Shot' Part II: Fertility Risks.* Truth for Health Foundation, LifeSite News, 2021. https://www.lifesitenews.com/conference-stop-the-shot/stop-the-shot-8-19/.

257. McLoone, D., *Stop the Shot conference doctors: pregnant women should never take the vaccine,* LifeSite News, 2021. https://www.lifesitenews.com/news/stop-the-shot-conference-doctors-pregnant -women-should-never-take-the-vaccine/.

258. Palmer, M., S. Bhakdi, and S. Hockertz, *Expert statement regarding Comirnaty—COVID-19- mRNA-Vaccine for children.* Children's Health Defense, 2021. https://childrenshealthdefense .org/wp-content/uploads/expert-evidence-pfizer-children.pdf.

259. FDA, *Coronavirus (COVID-19) Update: June 25, 2021.* U.S. Food & Drug Administration, 2021. https://www.fda.gov/news-events/press-announcements/coronavirus-covid-19-update-june -25-2021.

260. Lazurus, R., *Electronic Support for Public Health–Vaccine Adverse Event Reporting System (ESP:VAERS). .* The Agency for Healthcare Research and Quality (AHRQ), 2010. https: //digital.ahrq.gov/sites/default/files/docs/publication/r18hs017045-lazarus-final-report-2011 .pdf.

261. Cavallaro, O., *Number Of Vaccine-Related Deaths, Injuries Rising, Proving Jabs 'As Deadly As COVID-19 Itself.* Christianity Daily, 2021. https://www.christianitydaily.com/articles/12295/20210619

/number-of-vaccine-related-deaths-injuries-rising-proving-jabs-as-deadly-as-covid-19-itself
.htm.

262. Roan, S., *Swine flu 'debacle' of 1976 is recalled.* Los Angeles Times, 2009. https://www.latimes
.com/archives/la-xpm-2009-apr-27-sci-swine-history27-story.html#:~:text=Waiting%20in%20
long%20lines%20at,receiving%20the%20vaccine%3B%2025%20died.

263. Johnson, R., *Senator Ron Johnson to Francis Collins, Rochelle Wallensky, and Janet Woodcock.* U.S.
Senate, 2021. https://www.ronjohnson.senate.gov/services/files/17788FED-A947-4143-8C1B
-95C59E60EE87.

264. Paardekooper, C., *Variation in Toxicity of COVID Vaccine Batches.*, Bitchute, 2021. https://www
.bitchute.com/video/6xIYPZBkydsu/.

265. Hart, T., *Recent deaths in young people in England and Wales.* HART: Health Advisory & Recovery
Team, 2021. https://www.hartgroup.org/recent-deaths-in-young-people-in-england-and-wales/.

266. Johnson, R., *Senator Ron Johnson to Lloyd J. Austin III, F (letter).* U.S. Senate, 2021. https://www
.ronjohnson.senate.gov/2022/2/sen-johnson-to-secretary-austin-has-dod-seen-an-increase-in
-medical-diagnoses-among-military-personnel.

267. Menge, M., *BREAKING: Fifth largest life insurance company in the US paid out 163% more for deaths
of working people ages 18–64 in 2021 - Total claims/benefits up $6 BILLION.* Crossroads Report,
2022. https://crossroadsreport.substack.com/p/breaking-fifth-largest-life-insurance?s=r.

268. CDC, *COVID-19 Pandemic Planning Scenarios.* Centers for Disease Control and Prevention,
2021. https://www.cdc.gov/coronavirus/2019-ncov/hcp/planning-scenarios.html.

269. Pierce, C.A., et al., *Immune responses to SARS-CoV-2 infection in hospitalized pediatric and adult
patients.* Sci Transl Med, 2020. 12(564). https://www.ncbi.nlm.nih.gov/pubmed/32958614.

270. Dowell, A.C., et al., *Children develop robust and sustained cross-reactive spike-specific immune
responses to SARS-CoV-2 infection.* Nat Immunol, 2022. 23(1): p. 40–49. https://www.ncbi.nlm
.nih.gov/pubmed/34937928.

271. Fink, J., *Trump Official Ben Carson Calls COVID Vaccine 'Giant Experiment,' Questions Shot for
Kids.* Newsweek, 2021. https://www.newsweek.com/trump-official-ben-carson-calls-covid
-vaccine-giant-experiment-questions-shot-kids-1645653.

272. Krug, A., J. Stevenson, and T.B. Hoeg, *BNT162b2 Vaccine-Associated Myo/Pericarditis in
Adolescents: A Stratified Risk-Benefit Analysis.* Eur J Clin Invest, 2022. 52(5): p. e13759. https:
//www.ncbi.nlm.nih.gov/pubmed/35156705.

273. Le Vu, S., et al., *Age and sex-specific risks of myocarditis and pericarditis following Covid-19 messenger
RNA vaccines.* Nat Commun, 2022. 13(1): p. 3633. https://www.ncbi.nlm.nih.gov/pubmed
/35752614.

274. Sun, C.L.F., E. Jaffe, and R. Levi, *Increased emergency cardiovascular events among under-40
population in Israel during vaccine rollout and third COVID-19 wave.* Sci Rep, 2022. 12(1): p. 6978.
https://www.ncbi.nlm.nih.gov/pubmed/35484304.

275. Cohen, K.W., et al., *Longitudinal analysis shows durable and broad immune memory after SARS-
CoV-2 infection with persisting antibody responses and memory B and T cells.* Cell Rep Med, 2021.
2(7): p. 100354. https://www.ncbi.nlm.nih.gov/pubmed/34250512.

276. Gazit, S., et al., *Severe Acute Respiratory Syndrome Coronavirus 2 (SARS-CoV-2) Naturally
Acquired Immunity versus Vaccine-induced Immunity, Reinfections versus Breakthrough Infections: A
Retrospective Cohort Study.* Clin Infect Dis, 2022. 75(1): p. e545-e551. https://www.ncbi.nlm.nih
.gov/pubmed/35380632.

277. Mathioudakis, A.G., et al., *Self-Reported Real-World Safety and Reactogenicity of COVID-19
Vaccines: A Vaccine Recipient Survey.* Life (Basel), 2021. 11(3). https://www.ncbi.nlm.nih.gov
/pubmed/33803014.

278. Mercola, J., *Leaked Document Reveals 'Shocking' Terms of Pfizer's International Vaccine Agreements..*
The Defender, 2021. https://childrenshealthdefense.org/defender/leaked-document-terms-pfizers
-international-vaccine-agreements/.

279. TrialSite, *Did Pfizer Fail to Perform industry Standard Animal Testing Prior to Initiation of mRNA
Clinical Trials?* TrialSite News, 2021.

280. Gutschi, M.L., *An Independent Analysis of the Manufacturing and Quality Issues of the
BNT162b BioNTech/Pfizer Vaccine Identified by the European Medicines Agency.* Canadian

Covid Care Alliance, 2022. https://www.canadiancovidcarealliance.org/wp-content/uploads
/2022/10/22OC22_EMA-Analysis-of-BNT162b-Manufacture.pdf.

281. Pfizer, *5.3.6 CUMULATIVE ANALYSIS OF POST-AUTHORIZATION ADVERSE EVENT REPORTS OF PF-07302048 (BNT162B2) RECEIVED THROUGH 28-FEB-2021.* Pfizer 2021. https://phmpt.org/wp-content/uploads/2021/11/5.3.6-postmarketing-experience.pdf.

282. DOJ, S., *Justice Department Announces Largest Healthcare Fraud Settlement in Its History.* U.S. Department of Justice, 2009. https://www.justice.gov/opa/pr/justice-department-announces-largest -health-care-fraud-settlement-its-history.

283. Driver, J., *The History of Utilitarianism.* 2009, Stanford Encyclopedia of Philosophy. https: //plato.stanford.edu/entries/utilitarianism-history/.

284. Shermer, M., *Why Malthus Is Still Wrong. Why Malthus makes for bad science policy.* Scientific American, 2016. https://www.scientificamerican.com/article/why-malthus-is-still-wrong/.

285. Hayashi, C., *How hubris put our health at risk.* World Economic Forum, 2013. https://www .weforum.org/agenda/2013/01/how-hubris-put-our-health-at-risk/.

286. Zhiyong, F., *China forcefully harvests organs from detainees, tribunal concludes.* NBC News, 2019. https://www.nbcnews.com/news/world/china-forcefully-harvests-organs-detainees-tribunal -concludes-n1018646.

287. Magness, P.W., *The Failure of Imperial College Modeling Is Far Worse than We Knew.* The American Institute for Economic Research, 2021. https://www.aier.org/article/the-failure-of -imperial-college-modeling-is-far-worse-than-we-knew/.

288. Magness, P.W., *The Failures of Pandemic Central Planning.* SSRN, 2021. https://papers.ssrn.com /sol3/papers.cfm?abstract_id=3934452.

289. Niemiec, E., *COVID-19 and misinformation: Is censorship of social media a remedy to the spread of medical misinformation?* EMBO Rep, 2020. 21(11): p. e51420. https://www.ncbi.nlm.nih.gov /pubmed/33103289.

290. WMA, *WMA Declarion of Helsinki—Ethical Principles for Medical Research Involving Human Subjects.* . 1964, World Medical Association. https://www.wma.net/policies-post/wma -declaration-of-helsinki-ethical-principles-for-medical-research-involving-human-subjects/.

291. Commission, *The Belmont Report,* E.P.a.G.f.t.P.o.H.S.o.R.a.t.N.C.f.t.P.o.H.S.o.B.a.B.R. Office of the Secretar, Editor. 1979, US Government: Washington, D.C. https://www.hhs.gov/ohrp /sites/default/files/the-belmont-report-508c_FINAL.pdf.

292. Jureidini, J., and L.B. McHenry, *The illusion of evidence based medicine.* BMJ, 2022. 376: p. o702. https://www.ncbi.nlm.nih.gov/pubmed/35296456.

293. Dall, C., *Authors retract controversial hydroxychloroquine study.* CIDRAP Center for Infectious Disease Research and Policy, 2020. https://www.cidrap.umn.edu/news-perspective/2020/06 /authors-retract-controversial-hydroxychloroquine-study.

294. Leef, G., *How our Federal overseers do science,* in *National Review.* 2021. https://www .nationalreview.com/corner/how-our-federal-overseers-do-science/.

295. Prasad, V., *At a time when the U.S. needed Covid-19 dialogue between scientists, Francis Collins moved to shut it down.* STAT News, 2021. https://www.statnews.com/2021/12/23/at-a-time -when-the-u-s-needed-covid-19-dialogue-between-scientists-francis-collins-moved-to-shut-it -down/.

296. NIH, *News Release: Lander, Collins set forth a vision for ARPA-H,* NIH, Editor. 2021, USG: DC. https://www.nih.gov/news-events/news-releases/lander-collins-set-forth-vision-arpa-h.

297. NIH, *ARPHA-H Mission,* NIH, Editor. 2022, HHS: DC. https://www.nih.gov/arpa-h/mission.

298. Koutsobinas, N., *Pfizer to Court: Toss Lawsuit That Revealed COVID-19 Vaccine Testing Issues,* in *NewsMax.* 2022, NewsMax. https://www.newsmax.com/us/pfizer-fda-ventavia-icon /2022/05/22/id/1070944/.

299. Baumann, J., *More Vaccines in 'Record Time' Seen for $6.5 Billion Medical Lab.* Boomburg Law, 2021. https://news.bloomberglaw.com/pharma-and-life-sciences/nih-chief-collins-on-bidens -6-5-billion-research-center-plan.

300. ARPA-H, *ADVANCED RESEARCH PROJECTS AGENCY FOR HEALTH (ARPA-H),* ARPA-H, Editor. 2022, NIH: DC. https://www.nih.gov/arpa-h.

301. NIAID, *About NIAID: Vaccine Research Center,* NIAID, Editor. 2022, HHS: DC. https://www .niaid.nih.gov/about/vrc.

302. Facher, L., *In the search for an ARPA-H director, the White House is zeroing in on DARPA veterans.* STAT News, 2022. https://www.statnews.com/2022/07/07/arpa-h-director-search-darpa-vets/.

303. Facher, L., *Former DARPA director: Biden's new science agency should be independent, not an NIH office.* STAT News, 2021. https://www.statnews.com/2021/11/18/former-darpa-director -bidens-new-science-agency-should-be-independent-not-an-nih-office/.

304. Fabb, S., *Dr. Renee Wegrzyn Appointed as First Director of New ARPA-H Agency.* Federation of Associations in Behavioral and Brain Sciences (FABB), 2022. Dr. Renee Wegrzyn Appointed as First Director of New ARPA-H Agency.

305. Wolin, S.S., *Democracy Incorporated: Managed Democracy and the Specter of Inverted Totalitarianism.* 2017: Princeton University Press. 400.

306. Emmons, L., *AP source who 'fact checked' Mass Formation Psychosis theory encouraged 'behavioral nudging' people into Covid compliance, quoted Goebbels,* in *PM.* 2022. https://thepostmillennial. com/ap-writer-fact-checked-mass-formation-psychosis-theory-encouraged-cajoling-covid -compliance.

307. Yale, *COVID-19 Vaccine Messaging, Part 1.* 2020, Yale University, Clinical Trials .gov: DC. https://clinicaltrials.gov/ct2/show/NCT04460703.

308. James, E.K., et al., *Persuasive messaging to increase COVID-19 vaccine uptake intentions.* Vaccine, 2021. 39(49): p. 7158–7165. https://www.ncbi.nlm.nih.gov/pubmed/34774363.

309. Rayner, G., *Use of fear to control behaviour in Covid crisis was 'totalitarian', admit scientists.* The Telegraph, 2021. https://www.telegraph.co.uk/news/2021/05/14/scientists-admit-totalitarian -use-fear-control-behaviour-covid/.

310. Desmet, M., *The Psychology of Totalitarianism.* 2022.

311. Freud, S., *Group Psychology and the Analysis of the Ego* 1989 (english version) ed. Part of Freud library. 1921: Mass Market Paperback.

312. Arendt, H., *The Origins of Totalitarianism* 1966: Harcourt Brace & World.

313. Boon, G., *The Crowd A Study of the Popular Mind* 1921.

314. Thaler, R.H., and C.R. Sunstein, *Nudge Hardcover* 2021: Yale University Press. 384.

315. Abaluck, J., et al., *Impact of community masking on COVID-19: A cluster-randomized trial in Bangladesh.* Science, 2022. 375(6577): p. eabi9069. https://www.ncbi.nlm.nih.gov/pubmed /34855513.

316. Bundgaard, H., et al., *Effectiveness of Adding a Mask Recommendation to Other Public Health Measures to Prevent SARS-CoV-2 Infection in Danish Mask Wearers: A Randomized Controlled Trial.* Ann Intern Med, 2021. 174(3): p. 335–343. https://www.ncbi.nlm.nih.gov/pubmed/33205991.

317. Alexander, P.E., *More than 150 Comparative Studies and Articles on Mask Ineffectiveness and Harms.* Brownstone Institute, 2021. https://brownstone.org/articles/more-than-150-comparative -studies-and-articles-on-mask-ineffectiveness-and-harms/.

318. Pilgner, J., *Silencing the Lambs — How Propaganda Works.* . Consortium News, 2022. 27(279).

319. Haskins, J., *Are Financial Institutions Using Esg Social Credit Scores To Coerce Individuals, Small Businesses? Five reasons to believe banks, financial institutions are expanding ESG.* The Heartland Institute, 2022. https://www.heartland.org/publications/publications-resources/publications/financial-institutions -are-expanding-esg-social-credit-scores-to-target-individuals-small-businesses.

320. Mountbatten-Windsor, C.P.A.G.K.C., *The Great Reset: Today, through HRH's Sustainable Markets Initiative and the World Economic Forum, The Prince of Wales launched a new global initiative, The Great Reset,* in *Prince of Wales.* 2020. https://www.princeofwales.gov.uk/thegreatreset.

321. Raynor, G., *Use of fear to control behaviour in Covid crisis was 'totalitarian', admit scientists Members of Scientific Pandemic Influenza Group on Behaviour express regret about 'unethical' methods."* Telegraph, 2021.

322. Schaefer, K., *China's Corporate Social Credit System,* in *Trivium,* U.S.-C.E.a.S.R. Commission, Editor. 2020, U.S.-China Economic and Security Review Commission.

323. Betz, B., *What is China's social credit system?,* in *Fox News.* 2020, Fox. https://www.foxnews.com /world/what-is-china-social-credit-system.

324. Sundance, *Trudeau Reversal Motive Surfaces - Canadian Banking Association Was Approved by World Economic Forum To Lead the Digital ID Creation.* The Last Refuge, 2022. https: //theconservativetreehouse.com/blog/2022/02/23/boom-trudeau-reversal-motive-surfaces -canadian-banking-association-was-approved-by-world-economic-forum-to-lead-the-digital -id-creation/.

325. FoxNews, *Veteran rips Trudeau's treatment of truckers: 'Treating Canadians like terrorists'.* Fox News, 2022. https://www.youtube.com/watch?v=CoBPDKnz1DY.

326. Miltimore, J., *Did Trudeau's Crackdown Spark a Bank Run in Canada? Five of Canada's largest banks mysteriously went offline this week after the government announced it was freezing assets of people supporting the Freedom Convoy, sparking rumors of a bank run or computer hacking.* Catalyst, 2022. https://catalyst.independent.org/2022/02/23/trudeaus-canada-bank-run/.

327. WEF, *Young Global Leaders: Community.* 2022 [cited 2022]; Available from: https://www.younggloballeaders.org/community?page=9&region=a0Tb00000000DC9EAM&x=4&y=23.

328. FamousPeople, *Chrystia Freeland Biography (10th Deputy Prime Minister of Canada).* The Famous People 2022; Available from: https://www.thefamouspeople.com/profiles/chrystia-freeland-14573.php.

329. Financial Samurai, *Foreign Real Estate Investors Are Coming To Buy Up American Homes.* Financial Samurai, 2022.

330. GCS, *17,000 Physicians and Medical Scientists Declare "COVID National Emergency Over" and Call on Congress to Restore Constitutional Democracy by Ending Emergency Powers,* Physicians and Medical Scientists, Editor. 2022, Global COVID Summit (GCS). https://globalcovidsummit.org/news/physicians-and-medical-scientists-declare-covid-national-emergency-over.

331. Biden, J., *Notice on the Continuation of the National Emergency Concerning the Coronavirus Disease 2019 (COVID-19) Pandemic.* 2022, White House, The Briefing Room: USA. https://www.whitehouse.gov/briefing-room/presidential-actions/2021/02/24/notice-on-the-continuation-of-the-national-emergency-concerning-the-coronavirus-disease-2019-covid-19-pandemic/.

332. Malone, R.W., *We the people, demand to see the data! CDC withholding evidence concerning COVID vaccine safety is scientific fraud.,* in *Substack @ rwmalonemd.* 2022. https://rwmalonemd.substack.com/p/we-the-people-demand-to-see-the-data?utm_source=url.

333. CCCA, *The Pfizer Inoculations For COVID-19 – More Harm Than Good – VIDEO. Watch this video of the Pfizer 6 month data which shows that Pfizer's COVID-19 inoculations cause more illness than they prevent. Plus, an overview of the Pfizer trial flaws in both design and execution.* 2022; Available from: https://www.canadiancovidcarealliance.org/media-resources/the-pfizer-inoculations-for-covid-19-more-harm-than-good-2/.

334. Vitals, A., *FDA authorizes Omicron boosters.* AXIOS, 2022. https://www.axios.com/2022/08/31/fda-retooled-covid-boosters-omicron.

335. Owens, C., *Omicron booster shots are right around the corner.* AXIOS, 2022. https://www.axios.com/2022/08/28/coronavirus-omicron-booster-vaccines-fda.

336. Marks, D., *NOW - FDA's CBER Director "hopes" the Updated "booster" Injection Will Hold and Not . . . Latest Tweet by Disclose.tv.* LATEST LY, 2022. https://www.latestly.com/socially/world/now-fdas-cber-director-hopes-the-updated-booster-injection-will-hold-and-not-latest-tweet-by-disclose-tv-4149760.html.

337. FDA, *Coronavirus (COVID-19) Update: FDA Authorizes Moderna, Pfizer-BioNTech Bivalent COVID-19 Vaccines for Use as a Booster Dose,* FDA, Editor. 2022, US Gov: DC. https://www.fda.gov/news-events/press-announcements/coronavirus-covid-19-update-fda-authorizes-moderna-pfizer-biontech-bivalent-covid-19-vaccines-use.

338. O''Brian, K., *Big Bird and Granny Bird Share Their Covid-19 Vaccine Experiences,* in *Adweek.* 2021. https://www.adweek.com/agencyspy/big-bird-and-granny-bird-share-their-covid-19-vaccine-experiences/174648/.

339. MaRS, *Advertising licensed healthcare products. For healthcare products, what counts as promotion or advertising?* MaRS Startup Toolkit, 2022. https://learn.marsdd.com/article/unlicensed-healthcare-products-no-advertising-or-promotion-allowed/.

340. MaRS, *Advertising licensed healthcare products. For healthcare products, what counts as promotion or advertising?* MaRs Startup Toolkit, 2012. https://learn.marsdd.com/article/advertising-licensed-healthcare-products/.

341. Gov, C., *Department of Justice Canada. Food and Drugs Act, Section 9.,* D. Canada, Editor. 2012, Canadian Gov. https://laws-lois.justice.gc.ca/eng/acts/F-27/page-4.html#docCont.

342. FDA, *United States Code. Title 21, Part 312, Investigational New Drug Application.,* FDA, Editor. 2022 US Gov. https://www.accessdata.fda.gov/scripts/cdrh/cfdocs/cfcfr/CFRSearch.cfm?fr=312.7.

343. FDA, *United States Code. (2012, April 1). Title 21, Part 812, Investigational Device Exemptions.*, FDA, Editor. 2022, USG: DC. https://www.accessdata.fda.gov/scripts/cdrh/cfdocs/cfcfr/CFRSearch.cfm?fr=812.7.

344. O'Brian, K., *Big Bird and Granny Bird Share Their Covid-19 Vaccine Experiences.* AdWeek, 2021. https://www.adweek.com/agencyspy/big-bird-and-granny-bird-share-their-covid-19-vaccine-experiences/174648/.

345. Drake, K.L., *Chapter 9. FDA regulation of the advertising and promotion of prescription drugs, biologics, and medical devices*, in *FDA regulatory affairs. A guide for prescription drugs, medical devices, and biologics*, D.J. Pisano and D.S. Mantus, Editors. 2009, Informa Healthcare: NY.

346. de Wet, C., *Chapter 12. Information and promotion*, in *The textbook of pharmaceutical medicine (6th ed.)*, J.P. Griffin, Editor. 2009, John Wiley & Sons Ltd.: West Sussex.

347. Malone, R.W., *ImmuneImprinting, ComirnatyandOmicron (part1).* Who is Robert Malone, Substack, 2022. https://rwmalonemd.substack.com/p/immune-imprinting-comirnaty-and-omicron.

348. Malone, R.W., *ImmuneImprinting, ComirnatyandOmicron (part2).* Who is Robert Malone, Subtack, 2022. https://rwmalonemd.substack.com/p/immune-imprinting-comirnaty-and-omicron-520.

349. Malone, R.W., *Testimony and Remarks, Pandemic Response. In preparation for Texas Senate hearings tomorrow.* Who is Robert Malone, Substack, 2022. https://rwmalonemd.substack.com/p/testimony-and-remarks-pandemic-response.

350. Harrington, M., *FDA tells Americans to 'install' booster update. The US is now treating its citizens like smartphones.* UnHerd: The Post, 2022. https://unherd.com/thepost/fda-tells-americans-to-install-booster-update/?mc_cid=e62af1ff61.

351. Schiffin, A. *Calling for coalitions: building partnerships between journalists and advocates.* 2020 Spring 2020; Available from: https://www.sipa.columbia.edu/academics/capstone-projects/calling-coalitions-building-partnerships-between-journalists-and.

352. Macleod, A., *Conflict of Interest? Bill Gates Gave $319 Million to Major Media Outlets, Documents Reveal*, in *The Defender.* 2021. https://childrenshealthdefense.org/defender/bill-melinda-gates-foundation-media-objectively/?utm_source=salsa&eType=EmailBlastContent&eId=3f69115a-411a-4d58-a05d-97dcbeefea4c.

353. Advanced Television, *TNI steps up fight against disinformation*, in *Advanced Television.* 2020. https://advanced-television.com/2020/07/13/tni-steps-up-fight-against-disinformation/.

354. Riley, N.S., and J. Piereson, *Donor-driven Journalism.* National Affairs, 2022. Fall, 2022(53). https://nationalaffairs.com/publications/detail/donor-driven-journalism.

355. DHS, *Sumary of Terrorism Threat to the U.S. Homeland*, DHS, Editor. 2022, USG. https://www.dhs.gov/ntas/advisory/national-terrorism-advisory-system-bulletin-february-07-2022.

356. Jay Bhattacharya, J., S. Gupta, and M. Martin Kulldorff, *The Great Barrington Declaration.* 2020. gbdeclaration.org.

357. BBC, *Trusted News Initiative (TNI) to combat spread of harmful vaccine disinformation and announces major research project. At a recent summit chaired by the BBC's new Director General, Tim Davie, the Trusted News Initiative (TNI) agreed to focus on combatting the spread of harmful vaccine disinformation.*, in *BBC.* 2020, BBC: UK. https://www.bbc.com/mediacentre/2020/trusted-news-initiative-vaccine-disinformation.

358. WSJ, *How Fauci and Collins Shut Down Covid Debate. They worked with the media to trash the Great Barrington Declaration.*, in *Wall Street Journal.* 2021. https://www.wsj.com/articles/fauci-collins-emails-great-barrington-declaration-covid-pandemic-lockdown-11640129116.

359. Achenbach, J., *Proposal to hasten herd immunity to the coronavirus grabs White House attention but appalls top scientists*, in *Washigngton Post.* 2020: DC. https://www.washingtonpost.com/health/covid-herd-immunity/2020/10/10/3910251c-0a60-11eb-859b-f9c27abe638d_story.html.

360. WSJ, *How Fauci and Collins Shut Down Covid Debate. They worked with the media to trash the Great Barrington Declaration*, in *WSJ.* 2021. https://www.wsj.com/articles/fauci-collins-emails-great-barrington-declaration-covid-pandemic-lockdown-11640129116.

361. Committee, S.H., *Senate HELP Committee Minority Oversight Staff Releases Interim Report Analyzing Origins of COVID-19 Pandemic.* 2022, U.S. Senate. https://www.help.senate.gov/ranking/newsroom/press/senate-help-committee-minority-oversight-staff-releases-interim-report-analyzing-origins-of-covid-19-pandemic.

362. Eban, K., and J. Kao, *COVID-19 Origins: Investigating a "Complex and Grave Situation" Inside a Wuhan Lab*. Propublica, 2022. https://www.propublica.org/article/senate-report-covid-19-origin-wuhan-lab.

363. CHD, *Herd Immunity: A False Rationale for Vaccine Mandates*. Children's Health Defense, 2019. https://childrenshealthdefense.org/news/herd-immunity-a-false-rationale-for-vaccine-mandates/.

364. Van dyke, T., *Fauci defends shifting herd immunity goal post based on polling*, in *Washington Examiner*. 2020: DC. https://www.washingtonexaminer.com/news/fauci-defends-shifting-herd-immunity-goal-post-based-on-polling.

365. Stolbert, S.G., *The Biotech Death of Jesse Gelsinger*. New York Times, 1999. https://archive.nytimes.com/www.nytimes.com/library/magazine/home/19991128mag-stolberg.html.

366. Tucker, J., *Did Economists Really Favor the Lockdowns?* The Brownstone Institute, 2021. https://brownstone.org/articles/did-economists-really-favor-the-lockdowns/.

367. Anthem, P., *Risk of hunger pandemic as coronavirus set to almost double acute hunger by end of 2020. New WFP figures indicate additional 130 million lives and livelihoods will be at risk*. 2020, World Health Organization: Geneva. https://www.wfp.org/stories/risk-hunger-pandemic-coronavirus-set-almost-double-acute-hunger-end-2020.

368. UN, *Policy Brief: The Impact of COVID-19 on Children*. United Nations Executive Summary, 2020. https://unsdg.un.org/sites/default/files/2020–04/160420_Covid_Children_Policy_Brief.pdf.

369. Gopinath, G., *The Great Lockdown: Worst Economic Downturn Since the Great Depression*, in *IMF Blog*. 2020, IMF https://www.imf.org/en/Blogs/Articles/2020/04/14/blog-weo-the-great-lockdown-worst-economic-downturn-since-the-great-depression.

370. Greenstone, M., and V. Nigam, *Does Social Distancing Matter?* SSRN, 2020. https://papers.ssrn.com/sol3/papers.cfm?abstract_id=3561244.

371. Hall, R.E., C.I. Jones, and P.J. Klenow, *Trading Off Consumption and COVID-19 Deaths*. NBER National Bureau of Economic Research, 2020. https://www.nber.org/papers/w27340.

372. Chin, V., et al., *Effect estimates of COVID-19 non-pharmaceutical interventions are non-robust and highly model-dependent*. J Clin Epidemiol, 2021. 136: p. 96–132. https://www.ncbi.nlm.nih.gov/pubmed/33781862.

373. Ioannidis, J.P.A., *Infection fatality rate of COVID-19 inferred from seroprevalence data*. Bull World Health Organ, 2021. 99(1): p. 19–33F. https://www.ncbi.nlm.nih.gov/pubmed/33716331.

374. Economist, *The risks of keeping schools closed far outweigh the benefits. Millions of young minds are going to waste*. The economist, 2020. https://www.economist.com/leaders/2020/07/18/the-risks-of-keeping-schools-closed-far-outweigh-the-benefits.

375. Mathews, Z., *Emily Oster: 'Schools Should Be Among First To Open And Last To Close'*. GBH, 2020. https://www.wgbh.org/news/local-news/2020/11/16/emily-oster-schools-should-be-among-first-to-open-and-last-to-close.

376. Chikina, M., and W. Pegden, *Modeling strict age-targeted mitigation strategies for COVID-19*. PLoS One, 2020. 15(7): p. e0236237. https://www.ncbi.nlm.nih.gov/pubmed/32706809.

377. Goolsbee, A., and C. Syverson, *Fear, Lockdown, and Diversion: Comparing Drivers of Pandemic Economic Decline 2020*. National Bureau of Ecominc Research, 2020. https://www.nber.org/papers/w27432.

378. IMF, *A Crisis Like No Other, An Uncertain Recovery*. International Monetary Fund, 2020. https://www.imf.org/en/Publications/WEO/Issues/2020/06/24/WEOUpdateJune2020.

379. Kulldorff, M., *COVID-19 Counter Measures Should be Age Specific*. Linked In, 2020. https://www.linkedin.com/pulse/covid-19-counter-measures-should-age-specific-martin-kulldorff/.

380. Fetzer, T., et al., *Coronavirus Perceptions and Economic Anxiety*. The Review of Economics and Statistics 2021. https://thepearsoninstitute.org/sites/default/files/2020–04/Coronavirus%20Perceptions%20And%20Economic%20Anxiety.pdf.

381. CDC, *COVID-19 Pandemic Planning Scenario*. CDC, 2021. https://www.cdc.gov/coronavirus/2019-ncov/hcp/planning-scenarios.html.

382. Abel, M., T. Byker, and J. Carpenter, *Socially Optimal Mistakes? Debiasing COVID-19 Mortality Risk Perceptions and Prosocial Behavior*. Institute of Labor Economics, 2020. https://docs.iza.org/dp13560.pdf.

383. Rothwell, J., *How misinformation is distorting COVID policies and behaviors*. Brookings, 2020. https://www.brookings.edu/research/how-misinformation-is-distorting-covid-policies-and-behaviors/.

384. Leonhardt, D., *Covid's Partisan Errors*. New York Times, 2021. https://www.nytimes.com/2021/03/18/briefing/atlanta-shootings-kamala-harris-tax-deadline-2021.html.

385. Sayers, F., *Neil Ferguson interview: China changed what was possible*. UnHerd, 2020. https://unherd.com/thepost/neil-ferguson-interview-china-changed-what-was-possible/.

386. Rosenthal, E., *It's Time to Scare People About Covid*. New York Times, 2020. It's Time to Scare People About Covid.

387. Arnon, A., J. Ricco, and K. Smetters, *Epidemiological and Economic Effects of Lockdown*. BPEA Conference Drafts, 2020. https://www.brookings.edu/wp-content/uploads/2020/09/Arnon-et-al-conference-draft.pdf.

388. Gupta, S., et al., *Effects of Social Distancing Policy on Labor Market Outcomes*. National Bureau of Ecominc Research, 2020. https://www.nber.org/papers/w27280.

389. Givetach, L., *U.N. warns of 'hunger pandemic' amid threats of coronavirus, economic downturn*. NBC News, 2020. https://www.nbcnews.com/news/world/u-n-warns-hunger-pandemic-amid-threats-coronavirus-economic-downturn-n1189326.

390. BBC, *Covid-19 disruptions killed 228,000 children in South Asia, says UN report*. BBC News, 2021. https://www.bbc.com/news/world-asia-56425115.

391. Crawford, M., *How science has been corrupted. The pandemic has revealed a darkly authoritarian side to expertise*. UnHerd, 2021. https://unherd.com/2021/05/how-science-has-been-corrupted/.

392. Perskey, J., *Retrospectives A Dismal Romantic*. Journal of Economic Perspectives, 1990. https://pubs.aeaweb.org/doi/pdfplus/10.1257/jep.4.4.165.

393. Salter, A.W., *How Economics Lost Itself in Data*. Wall Street journal, 2021. https://www.wsj.com/articles/how-economics-lost-itself-in-data-11611775849.

394. Ruhm, C., *Shackling the Identification Police?* National Bureau of Ecominc Research, 2018. https://www.nber.org/system/files/working_papers/w25320/w25320.pdf.

395. Rodrik, D., *How Economists and Non-Economists Can Get Along*. Project Syndicate, 2021. https://www.project-syndicate.org/commentary/economists-other-social-scientists-and-historians-can-get-along-by-dani-rodrik-2021-03.

396. Harris, C., et al., *The Binding Force of Economics*. Contemporary Methods and Austrian Economics, 2022. https://www.emerald.com/insight/content/doi/10.1108/S1529-213420220000026006/full/html.

397. Akeriof, G.A., *Sins of Omission and the Practice of Economics*. American Economic Association, 2020. https://www.aeaweb.org/articles?id=10.1257/jel.20191573.

398. Gupta, S., *Avoiding ambiguity*. Nature, 2001. 412(6847): p. 589. https://www.ncbi.nlm.nih.gov/pubmed/11493898.

399. Loh, M., *'Freedom Convoy' truckers who continue their anti-vaccine mandate blockades*. Business Insider, 2022. https://www.businessinsider.com/trudeau-canada-freeze-bank-accounts-freedom-convoy-truckers-2022-2.

400. Morse, J., *The Federal Reserve Cartel: The Rothschild, Rockefeller and Morgan Families*. The Event Chronicle, 2015. https://theeventchronicle.com/the-federal-reserve-cartel-the-rothschild-rockefeller-and-morgan-families/.

401. Quigley, C., *Tragedy & Hope: A History of the World in Our Time (The uncensored edition)*. 2nd (first printed in 1966) ed. 2004 (first printed in 1966).

402. K. Schwab and T. Malleret, *The Great Narrative (The Great Reset)*. 2021: Amazon.

403. Schwab, K., and T. Malleret, *COVID-19: The Great Reset*. 2020: Amazon. 280. https://2009–2017.state.gov/r/gec/index.htm.

404. McClaughry, J., *Thomas Jefferson and Public Debt*. Ethan Allen, 2021. https://www.ethanallen.org/thomas_jefferson_and_public_debt.

405. Malone, R.W., and J. Glasspool-Malone, *The World Economic Forum Young Leaders Program has over 3800 graduates*. The Malone Institute, 2022. https://maloneinstitute.org/wef.

406. Mercola, J., *Who Owns Big Pharma + Big Media? You'll Never Guess*. The Defender, 2021. https://childrenshealthdefense.org/defender/blackrock-vanguard-own-big-pharma-media/.

407. Stuffaford, A., *A Useful Pandemic: Davos Launches New 'Reset,' this Time on the Back of COVID*. National Review, 2020. https://www.nationalreview.com/2020/10/a-useful-pandemic-davos-launches-new-reset-this-time-on-the-back-of-covid/.

408. Stuffaford, A., *Larry Fink, 'Emperor'?* National Review, 2022. https://www.nationalreview.com/2022/02/larry-fink-emperor/.

409. Stuffaford, A., *The Great Reset: If Only It Were Just a Conspiracy.* Nat Neurosci, 2020. https://www.nationalreview.com/2020/11/the-great-reset-if-only-it-were-just-a-conspiracy/.

410. Stuffaford, A., *'Stakeholder Capitalism' A Sham? Unfortunately Not.* National Review, 2021. https://www.andrewstuttaford.com/archive/2022/2/2/stakeholder-capitalism-a-sham-unfortunately-not.

411. BBC, *Elon Musk warned he must protect Twitter users.* BBC News, 2022. https://www.bbc.com/news/business-61225355.

412. Obama, B., *Former President Obama speaks on disinformation's threat to democracy*, in *Youtube.* 2022. https://www.youtube.com/watch?v=ExEApwbhfqQ.

413. JHU, *Event 201: A Global Pandemic Exercise.* 2019, JHU Bloomberg School of Public Health: Center for Health Security: Baltimore. https://www.centerforhealthsecurity.org/our-work/exercises/event201/.

414. Baker, N., *The Lab-Leak Hypothesis For decades, scientists have been hot-wiring viruses in hopes of preventing a pandemic, not causing one. But what if ...?* Intelligencer, 2021. https://nymag.com/intelligencer/article/coronavirus-lab-escape-theory.html.

415. Glass, A., *Senate creates Truman committee, March 1, 1941.* Politco, 2018. https://www.politico.com/story/2018/03/01/senate-creates-truman-committee-march-1-1941-430139.

416. Halderstam, D., *The Best and the Brightest.* 1992: Random House.

417. BBC, *Terror and persuasion: Goebbels and the Ministry of Propaganda.* BBC News: Bitesize, 2022. https://www.bbc.co.uk/bitesize/guides/zpb9fcw/revision/2.

418. Wolin, S.S., *Democracy Incorporated: Managed Democracy and the Specter of Inverted Totalitarianism.* 2008: Princeton University Press. 376.

419. CDC, *CDC Foundation* CDC, 2022. https://www.cdc.gov/about/business/cdcfoun.htm.

420. Herman, B., *The NIH claims joint ownership of Moderna's coronavirus vaccine.* Axios, 2020. https://www.axios.com/2020/06/25/moderna-nih-coronavirus-vaccine-ownership-agreements.

421. Tanne, J.H., *Royalty payments to staff researchers cause new NIH troubles.* BMJ, 2005. 330(7484): p. 162. https://www.ncbi.nlm.nih.gov/pubmed/15661767.

422. Andrzejewski, A., *NIH Scientists Pocketed $350 Million in Royalties—Agency Won't Say How Much Went to Fauci.* The Defender, 2022. https://childrenshealthdefense.org/defender/nih-scientists-millions-royalties/.

423. HHS, *HHS FY 2022 Budget in Brief*, HHS, Editor. 2022, HHS: DC. https://www.hhs.gov/about/budget/fy2022/index.html.

424. McAndrew, A., *DoD Budget Request*, D.o. Defense, Editor. 2022, DoD: DC. https://comptroller.defense.gov/Budget-Materials/Budget2022/.

425. Brust, A., *Proposed NIH budget boost spending for experimentation, examines lower grant funding for minority scientists.* Federal News Network, 2022. https://federalnewsnetwork.com/budget/2022/05/proposed-nih-budget-boost-spending-for-experimentation-examines-lower-grant-funding-for-minority-scientists/.

426. Tabak, D.S.D., *Congressional Justification of the NIH request for the fiscal year (FY) 2023 budget,* N. Office of Budget, Editor. 2022, NIH: DC. https://officeofbudget.od.nih.gov/pdfs/FY23/br/Overview%20of%20FY%202023%20Presidents%20Budget.pdf.

427. Azar, A., *Determination that a Public Health Emergency Exists*, HHS, Editor. 2020, HHS: DC. https://www.phe.gov/emergency/news/healthactions/phe/Pages/2019-nCoV.aspx.

428. Cochran, N., *The Secretary of Helth and Human Services Letter to Governors*, ASPR, Editor. 2021, ASPR: DC. https://aspr.hhs.gov/legal/PHE/Pages/Letter-to-Governors-on-the-COVID-19-Response.aspx.

429. ASPR, *Public Health Emergency Determinations to Support an Emergency Use Authorization*, ASPR, Editor. 2022, ASPR: DC. https://aspr.hhs.gov/legal/Section564/Pages/default.aspx.

430. FDA, *MCM Emergency Use Authorities FDA can allow the emergency use of MCMs through Emergency Use Authorization (EUA) and authorities related to emergency use of approved MCMs,* FDA, Editor. 2022, FDA: DC. https://www.fda.gov/emergency-preparedness-and-response/mcm-legal-regulatory-and-policy-framework/mcm-emergency-use-authorities.

431. OPM, *Policy, Data, Oversight: Senior Executive Service*, OPM, Editor. 2022, OPM: OPM.gov. https://www.opm.gov/policy-data-oversight/senior-executive-service/.

432. Tucker, J., *The Astonishing Implications of Schedule F*. The Brownstone Institute, 2022. https://brownstone.org/articles/the-astonishing-implications-of-schedule-f/.

433. Washington Post, *Trump's newest executive order could prove one of his most insidious*. Washington Post, 2020. https://www.washingtonpost.com/opinions/trumps-newest-executive-order-could-prove-one-of-his-most-insidious/2020/10/23/c8223cac-1561-11eb-bc10-40b25382f1be_story.html.

434. LibertyCounsel, *Biden's Amendments Hand U.S. Sovereignty to the WHO*. Liberty Counsel, 2022. https://lc.org/newsroom/details/051222-bidens-amendments-hand-us-sovereignty-to-the-who-1.

435. Koehli, H., *The Psychology of Totalitarianism Part 4, On the Fractal Nature of Conspiracy*. Political Ponerology, Substack, 2022. https://ponerology.substack.com/p/the-psychology-of-totalitarianism-8a8.

436. Crawford, M., *Covid was liberalism's endgame. Liberal individualism has an innate tendency towards authoritarianism*. UnHerd, 2022. https://unherd.com/2022/05/covid-was-liberalisms-endgame/.

437. Malone, R.W., *Ivermectin: Why is the Administrative State willing to kill you? Liberalism's endgame, public health despotism, Scientism, Big Tech*. RWMaloneMD Substack, 2022. https://rwmalonemd.substack.com/p/Ivermectin-why-is-the-administrative.

438. Janus, I.L., *Victims of Groupthink Psychologicat Study of Foreign Policy*. 1972: Houghton Mifflin.

439. Rickards, J., *The War on Cash Entering Bold New Phase Daily Reckoning by James Rickards, February 15, 2022*. Daily Reckoning 2022. https://dailyreckoning.com/the-war-on-cash-entering-bold-new-phase/.

440. Metzgar, E., *Smith-Mundt reform: In with a whimper?*, in *Columbia Journalism Review*. 2013, Columbia University. https://archives.cjr.org/behind_the_news/smith-mundt_modernization_pass.php.

441. Spiering, C., *Barack Obama Calls for More Censorship: First Amendment 'Does Not Apply to Facebook and Twitter'*, in *Breitbart*. 2022. https://www.breitbart.com/politics/2022/04/21/barack-obama-calls-more-censorship-first-amendment-does-not-apply-facebook-twitter/).

442. Bloomberg, *Bloomburg. Amazon, Alphabet among tech firms meeting with White House on coronavirus response*, in *Los Angeles Times*. 2020. https://www.latimes.com/business/technology/story/2020-03-11/white-house-amazon-microsoft-alphabet-coronavirus-response.

443. Romm, T., *White House asks Silicon Valley for help to combat coronavirus, track its spread and stop misinformation.*, in *Washington Post*. 2020. https://www.washingtonpost.com/technology/2020/03/11/white-house-tech-meeting-coronavirus/.

444. DeSantis, R., and Staff, *Governor Ron DeSantis to Biden Administration: Release Stranglehold on Life-saving Monoclonal Antibodies (Jan 3, 2022)*. 2022, State of Florida: Florida. https://www.flgov.com/2022/01/03/governor-ron-desantis-to-biden-administration-release-stranglehold-on-life-saving-monoclonal-antibodies/.

445. FSMB, *Federation of State Medical Boards*. 2022, FSMB. https://www.fsmb.org/u.s.-medical-regulatory-trends-and-actions/guide-to-medical-regulation-in-the-united-states/introduction/.

446. Bleau, H., *There is "no way" Florida will ever support the World Health Organization's (W.H.O.) global pandemic treaty, Florida Gov. Ron DeSantis (R) said on Monday.*, in *Breitbart*. 2022.

447. Hill-Gray, R., *Food Processing Plants Have Been Catching on Fire — Here's Why*. Market Realist, 2022. https://marketrealist.com/p/food-processing-plant-fires/.

448. Ng, J., *Chronic Food Shortages From Lettuce to Sriracha and the Rise of Chicken*. Bloomburg, 2022. https://www.bloomberg.com/news/newsletters/2022-06-10/supply-chain-latest-chicken-popularity-and-shortages-of-lettuce-and-sriracha.

449. Harrington, T., *Drs. Walensky and Offit: It's All in Good Fun*. The Brownstone Institute, 2022. https://brownstone.org/articles/drs-walensky-and-offit-its-all-in-good-fun/.